Edited by
Richard M. De La Rue, Siyuan Yu, and
Jean-Michel Lourtioz

Compact Semiconductor Lasers

Related Titles

Ohtsu, M. (ed.)

Frequency Control of Semiconductor Lasers

1996
Print ISBN: 978-0-471-01341-9

Physics of Optoelectronic Devices

1995
Print ISBN: 978-0-471-10939-6
eMobi-lite ISBN: 978-0-470-30140-1

Coldren, L.A., Corzine, S.

Diode Lasers and Photonic Integrated Circuits

1995
Print ISBN: 978-0-471-11875-6
eMobi-lite ISBN: 978-0-470-30154-8

Fukuda, M.

Optical Semiconductor Devices

1999
Print ISBN: 978-0-471-14959-0
eMobi-lite ISBN: 978-0-470-29851-0

Saleh, B.E., Teich, M.C.

Fundamentals of Photonics, Online Version

2nd Edition

2001
Print ISBN: 978-0-471-21374-1

Choi, H.K. (ed.)

Long-Wavelength Infrared Semiconductor Lasers

2004
Print ISBN: 978-0-471-39200-2
eMobi-lite ISBN: 978-0-470-31334-3
Adobe PDF ISBN: 978-0-471-64980-9
ISBN: 978-0-471-64981-6

Luryi, S., Xu, J., Zaslavsky, A. (eds.)

Future Trends in Microelectronics

The Nano, the Giga, and the Ultra

2004
Print ISBN: 978-0-471-48405-9

May, G.S., Spanos, C.J.

Fundamentals of Semiconductor Manufacturing and Process Control

2006
Print ISBN: 978-0-471-79028-0

Korvink, J.G., Greiner, A.

Semiconductors for Micro- and Nanotechnology

An Introduction for Engineers

2002
Print ISBN: 978-3-527-30257-4
ISBN: 978-3-527-60022-9
Adobe PDF ISBN: 978-3-527-61625-1

Epstein, R., Sheik-Bahae, M. (eds.)

Optical Refrigeration
Science and Applications of Laser Cooling of Solids

2009
Print ISBN: 978-3-527-40876-4
ISBN: 978-3-527-62804-9
Adobe PDF ISBN: 978-3-527-62805-6

Okhotnikov, O.G. (ed.)

Semiconductor Disk Lasers
Physics and Technology

2010
Print ISBN: 978-3-527-40933-4
ISBN: 978-3-527-63039-4
Adobe PDF ISBN: 978-3-527-63040-0

Khanh, T., Bodrogi, P., Vinh, Q., Winkler, H. (eds.)

LED Lighting
Technology and Perception

2016
Print ISBN: 978-3-527-41212-9
ISBN: 978-3-527-67014-7
eMobi ISBN: 978-3-527-67015-4
ePub ISBN: 978-3-527-67016-1
Adobe PDF ISBN: 978-3-527-67017-8

Edited by
Richard M. De La Rue, Siyuan Yu, and Jean-Michel Lourtioz

Compact Semiconductor Lasers

Verlag GmbH & Co. KGaA

Editors

Prof. Richard M. De La Rue
Optoelectronics Research Group
School of Engineering
University of Glasgow
Rankine Building
Oakfield Avenue
Glasgow G12 8LT
Scotland, U.K.

Prof. Siyuan Yu
Dept. of Elect. Engineering
University of Bristol
Merchant Venturers Building
Bristol, BS8 1UB
United Kingdom

Dr. (Prof.) Jean-Michel Lourtioz
Directeur de Recherche CNRS
Vice President of Université Paris-Sud
Mission Campus
Bâtiment 209E
91405 Orsay Cedex
France

Cover
Courtesy of Gabor Mezosi

All books published by **Wiley-VCH** are carefully produced. Nevertheless, authors, editors, and publisher do not warrant the information contained in these books, including this book, to be free of errors. Readers are advised to keep in mind that statements, data, illustrations, procedural details or other items may inadvertently be inaccurate.

Library of Congress Card No.: applied for

British Library Cataloguing-in-Publication Data
A catalogue record for this book is available from the British Library.

Bibliographic information published by the Deutsche Nationalbibliothek
The Deutsche Nationalbibliothek lists this publication in the Deutsche Nationalbibliografie; detailed bibliographic data are available on the Internet at <http://dnb.d-nb.de>.

© 2014 Wiley-VCH Verlag GmbH & Co. KGaA, Boschstr. 12, 69469 Weinheim, Germany

All rights reserved (including those of translation into other languages). No part of this book may be reproduced in any form – by photoprinting, microfilm, or any other means – nor transmitted or translated into a machine language without written permission from the publishers. Registered names, trademarks, etc. used in this book, even when not specifically marked as such, are not to be considered unprotected by law.

Print ISBN: 978-3-527-41093-4
ePDF ISBN: 978-3-527-65537-3
ePub ISBN: 978-3-527-65536-6
mobi ISBN: 978-3-527-65535-9
oBook ISBN: 978-3-527-65534-2

Cover-Design Adam-Design, Weinheim, Germany
Typesetting Laserwords Private Limited, Chennai, India
Printing and Binding Markono Print Media Pte Ltd, Singapore

Printed on acid-free paper

Contents

Preface and Introduction *XIII*
List of Contributors *XXI*
Color Plates *XXV*

1	**Nanoscale Metallo-Dielectric Coherent Light Sources** *1*
	Maziar P. Nezhad, Aleksandar Simic, Amit Mizrahi, Jin-Hyoung Lee, Michael Kats, Olesya Bondarenko, Qing Gu, Vitaliy Lomakin, Boris Slutsky, and Yeshaiahu Fainman
1.1	Introduction *1*
1.2	Composite Metallo-Dielectric-Gain Resonators *4*
1.2.1	Composite Gain-Dielectric-Metal Waveguides *5*
1.2.2	Composite Gain-Dielectric-Metal 3D Resonators *7*
1.3	Experimental Validations of Subwavelength Metallo-Dielectric Lasers for Operation at Room-Temperature *9*
1.3.1	Fabrication Processes for Subwavelength Metallo-Dielectric Lasers *10*
1.3.2	Characterization and Testing of Subwavelength Metallo-Dielectric Lasers *11*
1.4	Electrically Pumped Subwavelength Metallo-Dielectric Lasers *13*
1.4.1	Cavity Design and Modeling of Electrically Pumped Subwavelength Metallo-Dielectric Lasers *13*
1.4.2	Fabrication of Electrically Pumped Subwavelength Metallo-Dielectric Lasers *17*
1.4.3	Measurements and Discussion of Electrically Pumped Subwavelength Metallo-Dielectric Lasers *18*
1.5	Thresholdless Nanoscale Coaxial Lasers *20*
1.5.1	Design and Fabrication of Thresholdless Nanoscale Coaxial Lasers *22*
1.5.2	Characterization and Discussion of Thresholdless Nanoscale Coaxial Lasers *23*
1.6	Summary, Discussions, and Conclusions *27*
	Acknowledgments *29*
	References *29*

2		**Optically Pumped Semiconductor Photonic Crystal Lasers** *33*
		Fabrice Raineri, Alexandre Bazin, and Rama Raj
2.1		Introduction *33*
2.2		Photonic Crystal Lasers: Design and Fabrication *35*
2.2.1		Micro/Nano Cavity Based PhC Lasers *36*
2.2.1.1		Lasers Based on 2D PhC Cavities *36*
2.2.1.2		Lasers Based on 3D PhC Cavities *44*
2.2.2		Slow-Light Based PhC Lasers: DFB-Like Lasers *46*
2.2.2.1		2D PhC DFB-Like Lasers for In-Plane Emission *47*
2.2.2.2		2D PhC DFB- Like Lasers for Surface Emission *50*
2.3		Photonic Crystal Laser Characteristics *52*
2.3.1		Rate Equation Model and PhC Laser Parameters *52*
2.3.1.1		Linear Rate Equation Model *52*
2.3.1.2		PhC Laser Parameters *53*
2.3.2		The Stationary Regime in PhC Lasers *55*
2.3.3		Dynamics of PhC Lasers *60*
2.4		The Final Assault: Issues That Have Been Partially Solved and Others That Remain to Be Solved Before Photonic Crystal Lasers Become Ready for Application *65*
2.4.1		Room Temperature Continuous Wave Room Temperature Operation of Photonic Crystal Nano-Lasers *66*
2.4.1.1		CW Operation via Nonradiative Recombination Reduction *66*
2.4.1.2		CW Operation via Increased Heat Sinking *69*
2.4.2		Interfacing and Power Issues *74*
2.4.2.1		Interfacing an Isolated PhC Cavity-Based Device with the External World *74*
2.4.2.2		Interfacing Active-PhC Cavity-Based Devices within an Optical Circuit *77*
2.5		Conclusions *82*
		References *83*
3		**Electrically Pumped Photonic Crystal Lasers: Laser Diodes and Quantum Cascade Lasers** *91*
		Xavier Checoury, Raffaele Colombelli, and Jean-Michel Lourtioz
3.1		Introduction *91*
3.2		Near-Infrared and Visible Laser Diodes *93*
3.2.1		Photonic Crystal Microcavity Lasers *93*
3.2.1.1		Photonic Crystals in Suspended (or Sustained) Membrane *93*
3.2.1.2		Other Promising Designs *96*
3.2.2		Waveguide Lasers in the Substrate Approach: Weak Vertical Confinement *98*
3.2.2.1		DFB-Like Photonic Crystal Waveguide Lasers *100*
3.2.2.2		α DFB Lasers *102*
3.2.2.3		Ridge Waveguide Lasers with Photonic Crystal Mirrors *103*
3.2.3		Photonic Crystal Surface-Emitting Lasers (PCSELs) *106*

3.2.4	Nonradiative Carrier Recombination in Photonic Crystal Laser Diodes *109*	
3.3	Mid-Infrared and Terahertz (THz) Quantum Cascade Lasers *112*	
3.3.1	Microdisk QCLs *114*	
3.3.1.1	Microdisk QC Lasers at Mid-Infrared Wavelengths *114*	
3.3.1.2	THz Waves: Several Closely Spaced Demonstrations of Small Volume Micro-Lasers Exploiting Double-Sided Metallic Cavities *115*	
3.3.2	Photonic Crystal QCLs: Surface Emission and Small Modal Volumes *116*	
3.3.2.1	Mid-Infrared Quantum Cascade Lasers with Deeply Etched Photonic Crystal Structures *116*	
3.3.2.2	Mid-Infrared Quantum Cascade Lasers with Thin Metallic Photonic Crystal Layer *121*	
3.3.2.3	Terahertz (THz) 2D Photonic Crystal Quantum-Cascade (QC) Lasers *123*	
3.3.3	Toward THz QC Lasers with Truly Subwavelength Dimensions *131*	
3.4	Concluding Remarks and Prospects *135*	
	References *138*	

4 **Photonic-Crystal VCSELs** *149*
Krassimir Panajotov, Maciej Dems, and Tomasz Czyszanowski

4.1	Introduction *149*
4.2	Numerical Methods for Modeling Photonic-Crystal VCSELs *156*
4.3	Plane-Wave Admittance Method *158*
4.4	Impact of Photonic-Crystal depth on VCSEL Threshold Characteristics *166*
4.5	Top and Bottom-Emitting Photonic-Crystal VCSELs *170*
4.6	Enhanced Fundamental Mode Operation in Photonic-Crystal VCSELs *173*
4.7	Highly Birefringent and Dichroic Photonic-Crystal VCSELs *177*
4.8	Photonic-Crystal VCSELs with True Photonic Bandgap *181*
4.9	Summary and Prospects *185*
	References *188*

5 **III–V Compact Lasers Integrated onto Silicon (SOI)** *195*
Geert Morthier, Gunther Roelkens, and Dries Van Thourhout

5.1	Introduction *195*
5.2	Bonding of III–V Membranes on SOI *197*
5.2.1	Adhesive Bonding *199*
5.2.2	Direct Bonding *201*
5.2.3	Substrate Removal *203*
5.3	Heterogeneously Integrated Edge-Emitting Laser Diodes *204*
5.3.1	Fabry-Perot Lasers *205*

5.3.2	Mode-Locked Lasers	*207*
5.3.3	Racetrack Resonator Lasers	*207*
5.3.4	DFB and Tunable Lasers	*209*
5.3.4.1	Distributed Feedback Lasers	*209*
5.3.4.2	Distributed Bragg Reflector Lasers	*209*
5.3.4.3	Sampled Grating DBR Lasers	*209*
5.3.4.4	Ring-Resonator Based Tunable Laser	*210*
5.3.4.5	Heterogeneously Integrated Multiwavelength Laser	*210*
5.3.5	Proposed Novel Laser Architectures	*211*
5.3.5.1	Exchange Bragg Coupling Laser Structure	*211*
5.3.5.2	Resonant Mirrors	*211*
5.3.6	Heat Sinking Strategies for Heterogeneously Integrated Lasers	*212*
5.4	Microdisk and Microring Lasers	*213*
5.4.1	Design of Microdisk Lasers	*215*
5.4.2	Static Operation	*217*
5.4.3	Dynamic Operation and Switching	*219*
5.4.3.1	Direct Modulation	*219*
5.4.3.2	All-Optical Set-Reset Flip-Flop	*219*
5.4.3.3	Gating and Wavelength Conversion	*221*
5.4.3.4	Narrowband Optical Isolation and Reflection Sensitivity of Microdisk Lasers	*222*
5.4.3.5	Phase Modulation	*224*
5.4.3.6	Microwave Photonic Filter	*224*
5.5	Summary and Conclusions	*226*
	References	*226*
6	**Semiconductor Micro-Ring Lasers**	***231***
	Gábor Mezosi and Marc Sorel	
6.1	Introduction	*231*
6.2	Historical Review of Major Contributions to Research on SRL Devices	*232*
6.3	Waveguide Design of Semiconductor Ring Lasers	*235*
6.4	Bending Loss in Semiconductor Ring Lasers	*238*
6.5	Nonradiative Carrier Losses	*240*
6.6	Semiconductor Microring and Microdisk Lasers with Point Couplers	*242*
6.7	Junction Heating in Small SRL Devices	*246*
6.8	RIE-Lag Effects in Small SRL Devices	*248*
6.9	Racetrack Geometry Microring Lasers	*249*
6.10	Chapter Summary	*252*
	References	*253*
7	**Nonlinearity in Semiconductor Micro-Ring Lasers**	***257***
	Xinlun Cai, Siyuan Yu, Yujie Chen, and Yanfeng Zhang	
7.1	Introduction	*257*

7.2	General Formalism 260	
7.2.1	Fundamental Equations for Semiconductor Ring Lasers 260	
7.2.2	Third Order Susceptibility and Polarization 265	
7.2.3	Generalized Equations in Matrix Form 270	
7.3	Numerical Results for Micro-Ring Lasers 272	
7.3.1	L–I Characteristics 274	
7.3.2	Temporal Dynamics and Lasing Spectra 275	
7.3.3	Lasing Direction Hysteresis 277	
7.3.4	Experimental Results for Racetrack Shaped SRLs 281	
7.4	Numerical Results for Unidirectional Micro-Ring Lasers 283	
7.4.1	Unidirectional SRL 284	
7.4.2	Impact of the Feedback Strength on the Operation of Unidirectional SRL 287	
7.5	Summary and Conclusions 293	
	References 294	

Index *297*

Preface and Introduction

The title of this book is *"Compact Semiconductor Lasers"* and, in accordance with this title, it aims to provide a nearly comprehensive review of efforts in various research laboratories to create semiconductor lasers that are progressively smaller. In our view, efforts to make semiconductor lasers ever more compact are mainly driven by the need to integrate semiconductor lasers in photonic integrated circuits (PICs) or optoelectronic integrated circuits (OEICs). Two major technical advantages can be derived from being small – taking up less space on a chip, and consuming less power per laser – both of which are necessary for higher density and higher device-count integration.

The primary industrial relevance of the compact semiconductor laser possibly comes from its potential for application in advanced PICs aimed at communication applications. PICs, such as those integrating high quality laser sources, modulators, and on-chip monitoring devices, have already been applied commercially for long distance optical communications. While not yet a major constraint on their application, compactness, and power-efficiency are nevertheless high on the wish list for such components. Yet one needs to be aware that compactness may be an attribute that conflicts with other requirements. For instance, the spectral purity, frequency precision, and tunability that are essential for high precision sources such as distributed-feedback (DFB) and distributed Bragg reflector (DBR) waveguide lasers imply the need for device lengths that are specified as several hundred micrometers, if not several millimeters.

Other emerging important information transmission applications may impose very different requirements on laser sources. Optical interconnects have at their base the familiar silicon very-large-scale integration (VLSI) circuit, or more colloquially, the chip. The driving concept of the optical interconnect is that light waves rather than electronic wires should provide the primary means for the transport of information between silicon chips or, more speculatively, from one part of a silicon chip to another. Used properly, optical interconnects could be a vitally important way of bypassing and/or overcoming bottlenecks in data processing and transmission. Therefore, optical interconnects are one form of integrated circuit where photonics might play a role. The compact semiconductor laser could, for instance, simply be organized, in large or small arrays, around the edges of the silicon VLSI chip. Useful performance could mean that each compact laser was both dissipating and delivering sub-microwatt-scale power levels, together with signal bandwidths of 10 Gb s^{-1} and beyond in each channel. It is obvious that existing communication lasers cannot be suitable, because of their size and power consumption. Here compact semiconductor lasers in the 1300–1600 nm band are the best candidates, since their development is more mature and, being in the transparency window of silicon, they are more complementary metal-oxide semiconductor (CMOS)-compatible than their counterparts at other, shorter, wavelengths.

With an increasing amount of information carried by optical waves, there has also been strong interest in the realization of what we may call "digital" PICs/OEICs. One can envisage the use of compact semiconductor lasers to realize *photonic* digital logical functionalities that are direct counterparts of several of the electronic logical functionalities. These are routinely built into the silicon VLSI chip, utilizing the nonlinear optical response of compact semiconductor lasers such as optical bistability for binary logic functions including Boolean logic, buffering, and storage. Such nonlinearity stems from the fundamental photon–charge-carrier interaction dynamics of the semiconductor and depends strongly on the structure of the laser cavity. The compact semiconductor laser based PIC/OEICs that might emerge will probably be quite different from the PICs and OEICs that have been most well developed so far for the transmission of information.

Alternative forms of functional PIC/OEIC, such as those being developed for sensing applications, may require different wavelengths and hence different semiconductor materials and structures, yet will still benefit from reduced laser size. For example, quantum cascade lasers (QCLs) that operate in the far infrared can benefit from compactness in applications where low power consumption is required including, for instance, vehicle embarked systems. Finally, it is possible to envisage a number of other situations where compact near-infrared semiconductor lasers could play a role. Examples could include single and paired photon sources that are relevant to various quantum processing applications and novel optomechanical device structures.

The fundamental mechanism that supports both reduced size and low power consumption is the high optical gain provided by direct bandgap semiconductor active materials and the quantum structures (quantum wells, quantum dots, etc.) based on these materials, making it possible for lasing to take place even with a tiny volume of active material. In this perspective, the effort toward smaller size and higher energy efficiency is largely an effort to realize ever-tighter simultaneous confinement of both the photon and the electron in this active volume. The reader may infer that all the structures covered in this book are aimed at strong confinement, which optically benefits from the high refractive index of the semiconductor materials used and at its most extreme is exemplified by the use of metal clad structures to overcome the optical diffraction limit. Truly subwavelength semiconductor lasers are indeed attainable with the incorporation of metal, provided that the resulting absorption losses are sufficiently low as applies, for instance, in the long wavelength region of the spectrum. Whatever may be the structure, much effort must also be dedicated to the electronic confinement, since the reduced active volume typically means a much-increased surface-area to active-volume ratio that leads to higher levels of nonradiative carrier recombination.

Making semiconductor lasers more and more compact is, by itself, a worthwhile technological challenge at the level of targeted applications.

A broader commentary on semiconductor lasers and their applications at this point may help our reader to place the contents of this book in the wider context of a field that has been established for 50 years and yet is still evolving rapidly. The compact semiconductor lasers described in the different chapters of this book could enter progressively into the armory of contemporary optoelectronics and photonics technologies, where "conventional" semiconductor lasers are already commonplace items. The latter are routinely manufactured, en masse, with production volumes that are measured in hundreds of millions per year. In fact, almost all of the lasers in the world are semiconductor lasers. Furthermore, semiconductor laser *diodes* are the primary light source, that is, the primary converters between electrical power and optical power, for a large fraction of all other lasers, for example, where the characteristic gain medium of the laser is a doped fiber, a single-crystal slab, or a rod of glass. In terms of the conversion efficiency between electrical power (current) and optical power, no other laser comes close to the semiconductor laser.

It is salutary to contrast production volume numbers for individual lasers including, for example, the vertical cavity surface-emitting lasers (VCSELs) that are often organized in arrays with *financial* numbers. Approximately half of the financial volume for the production of "lasers," around the world, is made up of semiconductor diode lasers. In contrast, the remaining half is almost totally made up of the "big" lasers that are used for applications such as numerically controlled machining and welding. In the case of such large lasers, the actual laser is, of course, built into a physically large machine that is a complete system of electronics, precision mechanisms, and optics.

New applications for semiconductor lasers continue to appear and they may, for instance, even displace the conventional light-emitting diode (LED) in parts of the display technology market. But there is also the possibility that some large-scale applications will disappear or greatly diminish in importance, at least in their most typical form for example, the use of blue diode lasers in compact disk (CD) equipment. The optical spectrum covered by semiconductor lasers is vast, although wavelength tunability remains an important issue. The spectral range of semiconductor lasers that are already being exploited commercially – or will probably be exploited commercially in the near-future – extends from the ultra-violet (UV), through the visible and a long way on into the infra-red (IR). With the inclusion of the quantum-cascade (QC) laser, the IR spectral coverage possible with semiconductor lasers ranges all the way from the shortest IR wavelengths, through the mid-IR to the far-IR, which last spectral region itself extends down in frequency to the terahertz (THz) region.

The list of applications where semiconductor lasers are used is, as already mentioned, extensive and we shall do no more than mention some of the more obvious and important applications. One important area, with financial values (revenue, turnover, and capitalization) measured in tens of billions of dollars, is fiber-optical telecommunications. In this domain, DFB and DBR lasers are the source of choice for large information bandwidth transmission over long distances and seem likely to remain so. On the other hand, the rapidly growing short-reach optical fiber communications market, for example, in access networks ("fiber-to-the-home" or "last mile") and in the very large-scale data centers that support the Internet "cloud" services, demands large numbers of low-cost semiconductor lasers that are mainly conventional Fabry–Perot (FP) lasers and VCSELs. Newer types of integrated laser, such as those that can emit multiple wavelengths at a low cost, could prove to be vital in the future upgrade of such systems.

Line-of-sight, building-to-building communication using modulated semiconductor lasers, over kilometer scale distances, is of potential importance, although the atmospheric transmission is often impeded by adverse weather conditions. Useful communication in space over thousands of kilometers has already been demonstrated, which is more challenging because of other factors. Although the scale of the market and activity in the domain of free-space optical telecommunications has, so far, been much smaller than for fiber-optics based telecommunications, semiconductor lasers will surely form an important ingredient in this field.

The CD, as used in CD players for audio reproduction (e.g., for music) and in DVD (digital video/versatile disk) players, has been much the largest market in volume for the semiconductor laser, with annual demand amounting to several tens of millions of suitably packaged individual lasers. The semiconductor laser in a CD player is used to read the fine pattern of holes that has been created in the thin sheet of metal deposited on the surface of the plastic substrate. This quite specialized application of the semiconductor laser, despite its historical success, may yet diminish considerably in importance – or even disappear – because of trends in information transmission, storage, and delivery associated with the Internet. More generally, data storage by optical methods that include holographic techniques must compete with electronic and magnetic memory methods that have their own areas of strength, as well as weakness. Our view is that the intrinsic merits of the semiconductor laser are sufficiently strong that it will surely continue to play an important role in major markets, but it is not easy to predict what these will be in, say, 20 years time. For the CD player and DVD player, the optical wavelength used to read the pattern in the disk should in principle be as short as possible, since the resolution-determined hole-packing density (i.e., the information density) increases at least as strongly as the inverse of the wavelength squared. This requirement has led to the use of gallium nitride-based diode lasers that emit at wavelengths in the blue-violet region of the visible spectrum. Laser diodes with even shorter emission wavelengths, some way into the UV, are a credible future possibility.

At the other end of the spectrum from the UV and visible regions, penetrative imaging with electromagnetic waves at long enough wavelengths to be safe for moderate exposure of human beings, that is, THz imaging, since THz radiation is nonionizing, is clearly of emerging importance. But there is a need for sources of THz frequency electromagnetic waves that have much greater efficiency than has so far been possible, as well as other desirable properties such as rapid switching and modulation, compactness, and room temperature operation. This requirement indicates the desirability of a suitably scaled semiconductor-based source for the electromagnetic energy required, but so far this efficient semiconductor source for THz electromagnetic waves has largely been conspicuous by its absence.

For Lidar and optical radar applications in the mid- and near IR, semiconductor lasers, because of their characteristically high efficiency, are of potentially great importance in situations where the output light beam is required to propagate over large distances in "free-space," in gaseous atmospheres (in particular, the earth's atmosphere), and fluid environments such as the sea. Probing the atmosphere, for example, for gaseous or particle pollutants, using optical radar techniques and sources of coherent short light pulses provides a challenging domain of potential applications for the semiconductor laser. Issues of beam-quality and directional control and the special optical systems that are likely to be required become important. For some of these applications, QCLs that offer wide spectral coverage through much of the IR spectrum have been researched quite intensively. QCLs are still heterostructure semiconductor lasers but have a dramatically different basic epitaxial layer structure, as well as distinctly different physical principles of operation because they are based on unipolar transitions, instead of electron–hole recombination. Polarization of the emission is naturally TM (transverse magnetic), that is, with the predominant magnetic field component of the emitted light parallel to the defining device plane while it is mostly TE (transverse electric) in conventional diode lasers. The range of possible wavelengths obtainable with the QCL extends from the mid-IR, through the far-IR and down to THz frequencies, although the coherent emitted radiation that justifies the epithet "laser" becomes progressively more difficult to generate, as the wavelength increases. The range of emission wavelengths possible with the QCLs overlaps, at near-IR wavelengths, with what is obtainable from conventional diode heterostructure lasers that have a substantial fraction of antimony in the composition of the light emitting III–V semiconductor region.

We return to the important, indeed basic, question: "What is to be gained from the pursuit of compactness in the semiconductor laser?" – a pursuit that is at the center of the research that is analyzed in this compact book. A partial answer to the question of "why compactness?" or "what is compactness for?" comes from considerations that also apply for the classic and central device of modern electronics, the transistor. Transistors are, in their standard format, three-terminal devices that are capable of both *switching* and *amplification* and, with feedback, also of *oscillation*. With appropriate organization, semiconductor lasers are capable of providing the same three functions, and it should also be born in mind that semiconductor lasers are devices that involve both electronic and photonic functionality. For example, when used as an amplifier in the "semiconductor optical amplifier (SOA)," the semiconductor laser takes an optical input and produces a higher power optical output that replicates the input but the amount of amplification of the light depends on electrical control of the gain of the optical amplifier.

In modern integrated electronics, the transistor is firstly a high-speed electronic switch. In a transistor, a controlling data stream goes into the transistor and is processed onto or into another data stream. The maximum speed (i.e., the rate) at which the switching operations involved can be carried out is restricted by the device size, from "large" down to very small sizes. As is well known, the silicon integrated circuit that is at the heart of modern electronics typically has several million interconnected transistors on it, as well as other components such as resistors and capacitors.

Semiconductor lasers can be organized to behave like their electronic counterpart just mentioned, the transistor. In a standard configuration, a heterostructure diode laser has its light output level modulated, in the simplest case, in amplitude by varying the level of the injection current that drives the laser into oscillation. Reducing the dimensions of the laser reduces the lasing threshold and the power consumption needed to modulate the optical output level, just as the reduction of the transistor dimensions implies a smaller energy dissipation per bit in an electronic circuit. For both devices, local heating, crosstalk effects, and low on-state/off-state contrast must be palliated to maintain high performance as the device size becomes progressively smaller, in order to produce a higher integration density. The use of semiconductor lasers to realize logical functionality and their possible integration into multi-stage logic circuits naturally push the semiconductor laser toward compactness. Considerations of speed of operation together with propagation delay, switching energy, and power consumption also provide the pressures that dictate compactness.

The chapters of this book, in order are

Chapter 1: Nano-scale metallo-dielectric coherent light sources.
Chapter 2: Optically pumped semiconductor photonic crystal lasers.
Chapter 3: Electrically pumped photonic crystal lasers: laser diodes and quantum cascade lasers.
Chapter 4: Photonic crystal VCSELs.
Chapter 5: III–V compact lasers integrated onto silicon (SOI).
Chapter 6: Semiconductor microring lasers.
Chapter 7: Nonlinearity in semiconductor microring lasers.

Although we do not claim that the organization of the present contribution to the literature on semiconductor lasers is totally systematic – and even less do we claim that the book is encyclopedic – we believe that what the book contains is organized appropriately, with a logical evolution of topics, and that it provides a good sample of the subtopics that constitute the whole field. Examples of subtopics that might have been included are the pillar geometry VCSEL-like laser and the promising vertical nanowire.

The book begins, in Chapter 1, with the topic of the metal enclosed nanoscale semiconductor laser. Detailed and rigorous electromagnetic analysis, together with creative design, leads to coherent light sources that can be substantially smaller than a free-space wavelength in all three space dimensions, while exploiting essentially the same III–V semiconductor based heterostructure gain medium as that in the conventional "macrolaser." The combination of the intrinsically high gain available with III–V semiconductor structures and careful minimization of the potentially overwhelming propagation losses that occur for metals at optical frequencies produces the possibility of efficient coherent light sources that could viably be packed in million-scale numbers on a single, modest-area, wafer section.

Chapter 2 shares with the other chapters the basic aspect that the "natural" gain medium for coherent semiconductor light sources is invariably an epitaxial III–V semiconductor heterostructure. As in Chapter 1, the work described is primarily reliant on optical pumping processes for the gain medium. But the device structure is the radically different 2D photonic crystal (PhC) patterned membrane that provides strong confinement in all three space dimensions. The use of optical pumping has allowed a wide-ranging basic (i.e., fundamental) exploration of the characteristics and behavior of very compact lasers that exploit PhC principles for the generation of the feedback required for an optical oscillator.

Chapter 3 shares with Chapter 2 the fact that PhC structures are at the heart of the laser devices that have been investigated and that are described in detail in this chapter. The vital difference between the content of Chapter 3 and that of Chapter 2 is that the optical gain medium is pumped by means of electric charge-carrier injection. The research described in Chapter 3 is important because it directly addresses the technological issues that must be "solved" if the compact PhC-structured semiconductor laser is to be useful in a wide variety of situations – situations where the intrinsic efficiency of electrical pumping and the direct control that it provides are of paramount importance. Both PhC laser diodes and PhC QCLs are considered in this chapter, thereby covering a wide range of emission wavelengths from near infrared to THz waves. It is incidentally shown that the use of microwave- or electronics-inspired resonator structures provides the opportunity to design THz semiconductor lasers that can be much smaller than the emitted wavelength.

Chapter 4 continues with the PhC structuring theme, but with the clear alternative configuration of "vertical" emission that is mediated by the PhC structure. The title of the chapter immediately identifies the compact lasers involved as being a form of VCSEL but one in which the resonant cavity and laser performance stem from the use of a PhC structure on the lateral surface of the laser. The compact PhC VCSEL structure provides control of the single-mode power and polarization, while simultaneously optimizing the output power and coupling through the top surface. Using detailed modeling of the 3D structure, it is also shown that PhC-VCSELs with true photonic bandgap characteristics are feasible.

Chapter 5 departs radically from the previous chapters because it is centrally concerned with a hybrid situation in which the gain medium remains a III–V semiconductor epitaxial heterostructure, but the supporting substrate is a silicon-on-insulator (SOI) wafer section, to which the thinned-down III–V semiconductor laser structure is bonded and the III–V semiconductor substrate on which it has been grown is (almost) completely removed. This chapter suggests the strong plausibility of the transition from compact semiconductor lasers in the research lab to lasers in the real world of optical interconnect applications. The characteristic compact laser geometry of the work in Chapter 5 is the microdisk, which supports the so-called whispering gallery mode (WGM). The crucial – and so far not fully solved – challenge for this hybrid configuration is the need to drive the laser by electrical current injection.

Chapters 6 and 7 are concerned with ring-geometry semiconductor lasers and therefore share, in the simplest limiting case, the same circular geometry as that of the microdisk described in Chapter 5. Furthermore, there is considerable similarity between the mode structure of a disk and a ring with similar diameter and the ring can meaningfully be considered as a limiting case of a disk with the inner wall serving to limit the number of transverse (in this case meaning the radial direction) modes. By demonstrating intrinsically credible electrical pumping in such microring lasers and gaining an in-depth understanding of the photon–charge-carrier interaction dynamics in such structures, Chapters 6 and 7 also demonstrate – both from the experimental and the theoretical points of view – that compact semiconductor lasers could enter the real world of laser applications.

All three chapters – Chapters 5–7 – are substantially concerned with the bi-stable operating characteristics that make such lasers of interest as optical switches with a latching capability. If the electrical pumping and heat-dissipation issues associated with the hybrid configuration of disk lasers mounted on planar silicon waveguides can be addressed satisfactorily, it might eventually be that the compact hybrid semiconductor laser will become the preferred light source and optical switching device in high-density photonic integration based on silicon. Hybrid III–V/SOI integration might therefore win-out over III–V semiconductor monolithic integration.

As already mentioned, this book is by no means encyclopedic. We have not explicitly addressed the micropillar cavity geometry that has received considerable levels of attention for more than a decade. Only part of the research carried out on micropillar cavities was actually concerned with lasers as opposed, for example, to cavity quantum dynamics investigations. Issues such as the need to suppress nonradiative carrier recombination at dry-etch process-exposed surfaces that intersect with the quantum-well active region are arguably of particular importance in this case, as well as in the cases of the microring and microdisk lasers. The situation might well be different for vertical *nanowires* grown by the vapor–liquid–solid (VLS) method, because this approach generates structures with a much lower number of surface defects. Recent results [1] obtained using optical pumping on these laser structures, – including, for instance, spontaneous emission factors, β, close to unity over very wide bandwidths – have considerable promise for future developments of compact semiconductor lasers. Lasers based on vertical nanowires and micropillars could well be an appropriate additional topic in a future edition of this book.

Reference

1. Claudon, J., Bleuse, J., Malik, N.S., Bazin, M., Jaffrennou, P., Gregersen, N., Sauvan, C., Lalanne, P., and Gerard, J.-M. (2010) A highly efficient single-photon source based on a quantum dot in a photonic nanowire. *Nat. Photonics*, **4**, 174–177.

List of Contributors

Alexandre Bazin
Laboratoire de Photonique et de
Nanostructures
CNRS UPR20
Marcoussis
France

Olesya Bondarenko
University of California, San Diego
Department of ECE
9500 Gilman Drive
La Jolla
CA 92093
USA

Xinlun Cai
University of Bristol
Department of Electronic and
Electrical Engineering
Bristol
BS8 1UB
UK

Xavier Checoury
Université Paris Sud
Institut d'Electronique
Fondamentale
CNRS, UMR 8622
91405 Orsay
France

Yujie Chen
Sun Yat-sen University
State Key Laboratory of
Optoelectronic Materials and
technologies
School of Physics and
Engineering
Guangzhou
China

Raffaele Colombelli
Université Paris Sud
Institut d'Electronique
Fondamentale
CNRS, UMR 8622
91405 Orsay
France

Tomasz Czyszanowski
Technical University of Lodz
Institute of Physics, ul.
Wolczanska 219,93-005
Lodz
Poland

Maciej Dems
Technical University of Lodz
Institute of Physics, ul.
Wolczanska 219,93-005
Lodz
Poland

Yeshaiahu Fainman
University of California, San Diego
Department of ECE
9500 Gilman Drive
La Jolla
CA 92093
USA

Qing Gu
University of California, San Diego
Department of ECE
9500 Gilman Drive
La Jolla, CA 92093
USA

Michael Kats
University of California, San Diego
Department of ECE
9500 Gilman Drive
La Jolla
CA 92093
USA

Jin-Hyoung Lee
University of California, San Diego
Department of ECE
9500 Gilman Drive
La Jolla
CA 92093
USA

Vitaliy Lomakin
University of California, San Diego
Department of ECE
9500 Gilman Drive
La Jolla, CA 92093
USA

Jean-Michel Lourtioz
Directeur de Recherche CNRS
Vice President of Université
Paris-Sud
Mission Campus
Bâtiment 209E
91405 Orsay Cedex
France

Gábor Mezosi
Infineon Technologies Austria AG
High Voltage MOS Technology
Development
IFAT PMM DPC HVM TD

Amit Mizrahi
University of California, San Diego
Department of ECE
9500 Gilman Drive
La Jolla, CA 92093
USA

Geert Morthier
Ghent University –imec
Department of Information
Technology
Photonics Research Group
Belgium

Maziar P. Nezhad
University of California, San Diego
Department of ECE
9500 Gilman Drive
La Jolla
CA 92093
USA

Krassimir Panajotov
Vrije Universiteit Brussels
Department of Applied Physics
and Photonics
Pleinlaan 1050
Brussels
Belgium

and

Institute of Solid State Physics
72 Tzarigradsko Chaussee blvd
1784 Sofia
Bulgaria

Fabrice Raineri
Laboratoire de Photonique et de
Nanostructures
CNRS UPR20
Marcoussis
France

and

Université Paris Diderot
Physics Department
Sorbonne Paris Cité
75207 Paris Cedex 13
France

Rama Raj
Laboratoire de Photonique et de
Nanostructures
CNRS UPR20
Marcoussis
France

Gunther Roelkens
Ghent University –imec
Department of Information
Technology
Photonics Research Group
Belgium

Aleksandar Simic
University of California, San Diego
Department of ECE
9500 Gilman Drive
La Jolla
CA 92093
USA

Boris Slutsky
University of California, San Diego
Department of ECE
9500 Gilman Drive, La Jolla
CA 92093
USA

Marc Sorel
Optoelectronics Research Group
School of Engineering
University of Glasgow
Rankine Building
Oakfield Avenue
Glasgow G12 8LT
Scotland, U.K.

Dries Van Thourhout
Ghent University –imec
Department of Information
Technology
Photonics Research Group
Belgium

Siyuan Yu
University of Bristol
Department of Electronic and
Electrical Engineering
Bristol
BS8 1UB
UK

and

Sun Yat-sen University
State Key Laboratory of
Optoelectronic Materials and
technologies
School of Physics and
Engineering
Guangzhou
China

Yanfeng Zhang
Sun Yat-sen University
State Key Laboratory of
Optoelectronic Materials and
technologies
School of Physics and
Engineering
Guangzhou
China

Color Plates

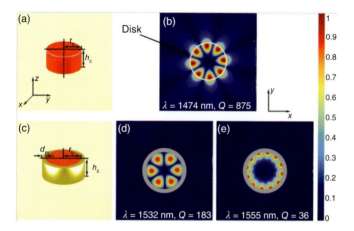

Figure 1.1 The $M = 4$ whispering gallery resonance for a thick semiconductor disk (a) is shown in (b) ($r_c = 460$ nm, $h_c = 480$ nm, and $n_{semi} = 3.4$). Note the spatial spread of the mode compared to the actual disk size. (c) The same disk encased in an optically thick ($d_m = 100$ nm) gold shield will have well-confined reflective (d) and plasmonic (e) modes but with much higher mode losses. $|E|$ is shown in all cases and the section plane is horizontal and through the middle of the cylinder. (From [7].) (This figure also appears on page 2.)

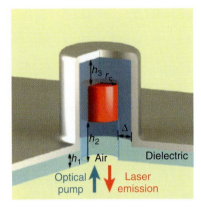

Figure 1.6 Schematic view of a practical realization of the laser cavity, compatible with planar fabrication techniques. The air gap at the bottom of the laser is formed after selective etch removal of the InP substrate. In the designed cavity the values for h_1, h_2, and h_3 are 200, 550, and 250 nm, respectively. (From [7].) (This figure also appears on page 9.)

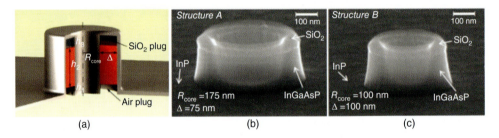

Figure 1.15 Nanoscale coaxial laser cavity. (a) Schematic of a coaxial laser cavity. (b,c) SEM images of the constituent rings in *Structure A* and *Structure B*, respectively. The side view of the rings comprising the coaxial structures is seen. The rings consist of SiO_2 on top, and quantum wells gain region underneath. (From [18].) (This figure also appears on page 21.)

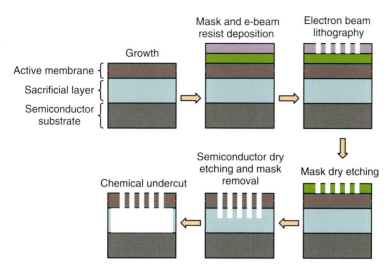

Figure 2.8 Process flow used in the fabrication of a 2D PhC made in an air-bridged membrane of III–V semiconductor. (This figure also appears on page 43.)

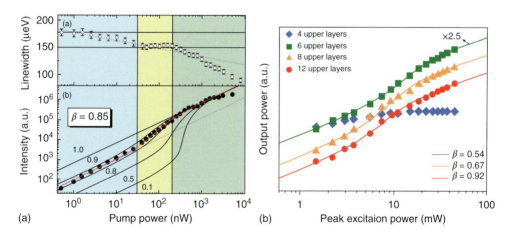

Figure 2.19 (a) Light–light curve measured at 4 K on a 2D PhC cavity patterned in GaAs based suspended membrane embedding one layer of InAs QDs as the active medium. (From [28].) (b) Light–light curve measured at 7 K on a GaAs-based 3D PhC cavity embedding three layers of InAs QDs as the active medium. (From [38].) (This figure also appears on page 57.)

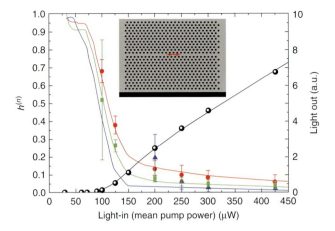

Figure 2.20 Emission statistics at room temperature of a 2D PhC laser made in an InP membrane containing InAsP QDs. Right-hand axis: Experimental (black dots) and calculated (black line) light-in–light-out curve. Left-hand axis: Experimental (red circles, green squares, and blue triangles for $n = 2$, 3, 4, respectively) and calculated (continuous lines of the corresponding colors) values of $h(n)$ as functions of pump power. Inset: SEM image of the sample. (From [78].) (This figure also appears on page 58.)

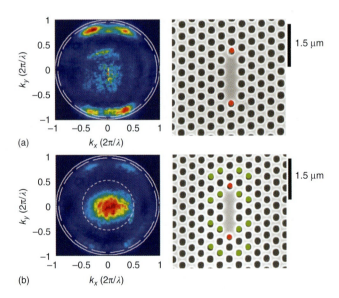

Figure 2.35 Right: SEM images of the unfolded (a) and folded (b) L3 nanocavities. Circles filled in red denote the shifted and shrunk-end holes to boost the Q-factor, those filled in green denote the size modulation at twice the period of the original lattice to achieve the band folding effect. Measured sizes are $a = 437$ nm, $s = 66$ nm, $r_0 = 122$ nm, $r_1 = 97$ nm, and $r_2 = 114$ nm. Left: Measured far-field of the unfolded (a) and folded (b) emission. The white line corresponds to the light line (emission at 90°), while the dashed white line corresponds to N.A. = 0.95, that is, the maximum angle collected in our set-up ($\sim 72°$). The short dashed line corresponds to 30° (N.A. = 0.5). (From [105].) (This figure also appears on page 75.)

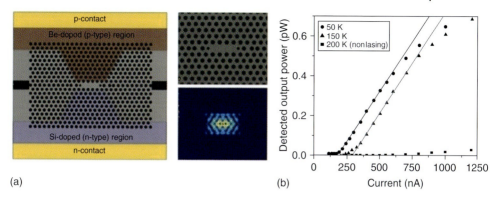

Figure 3.3 (a) Design of the electrically pumped photonic-crystal laser reported in [23]. Left: Schematic of the laser. The p-type doping region is indicated in red, and the n-type region in blue. The width of the intrinsic region is narrow in the cavity region to direct current flow to the active region of the laser. A trench is added to the sides of the cavity to reduce leakage current. Right: modified three-hole defect photonic-crystal cavity design (top) and a FDTD simulation of the E-field of the cavity mode in such a structure (bottom). (b) Experimental output power (detected on the spectrometer) as a function of the current through the laser at 50 K (blue points), 150 K (green points) and 200 K (nonlasing; black points). Red lines are linear fits to the above threshold output power of the lasers, which are used to define the threshold current levels. It is estimated, from the collected output power that the total power radiated above threshold is on the order of a few nanowatts. (This figure also appears on page 95.)

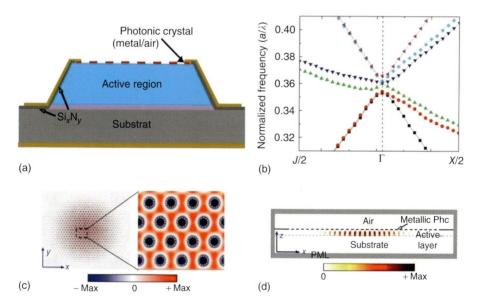

Figure 3.20 (a) Schematic cross section of the QC laser with thin metallic photonic crystal on top [121]. (b) Band diagram around the $a/\lambda = 0.36$ normalized frequency. (c) Electric field distribution (E_z component orthogonal to the semiconductor layers) of the monopole mode with an x/y section taken at the center of the laser active region. (d) Light intensity distribution for the monopole mode in an x/z section across the center, with the simulation domain indicated in gray. (This figure also appears on page 122.)

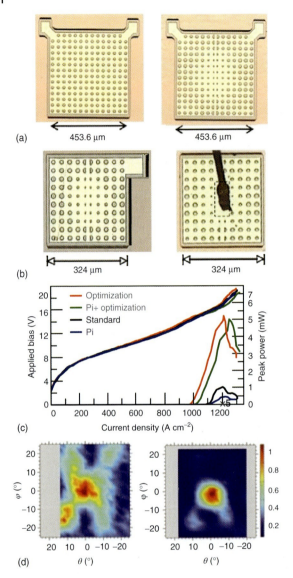

Figure 3.24 Optimized structures of 2D band-edge terahertz QC lasers with small device areas. (a) Microscope images of devices fabricated in [133]. Left: uniform with 8.1 μm hole radius. Right: graded with π-shift and hole radii in the 5.4–10.8 μm range. For both devices, the lattice period is 32.4 μm. Small regions in the two corners of each device are used for wire bonding. (b) Microscope images of devices fabricated in [135]. Left: 10-period graded with π-shift and one bonding pad on the device side. Right: 10 period graded with π-shift and central bonding pad. (c) Voltage–light–current density characteristics of devices fabricated in [133] with and without Q-factor optimization, with and without π-shift ($T = 78$ K). (d) Far-field emission patterns of devices shown in (b). (This figure also appears on page 130.)

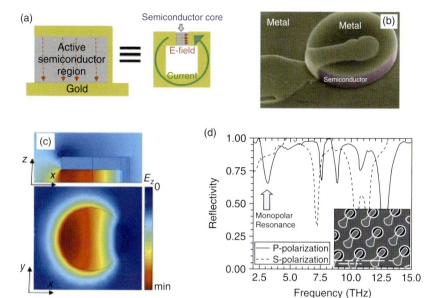

Figure 3.27 (a) Schematic of the split-ring resonator (SRR)-like structure proposed in [13]. (b) Scanning electron microscope image of a typical SRR-like terahertz resonator. (c) Side/top view of the field distribution (electric field component orthogonal to the metallic contacts, E_z, for the monopolar mode of the SRR-like resonator. (d) Typical reflectivity measurement for SRR-like resonators with 6 μm diameter. (The inset shows an SEM image of the resonator array). The solid line corresponds to the polarization which couples to the monopolar resonance, marked by a gray arrow. The dashed line corresponds to the 90°-rotated polarization, which does not couple to the monopolar mode. (This figure also appears on page 134.)

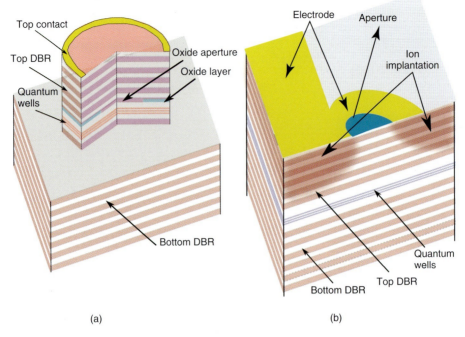

Figure 4.1 Typical vertical-cavity surface-emitting lasers (VCSELs). Transverse confinement of light is provided by either a partial lateral oxidation of an aluminum rich layer starting from the sides of the etched post (a) or ion implantation (b). (This figure also appears on page 150.)

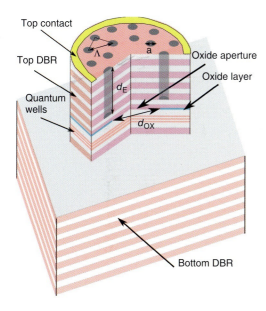

Figure 4.3 Photonic-crystal vertical-cavity surface-emitting laser (PhC-VCSEL). The PhC parameters: lattice pitch Λ, hole diameter a and etching depth d_E and the laser aperture diameter d_{Ox} are denoted. (This figure also appears on page 152.)

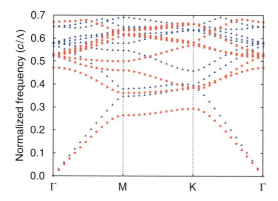

Figure 4.21 Band-diagram for in-plane modes. Red circles indicate TE-like modes and blue triangles TM-like ones. Here and in the rest of the paper we consider $n_H = 3.5$ and $n_L = 1.5$. (After [102].) (This figure also appears on page 182.)

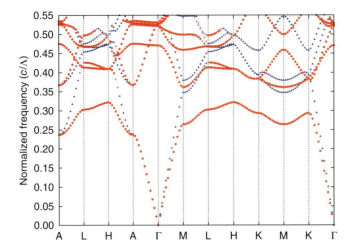

Figure 4.22 Three-dimensional band diagram. Red circles indicate the TE-like modes, blue triangles the TM-like modes and violet squares the modes with parity close to zero. (After [102].) (This figure also appears on page 183.)

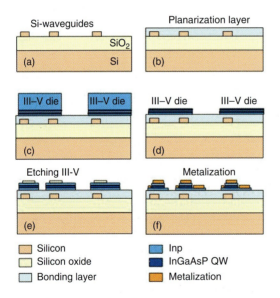

Figure 5.1 Schematics of heterogeneous integration process, with (a) processed SOI wafer or die, (b) planarized SOI (using oxide or BCB), (c) after bonding of unprocessed III–V material, (d) after III–V substrate removal, (e) after lithography and dry etching, and (f) after metalization. (This figure also appears on page 197.)

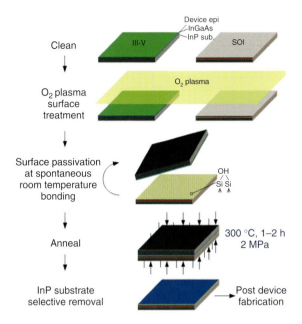

Figure 5.4 O_2-assisted low temperature bonding of III–V to Si. (After [13].) (This figure also appears on page 203.)

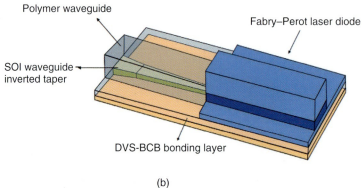

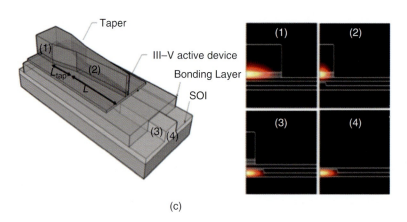

Figure 5.6 Overview of heterogeneously integrated Fabry–Perot type laser structures: (a) evanescent hybrid laser, (b) inverted taper coupler outside the laser cavity, and (c) intracavity spot-size converter. (This figure also appears on page 206.)

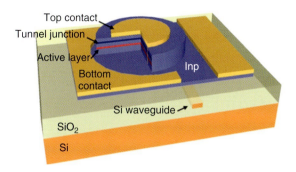

Figure 5.16 Schematic representation of the heterogeneous SOI-integrated microdisk laser. (This figure also appears on page 216.)

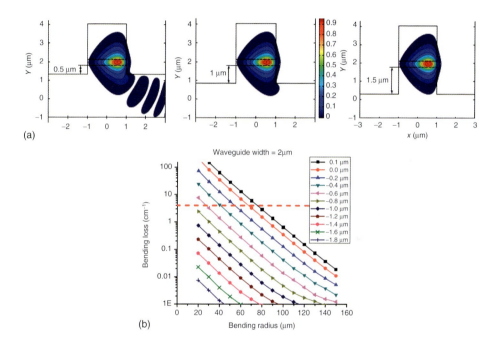

Figure 6.6 (a) Three examples of contour maps of the modal field distribution for a 20 μm radius and 2 μm wide waveguide for relative etch depths from the bottom of the core layer of 0.5, 1.0, and 1.5 μm, respectively. (b) Simulated bending losses of a deeply etched waveguide as a function of the bending radius, with the etch depth as a parameter. The internal propagation loss of a straight waveguide with the same waveguide dimensions is plotted as a reference (i.e., horizontal line at $\sim 4\,cm_{-1}$). (This figure also appears on page 238.)

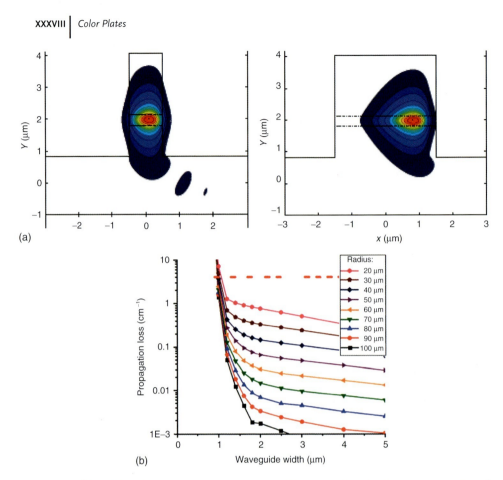

Figure 6.7 (a) The two contour maps illustrate the field distribution of a 20 μm radius waveguide for widths of 1 μm (left) and 3 μm (right) and a relative etch depth from the bottom of the core layer of 1.0 μm. (b) Simulated bending losses as a function of the width of a deeply etched waveguide, with the radius as a parameter. (This figure also appears on page 239.)

Color Plates | XXXIX

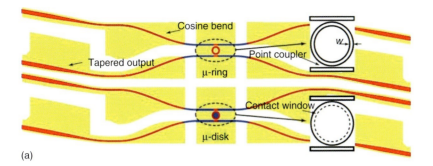

(a)

(b)

Figure 6.11 (a) Schematic of fabricated ring-shaped (top) and disk-shaped (bottom) laser devices. (b) SEM photograph of a 7 μm radius SDL. The inset shows a close up of a 6 μm diameter microdisk. (This figure also appears on page 244.)

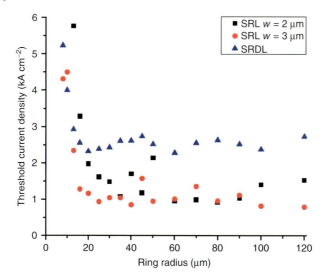

Figure 6.12 Measured threshold current density of micro-SRLs and SDLs – as a function of the device radius. Devices etched with the $Cl_2/Ar/N_2$ chemistry to an etch depth of 3.1 μm show very similar characteristics and are not described in this section. (This figure also appears on page 245.)

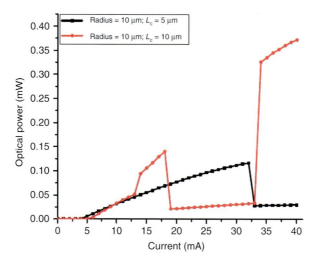

Figure 6.17 L–I characteristics measured on the CW direction for 10 μm radius racetrack SRLs with respective coupling lengths of 5 and 10 μm. The LI characteristics from the CCW direction (not shown in the figure) show an opposite trend that is a typical indication of directional bistability. (This figure also appears on page 250.)

1
Nanoscale Metallo-Dielectric Coherent Light Sources

Maziar P. Nezhad, Aleksandar Simic, Amit Mizrahi, Jin-Hyoung Lee, Michael Kats, Olesya Bondarenko, Qing Gu, Vitaliy Lomakin, Boris Slutsky, and Yeshaiahu Fainman

1.1
Introduction

Compact photonic components are important for the design and fabrication of integrated optical devices and circuits. In the case of light sources, reducing the size can result in improved metrics such as higher packing density and reduced power consumption and also may enhance cavity–emitter interactions such as the Purcell effect. Until recently it was commonly known that the minimum size for a laser is ultimately determined by the free-space wavelength, λ_0. For example, as the size of a conventional Fabry–Perot semiconductor laser is scaled down in all three dimensions toward λ_0, three effects adversely influence the lasing process. Firstly, the roundtrip path of the optical wave in the gain medium is shortened. Secondly, radiative losses from the end mirrors have an increased effect. Thirdly, the lateral field confinement in the resonator waveguide is reduced, resulting in a smaller overlap of the optical mode with the gain medium. All these effects lead to a significant increase in the lasing threshold. As a result lasing cannot be achieved below a certain size limit. By allowing the laser size to increase in one or two dimensions, it is possible to reduce the physical size of the laser in the remaining dimension(s) to values below this limit. For example, the disk thickness in whispering-gallery-mode (WGM) lasers [1] can be reduced to a fraction of the free-space wavelength [2] but, to compensate for the small thickness, the disk diameter must be increased. It should be noted that, in addition to the optical mechanisms noted above, nonradiative surface recombination can have a non-negligible negative effect on the emitter efficiency and thus needs to be accounted for in the design and analysis of such sources. The ultimate challenge in this respect is concurrent reduction of the resonator size in all three dimensions, and, at the same time, satisfying the requirements for lasing action.

The size of an optical cavity can be defined using different metrics, for example, the physical dimensions of the cavity or the size of the optical mode. However, if the goal of size reduction is to increase the integration density (for example, in a

Compact Semiconductor Lasers, First Edition.
Edited by Richard M. De La Rue, Siyuan Yu, and Jean-Michel Lourtioz.
© 2014 Wiley-VCH Verlag GmbH & Co. KGaA. Published 2014 by Wiley-VCH Verlag GmbH & Co. KGaA.

laser array), the effective cavity size should account for both the overall physical dimensions of the resonator and the spread of the optical mode beyond the physical boundary of the resonator. By this token, most conventional dielectric laser cavities are not amenable to dense integration because they have either a large physical footprint or a large effective mode. For example, distributed Bragg resonators [3] and photonic-crystal cavities [4] (both of which can be designed to have very small mode volumes) have physical footprints that are many wavelengths in size, due to the several Bragg layers or lattice periods that are required for maintaining high finesse. On the other hand, it has been demonstrated that the diameter of thick (λ_0/n) micro-disk lasers can be reduced below their free-space emission wavelength [5]; however, the spatial spread of the resultant modes (which have low azimuthal numbers owing to the small disk diameters) into the surrounding space beyond the physical boundaries of the disks may lead to mode coupling and formation of "photonic molecules" in closely spaced disks [6]. For illustration purposes, an $M = 4$ WGM for a semiconductor disk with radius $r_c = 460$ nm and height $h_c = 480$ nm (Figure 1.1a) is shown in Figure 1.1b, clearly indicating the radiative nature of the mode and its spatial spread, which, as mentioned, can lead to mode coupling with nearby structures. (M is the azimuthal order of the resonance, corresponding to half the number of lobes in the modal plot of $|E|$.)

One approach to alleviate these issues is to incorporate metals into the structure of dielectric cavities, because metals can suppress leaky optical modes and effectively isolate them from their neighboring devices. The modes in these metallo-dielectric cavities can be grouped into two main categories: (i) surface bound (that is, surface

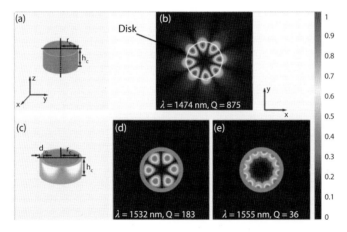

Figure 1.1 The $M = 4$ whispering gallery resonance for a thick semiconductor disk (a) is shown in (b) ($r_c = 460$ nm, $h_c = 480$ nm, and $n_{semi} = 3.4$). Note the spatial spread of the mode compared to the actual disk size. (c) The same disk encased in an optically thick ($d_m = 100$ nm) gold shield will have well-confined reflective (d) and plasmonic (e) modes but with much higher mode losses. $|E|$ is shown in all cases and the section plane is horizontal and through the middle of the cylinder. (From [7].). (Please find a color version of this figure on the color plates.)

plasmon polariton (SPP)) resonant modes and (ii) conventional resonant modes (called *photonic modes*), resulting purely from reflections within the metal cavity. Although they are highly confined, the disadvantage of plasmonic modes is their high loss, which is caused by the relatively large mode overlap of the optical field with the metal (compared to the reflective case). Owing to the high Joule loss at telecommunication and visible wavelengths, the lasing gain threshold for such cavities can be very large. On the other hand, the negative permittivity of metals not only allows them to support SPP modes, but also enables them to act as efficient mirrors. This leads to the second class of metallo-dielectric cavity modes, which can be viewed as lossy versions of the modes in a perfectly conducting metal cavity. Because the mode volume overlap with the metal is usually smaller than in the plasmonic case, in a cavity supporting this type of mode it is possible to achieve higher resonance quality-factors (Q-factors) and lower lasing gain thresholds, albeit at the expense of reduced mode confinement (compared to plasmonic modes). In general, both types of modes can exist in a metal cavity. Embedding the gain disk mentioned earlier in a gold shield (Figure 1.1c) effectively confines the resonant modes while increasing Joule losses. As discussed, the surface bound plasmonic mode (Figure 1.1e) has both a higher M number and higher losses ($M = 6$, $Q = 36$) compared to the non-plasmonic mode (Figure 1.1d, $M = 3$, $Q = 183$). It should be noted that even though the metal shield is the source of Joule loss, the large refractive index of the semiconductor core ($n_{semi} \approx 3.4$) aggravates the problem and increases both the plasmonic and Fresnel reflection losses. For SPP propagation on a (planar) semiconductor–metal interface, the threshold gain for lossless propagation is proportional to n_{semi}^3 [8]. This means that, even though plasmonic modes with relatively high Q can exist inside metal cavities with low-index cores (for example, silica, for which $n = 1.48$), using this approach to create a purely plasmonic, room-temperature semiconductor laser at telecommunication wavelengths becomes challenging, due to the order of magnitude increase in gain threshold. However, plasmonic modes also have an advantage in co-localizing the emitters with the resonant mode volume, thereby leading to a more efficient emission into the lasing mode. This mode of operation is discussed further below, but at this point we focus on novel composite metal-dielectric resonators and the resonant modes that they support.

One possible solution for overcoming the obstacle of metal loss is to reduce the temperature of operation, which will have two coinciding benefits: a reduction of the Joule losses in the metal and an increase in the amount of achievable semiconductor gain. Hill and colleagues [9] have demonstrated cryogenic lasing from gold-coated semiconductor cores with diameters as small as 210 nm. However, in this case the metal is directly deposited on the semiconductor core (with a 10-nm SiN electrical insulation layer between). As a result, owing to the large overlap of the mode with the metal, the estimated room-temperature cavity Q is quite low. The best case is ~180 for a silver coating (assuming the best reported value for the permittivity of silver [10]) which corresponds to an overall gain threshold of ~1700 cm^{-1} and is quite challenging to achieve at room temperature. Even though this device lases when cooled to cryogenic temperatures, it would be challenging to achieve room-temperature lasing with the same approach and a similar sized cavity, owing to the constraints imposed by the amount of available semiconductor gain and the metal losses. The gain coefficient for optically pumped bulk InGaAsP emitting at 1.55 μm is reported to be ∼200 cm^{-1} [11]. Electrically pumped multiple quantum wells (MQWs), on the other hand, have been reported to have higher material gain coefficients of over 1000 cm^{-1} [12]. Furthermore, recent results obtained from Fabry–Perot type metallic nanolasers at room temperature indicate that this level of gain is also achievable in bulk InGaAs [13]. However, even if the required gain is achievable at room temperature, efficient operation of the device would still be a challenge because of thermal heating and nonradiative recombination processes (for example, Auger recombination). In particular, to operate a densely packed array of such devices, thermal management would be a major concern, given the requisite intense pumping levels. Consequently, it is extremely important to optimize the resonator design so that the gain threshold is minimized. In this chapter we introduce novel, composite metal-dielectric, three-dimensional resonators, and lasers that are smaller than the wavelength in all three space dimensions (3D), can operate at room temperatures, and can even operate without a threshold [7, 14–18].

1.2
Composite Metallo-Dielectric-Gain Resonators

As indicated in the previous section, the drawback of using metals in optical resonators is their high dissipative loss. In this section, we show that the losses in metal-coated gain waveguides and in 3D laser resonators, can be significantly reduced by introducing a low-index "shield" layer between the gain medium and the metal [7, 14, 17].

Consider a composite gain waveguide (CGW) having a gain medium cylindrical core, a shield layer, and a metallic coating, as shown in Figure 1.2a [14]. For a given CGW cross-section size, the shield layer thickness is then tuned to maximize the confinement of the electric field in the gain medium and reduce the field penetration into the metal. By doing that, we increase the ability of the device to compensate for the dissipated power with power generated in the gain medium. A direct measure of that ability is the threshold gain, that is, the gain required for lossless propagation [8] in the CGW. The field attenuation in the shield layer resembles that of Bragg fibers [19]. The layer adjacent to the core, in particular, is of high importance [20] and has also been used, for example, to reduce losses in infrared hollow metallic waveguides [21].

Subsequently, we use the CGW model for the design of subwavelength 3D resonators. To confine the light in the longitudinal direction, the CGW is terminated from both sides by a low-index "plug" region covered with metal, which forms the closed cylindrical structure shown in Figure 1.2b.

A more practical nanolaser configuration from a fabrication point of view is the open structure with a SiO$_2$ substrate shown in Figure 1.2c. The inherent radiation

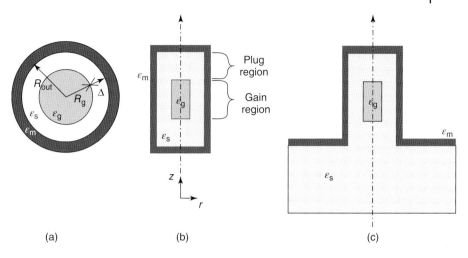

Figure 1.2 (a) Cross section of the metal-coated composite gain waveguide. (b) Cylindrical closed 3D resonator. (c) Cylindrical open 3D resonator. R_g is the radius of the gain core and R_{out} is the overall radius of the composite gain/dielectric core. Δ is the thickness of the dielectric shield. ϵ_g, ϵ_s, and ϵ_m are the relative permittivities of the gain, shield, and metal layers, respectively. (From [14].)

losses into the substrate provide means for collecting the laser light, in contrast to the closed structure, where extracting light requires modification of the metal coating, such as making an aperture in it. The *threshold gain* for the 3D resonators, defined as the gain required to compensate for the metal losses in the closed structure or to compensate for both the metal and radiation losses in the open structure, is shown in the following sections to be sufficiently low to allow laser action at room temperature.

1.2.1
Composite Gain-Dielectric-Metal Waveguides

We first consider the infinite CGW of Figure 1.2a, with relative permittivities $\varepsilon_g = \varepsilon'_g + j\varepsilon''_g$, ε_s, and $\varepsilon_m = \varepsilon'_m - j\varepsilon''_m$ of the gain medium, the shield layer, and the metal, respectively. Assuming a time dependence of $\exp(j\omega t)$, we have $\varepsilon''_g > 0$, $\varepsilon''_m > 0$. The radius of the gain medium is R_g, the shield layer thickness is $\Delta = R_{out} - R_g$, and the metallic coating layer begins at radius R_{out}. The eigenmodes of the CGW may be derived from the general solution of the longitudinal fields in each layer having the form:

$$U = [A J_m(k_r r) + B Y_m(k_r r)] f(m\varphi) e^{-j\beta z} \qquad (1.1)$$

where $U = E_z$ or H_z; J_m and Y_m are Bessel functions of the first and second kind, respectively; $k_r = \sqrt{k_0^2 \varepsilon - \beta^2}$, $k_0 = \omega/c$; ε is the relative permittivity of the layer; and $f(m\varphi)$ may be expanded by $\exp(\pm jm\varphi)$, where the integer m is the azimuthal index. The dispersion relation is found using the transfer matrix method [19]. For the threshold gain ε''_g, the propagation constant β is real, and the threshold gain

$$\varepsilon''_g = \varepsilon''_m \frac{\iint_{\text{Metal}} dA |E|^2}{\iint_{\text{Gain}} dA |E|^2} \quad (1.2)$$

where the integration in the numerator and denominator is over the cross section of the metal and each propagation mode may be found by imposing $\text{Im}\{\beta\} = 0$ in the dispersion relation and then finding the solutions in the plane $(\text{Re}\{\beta\}, \varepsilon''_g)$, similarly to [22].

The effect of the shield layer on the TE_{01} mode threshold gain is demonstrated in Figure 1.3, where ε''_g is plotted as a function of the shield thickness Δ for a given radius $R_{\text{out}} = 300$ nm (Figure 1.3a) and $R_{\text{out}} = 460$ nm (Figure 1.3b). In the simulations we assume a wavelength of $\lambda = 1550$ nm, $\varepsilon'_g = 12.5$ corresponding to an InGaAsP gain medium, $\varepsilon'_s = 2.1$ for a SiO_2 shield layer, and $\varepsilon_m = -95.9 - j11.0$ for a gold coating [23]. The rapid field decay in the gold layer permits us to assume that the metal extends to infinity, whereas in reality a coating layer of 100 nm would suffice. As the shield thickness increases, a lower percentage of the field penetrates into the metal, reducing the losses. On the other hand, the gain material occupies less of the CGW volume, which means that a higher gain is required to compensate for the dissipation losses in the metal. The trade-off between these two processes results in an optimal point at which the threshold gain is minimal. This typical behavior of low-order modes is seen in Figure 1.3 for the TE_{01} mode. For $R_{\text{out}} = 300$ nm, the improvement of the threshold gain from the $\Delta = 0$ (no shield layer) case is by a factor of 1.7, while for $R_{\text{out}} = 460$ nm, the improvement is by a factor of 6.1.

For larger radii, a lower threshold gain may be achieved, as shown in Figure 1.3 and further emphasized in Figure 1.4, where the minimal threshold gain ε''_g is depicted as a function of R_{out} for four low-order modes: TM_{01}, TE_{01}, HE_{11}, and HE_{21}. Having the highest confinement around the gain-medium core, the HE_{11}

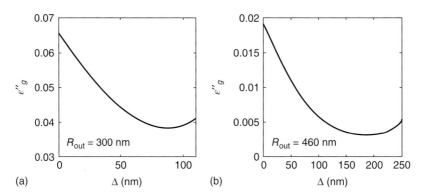

Figure 1.3 Threshold gain ε''_g as a function of the shield thickness Δ for the TE_{01} mode. (a) $R_{\text{out}} = 300$ nm. (b) $R_{\text{out}} = 460$ nm. (From [14].)

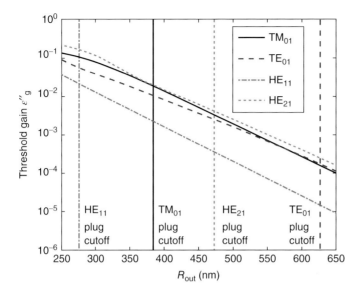

Figure 1.4 Minimum threshold gain as a function of R_{out}. The vertical lines show the cut-off of each mode in the 3D resonator plug region. (From [14].)

mode has the lowest threshold gain among the four modes. Generally, for small radii, the shield layer is less effective, since it quickly drives the mode below the cut-off. For large radii, the threshold gain is low, as the field penetration into the metal is small. The optimal shield layer thickness increases monotonically as a function of R_{out}. For the TE_{01} mode, it ranges between 80 and 330 nm, for a corresponding range of $R_{out} = 250$–650 nm.

1.2.2
Composite Gain-Dielectric-Metal 3D Resonators

The role of the metal coating, which is important in the infinite CGW model, becomes even more crucial for creating a 3D resonator. As explained above, the CGW facets are terminated by plug regions, which are short metallic waveguides filled with SiO_2 as seen in Figure 1.2b,c, in an approach similar to that of Hill *et al.* [9]. The plug ensures strong confinement of the field in the gain region, provided that the mode residing in it is below the cut-off, that is, decaying exponentially in the z direction.

For the plug region waveguide, the cut-off is not clearly defined since the modes are significantly different from those of the perfectly conducting cylindrical waveguide [24]. A reasonable definition for the cut-off situation is a waveguide with the radius R_{out} supporting a mode whose β is the closest to the origin on the complex β plane. That cut-off is shown for each one of the modes of Figure 1.4 by the vertical lines, providing a qualitative tool for choosing an operation mode for the entire 3D structure, as the chosen radius needs to be to the left of the vertical line corresponding to the operation mode. The smaller the device radius compared with the cut-off radius, the stronger the decay in the plug; consequently, the threshold gain is lower. While the HE_{11} mode achieves the lowest threshold gain for a given R_{out}, its cut-off in the plug region is at a small radius; working below this cut-off entails a relatively high threshold gain. It is therefore seen that the TE_{01} mode, which has the highest cut-off of the shown modes, is favorable. The result shows that modes corresponding to a larger R_{out} will have a significantly lower threshold gain. Another advantage of the TE_{01} mode is that in the gain region it couples only to symmetric TE modes in the plug region, whereas $m > 0$ modes are hybrid and may couple to all modes with the same azimuthal index.

Using the CGW model at the optimal point of Figure 1.3b as a starting point, a 3D closed resonator with $R_{out} = 460$ and a 100 nm thick gold coating was designed for the TE_{012} mode. 3D finite-element method (FEM) simulation results for the squared electric field magnitude $|E|^2$ normalized to its maximum value are shown in Figure 1.5. The overall height of the resonator is 1500 nm, and the overall diameter is 1120 nm, making it smaller than the vacuum wavelength in all three dimensions. The resonance was fine-tuned to a wavelength of 1550 nm by setting the gain cylinder height to be about 480 nm and the shield layer thickness to about 200 nm, which is close to the 190 nm predicted by the CGW model. The threshold gain, however, is in less good agreement with the CGW model; the value for the 3D resonator is $\varepsilon''_g \approx 0.011$, which corresponds to about 130 cm^{-1}, whereas the CGW model gives about 36 cm^{-1}. This discrepancy is due to the losses occurring in the plug region and the mode deformation at the interfaces between the plug and gain regions, two effects that are not taken into account in the CGW model. It is evident that the longer the resonator, the more accurately the CGW model describes the

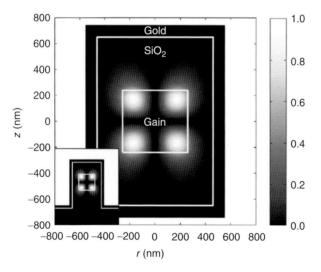

Figure 1.5 Cross section of a closed cylindrical 3D subwavelength laser resonator. The square of the electric field magnitude ($|E|^2$) normalized to its maximal value of the TE_{012} mode is shown. The inset shows a similar open structure.

behavior in the gain region. For instance, a longer resonator with the same radius and designed for the TE_{013} mode has a threshold gain of about 95 cm^{-1}.

If the structure shown in Figure 1.5 is designed with no shield layer in the gain region, but with the same overall radius and height, then the resulting threshold gain is about 420 cm^{-1}. The gain that may be achieved at room temperature by optical pumping of bulk InGaAsP is about 200 cm^{-1} [11]. It is therefore evident that a shield layer that lowers the threshold gain from 420 to 130 cm^{-1} is crucial to enable lasing at room temperature. Slightly modifying the structure for the open configuration, as shown in the inset of Figure 1.5, the field distribution remains nearly unchanged and the threshold gain increases only to about 145 cm^{-1}, owing to the radiation losses. The quality factor of this open resonator without gain is $Q = 1125$, whereas the values for the other 3D structures with a shield layer discussed above are even higher. Finally, we note that for electrical pumping, considerably higher gains may be reached [12] so that the structure, with appropriate changes, is expected to be even further reduced in size. In the following section we use the optimized thickness of a low-index shield layer between the gain medium and the metal coating of a 3D laser resonator for experimental demonstration of nanolasers.

1.3
Experimental Validations of Subwavelength Metallo-Dielectric Lasers for Operation at Room-Temperature

In continuation of the design methodology for 3D subwavelength lasers presented in the previous section, we now present the steps leading to actual implementation and characterization of the devices. The target device is shown in Figure 1.6, in

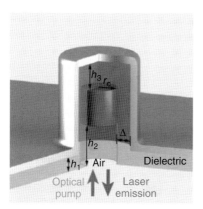

Figure 1.6 Schematic view of a practical realization of the laser cavity, compatible with planar fabrication techniques. The air gap at the bottom of the laser is formed after selective etch removal of the InP substrate. In the designed cavity the values for h_1, h_2, and h_3 are 200, 550, and 250 nm, respectively. (From [7].). (Please find a color version of this figure on the color plates.)

which a gain core is suspended in a bilayer shell of silica and metal. The device is pumped optically through the bottom aperture and the emitted light is also collected from the same aperture.

1.3.1
Fabrication Processes for Subwavelength Metallo-Dielectric Lasers

The metallo-dielectric laser structure was fabricated from an InGaAsP MQW stack grown on InP. Hydrogen silsesquioxane (HSQ) electron-beam resist was patterned into arrays of dots (Figure 1.7a) using a Raith 50 electron-beam writer, and the size of the dots was varied by changing the pattern size and/or the electron-beam dosage. Cylindrical structures were then etched using $CH_4/H_2/Ar$ reactive ion etching (RIE) (Figure 1.7b). Using an optimized and calibrated plasma-enhanced chemical vapor deposition (PECVD) process, the silica shield layer was grown to a thickness of ~200 nm (Figure 1.7c). Note that the outline of the embedded gain core is visible through the silica layer. In practice, the poor adhesion of gold to silica caused separation of the dielectric portion of the structure from the metal layer.

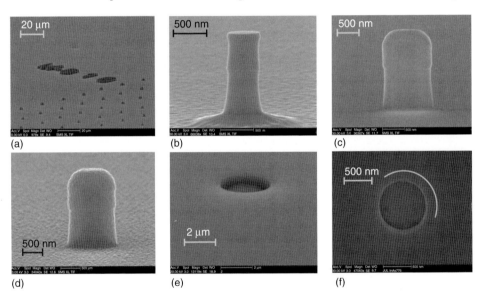

Figure 1.7 Various stages of the fabrication process: (a) Array of electron-beam patterned HSQ resist dots. (b) RIE etched pillar after oxygen plasma and BOE cleaning. The faint bump in the middle indicates the boundary between the InGaAsP and InP layers. (c) Etched pillar after PECVD of silica. The outline of the semiconductor pillar can be seen through the silica layer. (d) Silica covered pillar after undergoing aluminum sputtering (70 nm). (e) Tilted bottom view of one of the samples after selective InP etch with HCL. The surface is comprised of the PECVD deposited silica. The hole corresponds to the air hole shown in the diagram of Figure 1.6. (f) Contrast enhanced normal bottom view of a cavity. The circular outline around the air hole is due to the dielectric shield layer and agrees well with the target dielectric shield layer thickness of 200 nm. (From [7].)

Fortunately, aluminum exhibits better adhesion properties, and at the wavelength of interest its optical properties are very close to those of gold. (The cavity Q of the resonator with an aluminum coating ($\epsilon_m = -95.9 - j11$) [23] is 1004, which compares with 1030 for gold.) A layer of aluminum with a minimum thickness of 70 nm was sputtered over the silica covered pillars (Figure 1.7d). The sample was then bonded on its upper side to a glass slide using SU-8 resist, and the InP substrate was subsequently removed in a selective HCl etch, leaving an air void under the structure. Figure 1.7e shows the tilted bottom view of an air void, with the lower face of the gain core visible inside. Figure 1.7f shows the normal bottom view (with enhanced contrast levels) of a similar void. The faint outline of the silica shield layer is discernible in this image, verifying the 200 nm thickness of the shield.

1.3.2
Characterization and Testing of Subwavelength Metallo-Dielectric Lasers

For optical pumping we used a 1064 nm wavelength pulsed fiber laser operating at a repetition rate of 300 kHz and a pulse width of 12 ns. The pump beam was delivered to the samples using a ×20 or ×50 long working-distance objective, which also collected the emitted light. To estimate the amount of pump power absorbed by the core, a full three-dimensional finite-element analysis was carried out over a range of core sizes. Using a double 4f imaging system in conjunction with a pump optical filter (Semrock Razor Edge long wavelength pass filter), the samples were imaged onto either an IR InGaAs camera (Indigo Alpha NIR) or a monochromator (Spectral Products DK480) with a resolution of 0.35 nm and equipped with a cooled InGaAs detector in a lock-in detection configuration. Owing to the limitations of the electron-beam writing process, the samples were slightly elliptical. The major and minor diameters of the gain core for the particular sample under test were measured to be 490 and 420 nm, respectively. In Figure 1.8a, the light–light curve corresponding to a laser emitting at 1430 nm is shown and exhibits a slope change that indicates the onset of lasing at an external threshold pump intensity of about 700 W mm^{-2}. The same data set is shown in a log–log plot (Figure 1.8a, inset graph), with the slopes of different regions of operation indicated on the plot. The S-shaped curve clearly shows the transition from photoluminescence (PL) to amplified spontaneous emission (ASE) and finally into the lasing regime. Also shown in Figure 1.8a are the emission patterns of the defocused laser image captured with the IR camera, corresponding to continuous wave (CW) (Figure 1.8a-I) and pulsed (Figure 1.8a-II) pumping situations. The average pump intensity in each case was approximately 8 W mm^{-2}.

Only broad PL emission occurs in the CW case, owing to the low peak intensity. However, when the pump is switched to pulsed mode, lasing is achieved due to the 278-fold increase in peak power. At the same time, the defocused image forms a distinct spatial mode with increased fringe contrast, which is an indication of increased spatial coherence and is further evidence of lasing. The polarization of the emission has a strong linear component, which is attributable to the slight ellipticity

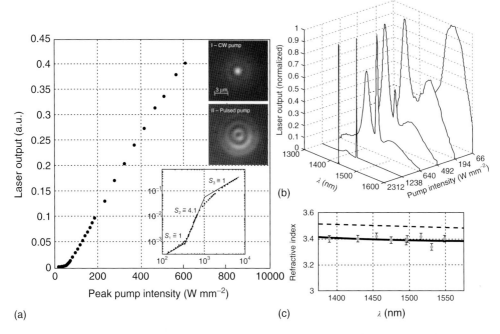

Figure 1.8 (a) Light–light curve for a nanolaser with major and minor core diameters of 490 and 420 nm (dotted curve). The same data set is shown as a log–log plot (dotted inset) together with the slopes for the PL, ASE, and lasing regions. Also shown are the images of the defocused emitted beam cross section (taken at about 10 μm away from the nanolaser exit aperture) for (I) CW pumping and (II) pulsed pumping. The appearance of the higher contrast fringes indicates increased coherence due to lasing. (b) Evolution of the emission spectra from PL to lasing. (c) Effective refractive indices (error bar data points) of the pumped MQW gain medium at lasing wavelengths, back-calculated from lasing spectra obtained from an array of nanolasers. Error bars were calculated assuming ±5 nm error in measuring the disk diameters from the SEMs. The dashed curve shows the effective refractive index of the unpumped MQW layer, as measured by a Filmetrics interferometric analyzer. The solid curve is offset down from the dashed one by a constant amount (0.102 RIU) that was chosen for best fit to the lasing data. The index reduction is consistent with the estimated free carrier effects. (From [7].)

of the gain core. Figure 1.8b shows the evolution of the emission, from a broad PL spectrum to a pair of competing ASE peaks and finally into a narrow lasing line at 1430 nm. The measured linewidth of this particular laser was 0.9 nm; however, linewidths as small as 0.5 nm were measured for other samples in the same size range.

Another way to verify the soundness and accuracy of the design and fabrication processes is to match the lasing wavelength with the target resonance of the cavity. However, owing to the high pump intensity, the refractive index of the gain core can vary substantially from its quiescent value which can shift the lasing line considerably from its target wavelength. Using the measured results from an array of lasers with slightly different sizes (which were measured individually using a scanning electron microscope, SEM) and exact three-dimensional finite-element modeling of each of the gain cores, the refractive index of the gain medium under specific pumping conditions was estimated at each lasing point (Figure 1.8c). Assuming a uniform drop in the refractive index over the spectrum of interest and using a least-squares fit of these data points, the estimated drop is ~0.102 refractive index units (RIUs) (least-squares fit) less than that obtained from interferometric multilayer measurements of the unpatterned wafers under low illumination intensity (Figure 1.8c, dashed line). We believe that this shift is mainly due to free carrier effects (a combination of band filling, bandgap shrinkage, and free carrier absorption), the net effect of which, at the estimated carrier density (about 1.2×10^{19} cm^{-3} for a 520 nm diameter core) is a refractive index drop of about 0.1 RIU. Furthermore a slight additional contribution (at most −0.004 RIU) may also be present due to compressive pressure on the gain cores, which is exerted by the thermal shrinkage of the aluminum layer after deposition in the sputtering chamber.

1.4
Electrically Pumped Subwavelength Metallo-Dielectric Lasers

The metallo-dielectric nanolaser described in the previous section is an optically pumped device. An electrically pumped device is more likely to lead to practical applications. In standard semiconductor fabrication technology, electrical contacts are usually implemented by using highly doped semiconductors. Apart from the increased optical losses that would result from the interaction of the optical mode with these highly doped regions, the vertical confinement of the resonator would also be compromised, since the higher index of the semiconductor would adversely affect the operation of the vertical "plugs." However, in this section we demonstrate that, through a careful design process and judicious selection of device parameters, both of these challenges can be surmounted.

1.4.1
Cavity Design and Modeling of Electrically Pumped Subwavelength Metallo-Dielectric Lasers

The platform for our devices is an InGaAs/InP double heterostructure grown on an InP substrate similar to the structure reported in [9]. The schematic of the laser structure is shown in Figure 1.9a. The intrinsic 300 nm thick (h_{core}) InGaAs bulk layer is the active layer and the upper (470 nm thick) and lower (450 nm thick) InP layers are the cladding layers through which the injected carriers flow into the active layer. Highly doped n-InGaAs on the top and p-InGaAsP in the lower cladding layer form the n- and p-contact layers, respectively. The width of the top and bottom InP cladding layers is intentionally reduced using selective wet etching to form a pedestal structure for enhancing the vertical optical

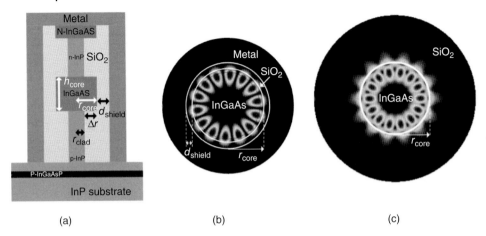

Figure 1.9 A schematic of subwavelength pedestal pillar laser shown in (a) where r_{core} is the radius of InGaAs gain layer, r_{clad} is the radius of InP cladding. Δr is the difference between r_{core} and r_{clad}. d_{shield} is the thickness of SiO$_2$ shield layer. h_{core} is the height of InGaAs gain medium. (b) The horizontal cross section of the electric field magnitude when r_{core} is 750 nm, r_{clad} is 690 nm ($\Delta r = 60$ nm), and d_{shield} is 150 nm with silver coating. (c) The horizontal cross section of the electric field magnitude in the same structure with only low-index dielectric (SiO$_2$) coating. r_{core} in (b,c) are the same value of 750 nm. (From [15].)

confinement. Thin dielectric and metal layers are coated on the pillar structure, thereby forming a metallo-dielectric cavity. As shown in the previous sections, the metallo-dielectric cavity is able to achieve efficient lateral mode confinement at the subwavelength scale due to the high reflectivity of the metal, while reducing the optical ohmic loss by minimizing the mode overlap with the metal by using a thin dielectric shield layer [7, 14]. The dielectric constant of bulk silver at room temperature ($\epsilon_{Ag} = -120.43 - j3.073$ at 1.55 μm) was used to simulate the metal coating [10]. We first designed a wavelength scale cavity in which the gain-core radius (r_{core}) is 750 nm ($2r_{core} \sim \lambda$), the cladding radius (r_{clad}) is 690 nm ($\Delta r = 60$ nm) and the thickness of the SiO$_2$ shield layer (d_{shield}) is 150 nm. The magnitude of the electric field (horizontal cross section) in the gain medium is shown in Figure 1.9b. A WGM with the azimuthal mode number (M) of 7 is supported and the electric field is strongly confined inside the metal cavity (TE mode). The thin active layer ($h_{core} = 300$ nm) allows only the lowest order mode in the vertical direction. The modal overlap with the metal coating is minimized by the dielectric shield, as shown in Figure 1.9b. As the dielectric shield layer becomes thinner, the field penetration into the metal coating increases, which results in a higher loss. In contrast, a thick dielectric shield layer could reduce the field penetration into the metal, resulting in a lower gain. Therefore, as discussed earlier, there should be an optimum thickness of the dielectric shield for a given gain-core radius.

For comparison, we numerically modeled the same pedestal structure coated only by a low-index dielectric ($n_{SiO_2} = 1.45$) and without a metal coating. The electric field magnitude of the resonant mode of this dielectric cavity is shown in Figure 1.9c. It shows significant modal spreading outside the gain medium, which results in low modal overlap with the gain medium, defined as a confinement factor. In our calculation, the confinement factor in the gain medium of the metallo-dielectric cavity (Figure 1.9b) is 0.39 compared to 0.19 in the purely dielectric cavity case (Figure 1.9c). The Q-factor of the metallo-dielectric cavity is 726, which is a factor of enhancement 5 on the Q-factor of the purely dielectric cavity ($Q = 151$). When the gain-core size is decreased below the wavelength scale, the effective confinement factor is significantly degraded, due to the radiation loss, which is something that has been studied for micro-disk resonators [25, 26]. Figure 1.10a shows the resonant mode of a purely dielectric cavity where $r_{core} = 350$ nm and $r_{clad} = 310$ nm ($\Delta r = 40$ nm) with a low-index dielectric (SiO$_2$) coating. This structure supports a WGM with $M = 3$ that exhibits a significant amount of mode spreading outside the gain medium and a low confinement factor in the gain medium (0.38).

However, by incorporation of the metallo-dielectric cavity (150 nm thick SiO$_2$ and Ag coating) as shown in Figure 1.10b, the resonant mode is strongly confined inside the subwavelength-scale cavity and the modal overlap with the gain is also significantly enhanced. In our calculation, the confinement factor in the gain medium for the metallo-dielectric cavity is 0.57. In consequence, for such small size resonators this structure produces 22 times increase in the Q-factor of the metallo-dielectric cavity ($Q = 468$), when compared to the Q-factor of the purely dielectric cavity ($Q = 21$). The metallo-dielectric cavity enables further reduction of the gain-core size while keeping efficient mode confinement in the gain medium. We numerically calculated the resonant mode of the metallo-dielectric cavity with $r_{core} = 220$ nm, $r_{clad} = 160$ nm ($\Delta r = 60$ nm), and $d_{shield} = 150$ nm. As shown in Figure 1.10c, for this case, the resonant mode is an axially symmetric mode (TE$_{011}$)

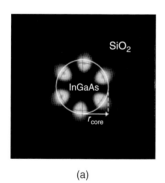 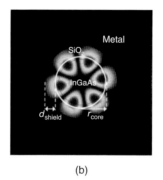 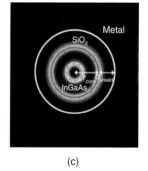

(a) (b) (c)

Figure 1.10 The electric field magnitude at the horizontal cross section of the active layer. (a) Pure dielectric cavity with $r_{core} = 350$ nm, $r_{clad} = 290$ nm ($\Delta r = 60$ nm), and SiO$_2$ coating. (b) Metallo-dielectric cavity with $r_{core} = 350$ nm, $r_{clad} = 290$ nm ($\Delta r = 60$ nm), $d_{shield} = 150$ nm, and Ag coating. The gain structure sizes, r_{core} and r_{clad}, in (a,b) are the same and the resonant mode is WGM with $M = 3$ for both cases. (c) Metallo-dielectric cavity with $r_{core} = 220$ nm, $r_{clad} = 160$ nm, $d_{shield} = 150$ nm, and Ag coating. The resonant mode is axially symmetric TE$_{011}$ mode. (From [15].)

where the electric field circulates around the gain core and is well-confined in the gain medium. The electric field is mostly contained inside the gain medium and has minimal mode overlap with the metal region. The Q-factor of this resonator is calculated to be 707 with a threshold gain of 236 cm^{-1}, which is lower than the bulk gain of InGaAs (400 cm^{-1}) at room temperature, at a wavelength of 1.5 µm [27]. The pedestal geometry is adopted in our metallo-dielectric cavity pillar laser structure to enhance the optical confinement in the vertical direction, while maintaining a conduit for carrier flow of both electrons and holes. To analyze the effect of the pedestal in our laser structure quantitatively, we have calculated the Q-factor and the threshold gain by varying r_{clad}.

The calculated Q-factor and threshold gain for a 750 nm core radius with various r_{clad} values is presented in Figure 1.11a. The shield thickness (d_{shield}) and the metal coating were kept constant. When r_{clad} is the same as r_{core} (cylinder type), the Q-factor is 163, and the threshold gain is 1505 cm^{-1}. As r_{clad} is reduced to 600 nm (pedestal type, $\Delta r = 150$ nm), the Q-factor is enhanced to 1731, which is about an order of magnitude improvement, and the threshold gain is decreased to 99 cm^{-1}, which represents a 93% reduction. As the pedestal undercut is made deeper (Δr is larger), the threshold gain is flattened and the resonant wavelength is shifted out of the optimal gain spectrum which is undesirable. We have also calculated the Q-factor and threshold gain when $r_{core} = 220$ nm as the pedestal size is changed, which is shown in Figure 1.11b. The Q-factor is enhanced from 152 (cylinder type, $\Delta r = 0$ nm) to 1572 (pedestal type, $\Delta r = 150$ nm), which is an order of magnitude improvement. The resonant mode is strongly confined inside the gain layer for the pedestal structure as shown in Figure 1.11d, – where $r_{core} = 220$ nm and $\Delta r = 120$ nm – and may be compared with the cylinder type structure shown in

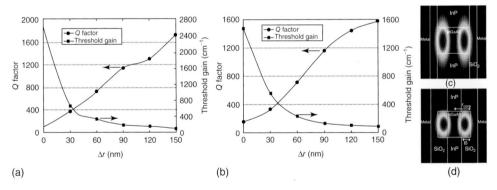

Figure 1.11 Numerical simulation results of the cavity Q-factor and threshold gain for various pedestal sizes. Δr (= $r_{core} - r_{clad}$) is pedestal undercut depth. (a) $r_{core} = 750$ nm, $d_{shield} = 150$ nm, and r_{clad} is varied from 750 to 600 nm ($\Delta r = 0$–150 nm). (b) $r_{core} = 220$ nm, $d_{shield} = 150$ nm, and r_{clad} is varied from 220 to 70 nm ($\Delta r = 0$–150 nm). (c) Vertical cross section of the resonant mode field (TE$_{011}$) magnitude when r_{core}, $r_{clad} = 220$ nm ($\Delta r = 0$ nm, cylinder type), and $d_{shield} = 150$ nm. (d) The resonant mode field (TE$_{011}$) magnitude when $r_{core} = 220$ nm, $r_{clad} = 100$ nm ($\Delta r = 120$ nm, pedestal type), and $d_{shield} = 150$ nm. (From [15].)

Figure 1.11c. The threshold gain is reduced from 1473 to 89 cm^{-1}, which is a 94% reduction. The threshold gain of 89 cm^{-1} is a promising result for the possible room temperature operation of this laser structure. As shown in both cases, the threshold gain is significantly reduced when a minimal pedestal undercut is employed. At $\Delta r = 60$ nm, the threshold gain for 750 and 220 nm r_{core} values are 338 and 236 cm^{-1}, respectively, and are therefore still lower than our target threshold gain of 400 cm^{-1} [27]. This reduction in the threshold gain has other advantages, since heat dissipation and carrier diffusion in the active layer have been identified as critical issues for most pedestal-type micro-disk lasers [28, 29].

1.4.2
Fabrication of Electrically Pumped Subwavelength Metallo-Dielectric Lasers

Wavelength scale (750 nm radius) and subwavelength scale (355 nm radius) circular masks were patterned on an InGaAs/InP heterostructure wafer by means of electron-beam lithography (EBL), using spin-coated HSQ resist. Subsequent dry etching was performed using CH$_4$/H$_2$/Ar gas chemistry to form subwavelength-scale pillar structures (as in the SEM image shown in Figure 1.12a). The selective etching of the cladding InP layers was performed using HCl:H$_3$PO$_4$ (1:3) wet etching and the result is shown in Figure 1.12b. 160 nm of InP was etched away on both sides through the wet etching process while the gain layer was preserved. A 150 nm thick layer of SiO$_2$ was conformally deposited on the pedestal pillar surface by means of a PECVD process (Figure 1.12c), which provides the low-index shield layer required to minimize the mode–metal overlap of the modal field with the metal and also passivates the InGaAs surface.

The SiO$_2$ layer on top of the subwavelength pillar structure was removed through the photoresist planarization process and SiO$_2$ dry etching, in order to access the n-side contact layer (n-InGaAs). Metal contacts (Ti/Pd/Au) were formed on the top

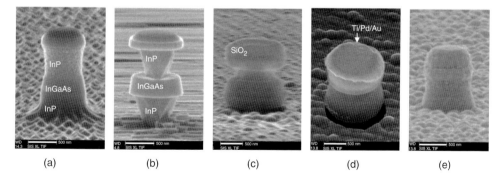

Figure 1.12 SEM images of subwavelength pillar laser structure during fabrication procedure. (a) Subwavelength pillar ($r_{core} = 395$ nm) structure after dry etching. (b) Pedestal pillar is formed by selective InP wet etching. (c) A thin SiO$_2$ layer (140 nm) is deposited on the pillar structure by PECVD. (d) N-contact metal (Ti/Pd/Au) layer deposited on the top of subwavelength pillar. (e) Silver is deposited on whole pillar structure. The scale bar in each image represents 500 nm. (From [15].)

of the pillar structure by electron-beam evaporation and liftoff (Figure 1.12d). After n-contact formation, a 200 nm thick silver layer was deposited so as to completely cover the pillar structures including the top and side wall of the pillar, and the n-type contact pad (Figure 1.12e). A 20 nm chromium (Cr) layer was deposited prior to the silver deposition for better adhesion. Since the high optical loss of Cr could degrade the Q-factor of the cavity and therefore increase the threshold gain, the unintentionally deposited Cr on the side wall of the pillar structure was subsequently removed by Cr wet etching while protecting the adhesion layer on the substrate by masking with photoresist. The p-contact was processed separately using photolithographic patterning and wet etching of the SiO_2 and InP layers, in order to access the underlying highly doped InGaAsP layer. The sample was then annealed to 400 °C for 60 s to reduce the contact resistance. Finally, the sample was mounted on the device package (TO 8) and wire-bonded.

1.4.3
Measurements and Discussion of Electrically Pumped Subwavelength Metallo-Dielectric Lasers

The devices were forward biased and the CW emission from the device was collected through a ×20 objective lens, and then imaged using a CCD camera. The spectral characteristics were analyzed by a monochromator with a maximum spectral resolution of 0.35 nm (with a 100 µm slit opening). The lasing characteristics of electrically pumped pedestal pillar lasers with two different gain-core radii (750 and 355 nm) were measured and analyzed. Figure 1.13a shows a SEM image of the pedestal pillar, in which $r_{core} = 750$ nm, $r_{clad} = 710$ nm, and the pillar height is 1.3 µm. The shield layer thickness (d_{shield}) was 140 nm and silver was coated to form a metal cavity. In the numerical simulation, the Q-factor was estimated to be 458 and the threshold gain was 534 cm^{-1} at the resonant wavelength of 1.50 µm. The lasing characteristics of this device at 77 K are shown in Figure 1.13b. Electroluminescence (EL) around 1.55 µm was observed when the injected current was higher than 20 µA. As the injected current increased, the emission spectrum showed spectral narrowing and the lasing peak appeared at 1.49 µm which is very close to the calculated resonant wavelength of 1.50 µm. The light output–injection current (L–I) curve (Figure 1.13c) shows a kink around the threshold current (50 µA) which is also an indication of the onset of lasing. The linewidth narrowed to 0.9 nm at an injection current level of 300 µA. We also investigated the temperature dependence of the lasing characteristics of this device. A local heater inside the cryostat kept the target temperature constant during the measurement. Lasing behavior was observed at 100, 120, and 140 K at a constant pumping current. The spectral evolution and L–I curve at 140 K are shown in Figure 1.13e. The lasing wavelength remained in the vicinity of 1.49 µm and the linewidth was also less than 1 nm at 140 K (Figure 1.13d). However, the threshold current increased to 240 µA (inset in Figure 1.13e) which is five times higher than the threshold current at 77 K. At 160 K, spectral narrowing was still observed at a wavelength of 1.49 µm, but the

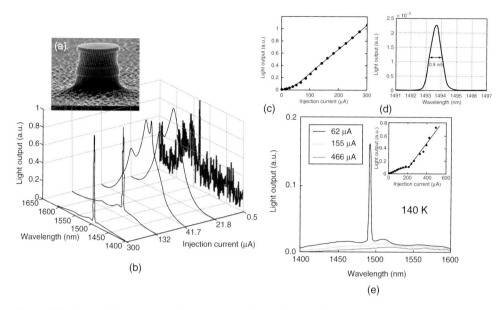

Figure 1.13 Lasing characteristics of 750 nm r_{core} pedestal pillar laser device. (a) An SEM image of the 750 nm r_{core} pedestal pillar structure. (b) Spectral evolution graphs with increasing injection current. (c) L–I curve of this device. (d) Linewidth measurement by a monochromator with 0.35 nm resolution. (e) Lasing spectrum measured at 140 K. The inset shows L–I curve at 140 K. (From [15].)

device failed to a reach lasing condition, primarily due to the heat generation inside the cavity and the higher optical losses in the metal cavity.

The lasing characteristics of 355 nm core radius subwavelength pillar laser were also investigated as shown in Figure 1.14. An SEM image of the pedestal pillar structure is shown in Figure 1.14a. The pillar structure had $r_{core} = 355$ nm, $r_{clad} = 310$ nm, $d_{shield} = 140$ nm, together with a 1.36 μm pillar height and silver coating. As discussed in the numerical simulation, this cavity structure supports the WGM with $M = 3$. From the simulation, the Q-factor was estimated to be 352 and the threshold gain was 692 cm^{-1} at the resonant wavelength of 1.38 μm. As shown in Figure 1.14b, spectral narrowing was observed as the injection current was increased and the lasing peak occurred at a wavelength of 1.41 μm. The threshold current was estimated to be around 540 μA which is 10 times higher than for the 750 nm r_{core} device due to the lower material gain at shorter wavelengths and the higher threshold gain value required. The resonant wavelength estimated using the simulation (1.38 μm) matches the measured results quite well. Higher resolution analysis of the lasing peak spectrum showed that the lasing peak at 1.41 μm is in fact a dual peak with 1.5 nm splitting (Figure 1.14d), which indicates the imperfect circular symmetry of the pillar structure because of inaccuracy in fabrication. When the temperature was increased, CW lasing operation was observed up to 100 K. As shown in Figure 1.14e, the onset of the lasing peak was clearly observed and the linewidth was about 6 nm at 100 K. At 120 K, the output

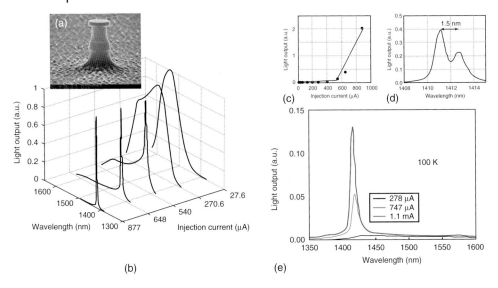

Figure 1.14 Lasing characteristics of 355 nm r_{core} pedestal pillar laser device. (a) An SEM image of 355 nm r_{core} pedestal pillar structure. (b) Spectral evolution graphs with increasing the injection current. (c) L–I curve of this device. (d) Linewidth measurement by a monochromator. (e) Lasing spectrum with difference injection currents measured at 100 K. (From [15].)

spectrum showed clear spectral narrowing with 8 nm linewidth, but failed to reach lasing, which could be due to the heating effect of the high driving current. It is to be expected that pulsed operation would reduce the impact of heating, enabling the device to operate at even higher temperatures. Based on the above numerical simulation, a smaller size ($r_{core} = 220$ nm) device should be able to lase with a low threshold gain (< 100 cm^{-1}). However, the fabrication difficulties, such as forming pedestal structures on a 200 nm wide InP cladding layer with the wet etching process, still pose a challenge and the thermal management issue becomes even more critical with subwavelength-scale pillar widths. We are currently working on resolving these issues by optimizing our fabrication processes and developing a low resistance contact design that could reduce the electrical power dissipation and self-heating in the device. Incorporation of quantum well or quantum dot gain structures in our laser devices could also allow for the building of highly efficient subwavelength-scale lasers [7, 30].

1.5
Thresholdless Nanoscale Coaxial Lasers

The devices described in previous sections, although smaller in all three dimensions than the free-space emitted wavelength, remain limited in a fundamental way: the lasing modes are essentially those of a hollow cylindrical metallic waveguide, and are subject to cut-off. Although, as observed in Section 1.2, the exact cut-off

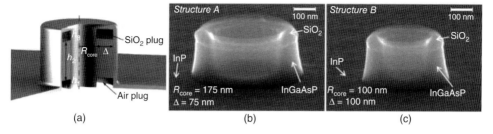

Figure 1.15 Nanoscale coaxial laser cavity. (a) Schematic of a coaxial laser cavity. (b,c) SEM images of the constituent rings in *Structure A* and *Structure B*, respectively. The side view of the rings comprising the coaxial structures is seen. The rings consist of SiO$_2$ on top, and quantum wells gain region underneath. (From [18].) (Please find a color version of this figure on the color plates.)

frequencies of the cylindrical structures discussed here differ somewhat from those of a cylinder with perfectly conducting walls, it is still the case that when the structure diameter is significantly smaller than the operating wavelength, modes such as those shown in Figure 1.5 cannot be sustained. One alternative design that could, in principle, permit size reduction beyond this limit is based on a coaxial waveguide geometry (Figure 1.15), which supports a TEM-like mode with no cut-off [18].

Coaxial nanolasers are also of theoretical interest because it is possible to design a cavity with the mode structure so sparse that only one mode falls within the gain bandwidth of the semiconductor core. A laser of this type exhibits a high spontaneous emission factor β (the ratio of spontaneous emission into the lasing mode to all spontaneous emission) that approaches unity and a pronounced Purcell effect [31] (enhancement or inhibition of spontaneous emission into the lasing mode compared to emission into free space, characterized by the Purcell factor). As a consequence, the characteristic kink in the log-scale light–light curve of the laser becomes smoothed out and less recognizable. Thresholdless laser behavior would imply a smooth curve throughout the pumping region, until saturation is reached. The theory of such lasers, particularly the modeling of subthreshold behavior and the definition of the lasing threshold, is an area of active research. It is also worth noting that, other things being equal, a large β value and strong Purcell enhancement tend to reduce the pump power required to obtain coherent radiation.

The cavity quantum electrodynamic (QED) effects caused by the interaction of matter and the electromagnetic field in subwavelength structures have been the subject of intense research in recent years [32]. The generation of coherent radiation in nanostructures has attracted considerable interest, both because of the QED effects that emerge in small volumes, and the potential of these devices for future applications, ranging from on-chip optical communication to ultrahigh resolution and high throughput imaging/sensing/spectroscopy [33, 34]. Current research efforts are directed at developing the "ultimate nanolaser": a scalable, low threshold, efficient source of radiation that operates at room-temperature, and occupies a small volume on a chip [35]. Coaxial nanostructures help to reach these goals, demonstrating the smallest volume laser device that operates at room temperature in CW mode and at telecom wavelengths. More importantly, for the first time, we have demonstrated thresholdless lasing with a broadband gain medium. Besides their application in the realization of deep subwavelength lasers, nanoscale coaxial resonators are the first step toward unveiling the potential of QED objects and meta-materials in which atom–field interactions generate new functionalities [36, 37].

1.5.1
Design and Fabrication of Thresholdless Nanoscale Coaxial Lasers

The miniaturization of laser resonators using dielectric or metallic material structures, faces two challenges: (i) the (eigen-) mode scalability, implying the existence of a self-sustained electromagnetic field, regardless of the cavity size and (ii) the unbalanced rate of decrease in the optical gain versus cavity loss, which results in a large and/or unattainable lasing threshold, as the volume of the resonator is reduced [25]. In this section, we describe and demonstrate a new approach to nanocavity design that resolves both challenges: firstly, subwavelength size nanocavities with modes far smaller than the free space operating wavelength are realized by designing a plasmonic coaxial resonator that supports the cut-off-free transverse electromagnetic (TEM) mode. Secondly, the high lasing threshold for small resonators is reduced by utilizing cavity QED effects, causing high levels of coupling of spontaneous emission into the lasing mode [38, 39]

The coaxial laser cavity is portrayed in Figure 1.15a. At the heart of the cavity lies a coaxial waveguide that supports plasmonic modes and is composed of a metallic rod enclosed by a metal-coated semiconductor ring [40, 41]. The impedance mismatch between a free-standing coaxial waveguide mode and free-space creates a resonator. However, our design uses additional metal coverage on top of the device, thin, low-index dielectric plugs of silicon dioxide at the top end, and air at the bottom end of the coaxial waveguide in order to improve the modal confinement. The role of the top SiO_2 plug is to prevent the formation of undesirable plasmonic modes at the top interface between the metal and the gain medium. The lower air plug enables coupling of the pump light into the cavity and also the coupling out of the light generated in the coaxial resonator. The metal in the sidewalls of the coaxial cavity is placed in direct contact with the semiconductor to ensure the support of plasmonic modes, providing a large overlap between the modes of the resonator and the emitters distributed in the volume of the gain medium. In addition, the metallic coating serves as a heat sink that facilitates room temperature and CW operation.

To reduce the lasing threshold, the coaxial structures are designed to maximize the benefits from the modification of the spontaneous emission due to the cavity QED effects [38, 39]. Because of their small size, the modal content of the nanoscale coaxial cavities is sparse, which is a key requirement for obtaining coupling of a high level of spontaneous emission into the lasing mode of the resonator. The modal content can be further modified by tailoring the geometry, that is, the radius of the core, the width of the ring, and the height of the gain and low-index plugs. Note that the number of modes supported by the resonator that can participate in the lasing process is ultimately limited to those that occur at frequencies that coincide with the gain bandwidth of the semiconductor gain material. In this work we have used a semiconductor gain medium composed of six quantum wells of $In_{x=0.56}Ga_{1-x}As_{y=0.938}P_{1-y}$ (10 nm thick) between barrier layers of $In_{x=0.734}Ga_{1-x}As_{y=0.57}P_{1-y}$ (20 nm thick), resulting in a gain bandwidth that spans frequencies corresponding to wavelengths in vacuum that range from 1.26 to 1.59 µm at room temperature, and from 1.27 to 1.53 µm at 4.5 K [42].

We consider two different versions of the geometry shown in Figure 1.15a, the first, referred to as *Structure A*, has an inner core radius of $R_{core} = 175$ nm, gain medium ring with a thickness of $\Delta = 75$ nm, lower plug height of $h_1 = 20$ nm, and quantum wells height of 200 nm covered by a 10 nm overlayer of InP resulting in a total gain medium height of $h_2 = 210$ nm, and upper plug height of $h_3 = 30$ nm. The second, *Structure B* is smaller in diameter, having $R_{core} = 100$ nm, and $\Delta = 100$ nm. The heights of the plugs and the gain medium are identical to those of *Structure A*. Figure 1.15b,c show the SEM images of the constituent rings in *Structure A* and *Structure B*, respectively. The two structures are fabricated using standard nanofabrication techniques.

Figure 1.16 shows the modal content of the two structures at 4.5 K, modeled using the 3D FEM eigenfrequency solver in the RF package of COMSOL Multiphysics. Figure 1.16a shows that for *Structure A* the fundamental TEM-like mode and the two degenerate HE_{11} modes are supported by the resonator and fall within the gain bandwidth of the gain material. This simulation was also repeated for *Structure A* with room temperature material parameters, showing that the two degenerate HE_{11} modes are redshifted to 1400 nm, and exhibit a reduced resonance quality factor of $Q \approx 35$, compared to $Q \approx 47$ at 4.5 K. The TEM-like mode is red shifted to 1520 nm with $Q \approx 53$, compared to $Q \approx 120$ at 4.5 K.

Structure B, shown in Figure 1.16b, supports only the fundamental TEM-like mode at a temperature of 4.5 K. The quality factor $Q \approx 265$ for this mode is higher than that of *Structure A*. In general, the metal coating and the small aperture of the nanoscale coaxial cavity inhibit the gain emitters from coupling into the continuum of free-space radiation modes [43]. Hence, the single-mode cavity of *Structure B* exhibits a very high spontaneous emission coupling factor ($\beta \approx 0.99$), approaching the condition for an ideal thresholdless laser [38, 39]. The spontaneous emission factor is calculated by placing randomly oriented and randomly positioned dipoles in the active area of the cavity, and then computing their emitted power at different wavelengths. The β-factor is given by the emitted power that spectrally coincides with the lasing mode, divided by the total emitted power [44].

1.5.2
Characterization and Discussion of Thresholdless Nanoscale Coaxial Lasers

Characterization of the nanoscale coaxial lasers was performed under optical pumping with a $\lambda = 1064$ nm laser pump beam, in both CW and pulsed regimes.

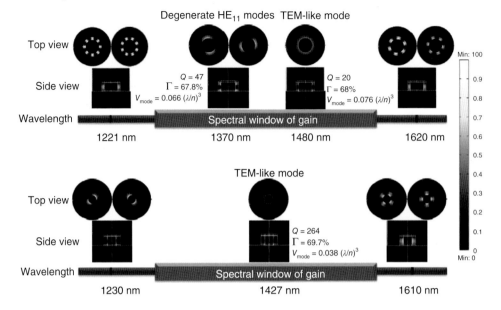

Figure 1.16 Electromagnetic simulation of nanoscale coaxial cavities. (a) The modal spectrum of the cavity of *Structure A* at a temperature of 4.5 K. This cavity supports a pair of HE_{11} degenerate modes and the fundamental TEM-like mode, within the gain bandwidth. (b) The modal spectrum of the cavity of *Structure B*. This cavity only supports the fundamental TEM-like mode within the gain bandwidth of the quantum wells. In the figures, Q is the quality factor of the mode, Γ is the energy confinement factor of the semiconductor region [21] and V_{mode} is the effective modal volume [22]. The color bar shows normalized $|E|^2$. Nominal permittivity values are used in this simulation. (From [18].)

Excitation of the cavity modes is confirmed by measurements of the far-field emission from the devices.

Figure 1.17 shows the emission characteristics of the nanoscale coaxial laser of *Structure A* operating at 4.5 K (light–light curve in frame (a), spectral evolution in frame (b), linewidth in frame (c)), and at room temperature (light–light curve in frame (d), spectral evolution in frame (e), and linewidth in frame (f)). The light–light curves of Figure 1.17a,d show standard laser action behavior, where spontaneous emission dominates at lower pump powers (referred to as the *photoluminescence region*), and stimulated emission is dominant at higher pump powers (referred to as the *lasing region*). The PL and lasing regions are connected through a pronounced transition region, referred to as the *amplified spontaneous emission region*. The evolution of the spectrum shown in Figure 1.17b,e also confirms the three regimes of operation: PL, ASE, and lasing. The spectral profiles at low pump powers reflect the modification of the spontaneous emission spectrum by the cavity resonances depicted in Figure 1.16a. The linewidth of the lasers shown in Figure 1.17c,f narrows with the inverse of the output power at lower pump levels (the solid trend line), which is in agreement with the well-known Schawlow–Townes formula

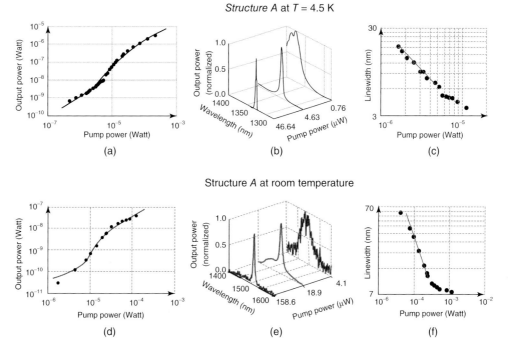

Figure 1.17 Optical characterization of nanoscale coaxial cavities, light–light curve, linewidth versus pump power, and spectral evolution diagram for lasers with threshold. Lasing in *Structure A*. (a) Light–light curve, (b) spectral evolution and (c) linewidth evolution at 4.5 K, (d) light–light curve, (e) spectral evolution, and (f) linewidth at room temperature. The pump power is calculated as the fraction of the power incident on the laser aperture. The solid curves in (a,d) are the best fit of the rate-equation model. The resolution of the monochromator was set to 3.3 nm. (From [18].)

for lasers operating below threshold [45]. Around threshold, the rapid increase in the gain–index coupling in semiconductor lasers slows down the narrowing of the linewidth, until carrier pinning resumes the modified Schawlow–Townes inverse power narrowing rate [46, 47]. In practice, only a few semiconductor lasers are shown to have above-threshold linewidth behavior that follows the modified Schawlow–Townes formula. In most of the lasers reported in the literature, the linewidth behavior is distinctly different from the inverse power narrowing rate. The mechanisms affecting the above-threshold linewidth, especially for lasers with high spontaneous emission coupling to the lasing mode, still constitute an important area of research [48–50].

A rate-equation model was adopted to study the dynamics of the photon-carriers interaction in the laser cavities. The light–light curves obtained from the rate-equation model for the laser of *Structure A* are shown as solid lines in Figure 1.17a,d. For the laser operating at 4.5 K, by fitting the rate-equation model to the experimental data, we found that almost 20% of the spontaneous emission is coupled into the lasing mode, which is assumed to be the mode with the highest quality factor (TEM-like mode). This assumption is validated by examining the far-field radiation pattern and the polarization state of the output beam. At room temperature, the surface and Auger nonradiative recombination processes dominate. As the carriers are lost through nonradiative channels, the ASE kink of the laser becomes more pronounced, and, as expected, the laser threshold shifts to higher pump powers levels.

Next, we examine the emission characteristics of *Structure B*. According to the electromagnetic analysis (Figure 1.16b), this structure is expected to operate as a thresholdless laser, since only one nondegenerate mode resides within the emission bandwidth of the gain medium. The emission characteristics of *Structure B* at 4.5 K are shown in Figure 1.18. The light–light curve of Figure 1.18a, which follows a straight line with no pronounced kink, agrees with the thresholdless lasing hypothesis. The thresholdless behavior is further manifested in the spectral evolution, seen in Figure 1.18b, where a single narrow, Lorentzian-like emission spectrum is obtained over the entire five-orders-of-magnitude range of pump power. This range spans from the first signal detected above the detection system noise floor at 720 pW pump power, to the highest pump power level of more than 100 μW. Since the homogeneously broadened linewidth of the gain medium is larger than the linewidth of the observed emission, the emission profile is attributed to the cavity mode. The measured linewidth at low pump power ($\Delta\lambda_{FWHM} \approx 5$ nm) which agrees with the transparency cavity Q-factor value of the TEM-like mode, as well as the radiation pattern, confirms the electromagnetic simulation given in Figure 1.16b. The evidence that the device indeed reaches lasing is further supported by the linewidth behavior. At low pump levels, the linewidth depicted in Figure 1.18c is almost constant, and does not narrow with output power, implying that the linewidth shows no subthreshold behavior [39, 47]. The lack of variation of linewidth with pump power is most probably the result of the increasing gain-index coupling, which is well-known as around-threshold behavior in semiconductor lasers [46, 47]. Another indication, and a more decisive proof, that *Structure B* does not exhibit subthreshold behavior is that the linewidth narrowing above the 100 nW pump power level does not follow the inverse power narrowing rate that is clearly identified in *Structure A*. This narrowing corresponds to the carrier-pinning effect, as further corroborated by the results of the rate-equation model for the carrier density. To the best of our knowledge, this linewidth behavior, though predicted in theory [48–50], has never been reported in any laser, and is unique to our thresholdless laser. The light–light curve obtained from the rate-equation model for the laser of *Structure B* at 4.5 K is shown by the solid line in Figure 1.18a. The best fit of our rate-equation model to the experimental data is achieved if 95% of the spontaneous emission is coupled to the lasing mode ($\beta = 0.95$). The deviation from the $\beta = 0.99$ value predicted by the electromagnetic simulation can be attributed to other nonradiative recombination processes that have not been considered in the rate-equation model, and to the spectral shift of the mode at higher pump levels that causes variations in the available gain for the mode. In summary, all the experimental observations, including output spectrum and beam profile, electromagnetic simulations, rate-equation model, and comparison with

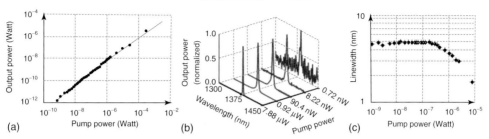

Figure 1.18 Optical characterization of nanoscale coaxial cavities, light–light curve, linewidth versus pump power, and spectral evolution diagram for thresholdless lasers. Thresholdless lasing in *Structure B*. (a) Light–light curve, (b) spectral evolution, and (c) linewidth evolution at 4.5 K. The pump power is calculated as the fraction of the power incident on the laser aperture. The solid curve in (a) is the best fit of the rate-equation model. The resolution of the monochromator was set to 1.6 nm. (From [18].)

the behavior of non-thresholdless lasers, suggest thresholdless lasing as the only plausible hypothesis that satisfactorily explains all aspects of the emission from the light emitting device, based on our results for *Structure B* at 4.5 K.

The thresholdless lasing that we have demonstrated in nanoscale coaxial cavities clearly differs from the situation in the state-of-the-art, high quality factor, photonic bandgap structures [51]. In the latter, near-thresholdless lasing is achieved in a quantum dot gain medium system with spectrally narrow band emission, and relies extensively on tuning of the cavity mode to the center of the quantum dot emission spectrum [51]. In the former, thresholdless lasing in a broadband gain medium is achieved with a low quality factor, single-mode metal cavity. Smaller size, a straightforward fabrication procedure, and better thermal properties are just a few of the advantages of nanoscale coaxial cavities for the realization of thresholdless lasing.

1.6
Summary, Discussions, and Conclusions

We have introduced and analyzed a new type of composite metallo-dielectric resonator that is confined in all three dimensions [7, 14–18]. We have investigated the effect of a low-index dielectric shield layer between the gain medium and the metal coating of a gain waveguide and a 3D laser resonator.

We have shown that the shield layer in the gain waveguide has an optimal thickness such that the threshold gain required to compensate for losses is minimal. The gain-waveguide results were used to design a novel 3D resonator that is smaller than the vacuum wavelength of the emitted light in all three dimensions, and has a sufficiently low threshold gain to allow lasing at room temperature. Using this design, we have demonstrated room-temperature lasing from subwavelength metallo-dielectric resonators. In addition to reducing the size below the emission wavelength in all three dimensions, the metal layer isolates the cavity from its outer environment.

The subwavelength nature of these emitters may lead to practical applications such as dense optical interconnects and laser arrays for optical trapping and manipulation of particles, both of which are currently limited by the larger size and lateral coupling effects in vertical-cavity surface-emitting laser (VCSEL) arrays. Also, the small size of the cavity may enable the implementation of high-speed directly modulated lasers through spontaneous emission enhancement by means of the Purcell effect.

We have also demonstrated a new design and fabrication of wavelength and subwavelength scale electrically driven lasers using metallo-dielectric cavities. In the design, the metal cavity, combined with a thin low-index dielectric layer, enabled a significant enhancement in modal confinement for both wavelength and subwavelength-scale cavities. Using a pedestal geometry improved the vertical mode confinement and showed a significant reduction in the threshold gain with increasing undercut in the pedestal cladding. Lowering the threshold gain below the target $400\,\text{cm}^{-1}$ requires a minimal undercut ($<60\,\text{nm}$) in the pedestal for 750 and 350 nm core radius lasers. This approach also provides an advantage for efficient heat transfer and carrier diffusion in the active region. Experimentally, we have presented the fabrication process for our designed structure, which is based on an InGaAs/InP double heterostructure. Laser devices were fabricated for 750 and 355 nm gain-core radii. We observed clear lasing operation at 77 K for both laser devices, with low threshold current values of 50 and 540 µA, respectively. For the $r_{core} = 750\,\text{nm}$ laser device, CW lasing operation at 1.49 µm was observed up to 140 K. The $r_{core} = 355\,\text{nm}$ laser device showed CW lasing operation up to 100 K. Numerical studies suggest that even smaller laser structures (core radius = 220 nm) could exhibit low threshold gain, making room temperature operation feasible.

Finally, with nanoscale coaxial resonant structures, we have successfully demonstrated room temperature, CW lasing and low-temperature, thresholdless lasing in a spectrally broadband semiconductor gain medium. Owing to the fundamental TEM-like mode with no cut-off, these cavities show ultrasmall mode confinement, offer a large mode-emitter overlap that results in optimal utilization of the pump power, and provide multifold scalability. Further developments toward electrical pumping of thresholdless nanoscale coaxial lasers that could operate at room temperature are in progress. The implications of our work are threefold. Firstly, the demonstrated nanoscale coaxial lasers have a great potential for future nanophotonic circuits on a chip. Secondly, thresholdless operation and scalability provide the first systematic approach toward the realization of QED objects and functionalities, specifically the realization of quantum meta-materials. Finally, this new family of resonators paves the way to in-depth study of the unexplored physics of emitter-field interaction.

It should be noted, however, that as the size of metallo-dielectric coherent sources is made smaller and smaller, the metallic losses become extremely high – due to the fact that a large portion of the mode overlaps with the metal in cases of extreme confinement – in contrast to the nanolasers discussed in this chapter, where the photonic modes are utilized. In addition, even though achieving short carrier lifetimes through large Purcell factors is desirable for increasing direct modulation rates, it can have detrimental effects on the optical or electrical injection levels required to achieve transparency and overcome losses [52]. This combination poses interesting challenges for practical implementation of extremely small CW coherent sources operating at room temperature. With this in mind, future research directions in the area of metallo-dielectric nanolasers include electrically pumped coaxial geometry devices, integration of nanolasers with silicon-on-insulator chip-scale waveguides, and gain material grown radially around a nanowire core. Theoretical analysis employing rate equations and the linewidth behavior of 3D nanolasers are also topics of great interest and importance.

Acknowledgments

The authors would like to acknowledge the support from the Defense Advanced Research Projects Agency (DARPA), the National Science Foundation (NSF), the NSF Center for Integrated Access Networks (CIAN), the Cymer Corporation, and the US Army Research Office.

References

1. McCall, S.L., Levi, A.F.J., Slusher, R.E., Pearton, S.J., and Logan, R.A. (1992) Whispering-gallery mode microdisk lasers. *Appl. Phys. Lett.*, **60**, 289–291.
2. Jin Shan, P., Po Hsiu, C., Tsin Dong, L., Yinchieh, L., and Kuochou, T. (1998) 0.66 μm InGaP/InGaAlP single quantum well microdisk lasers. *Jpn. J. Appl. Phys., Part 2*, **37**, L643–L645.
3. Md Zain, A.R., Johnson, N.P., Sorel, M., and De La Rue, R.M. (2009) High quality-factor 1-D-suspended photonic crystal/photonic wire silicon waveguide microcavities. *IEEE Photon. Technol. Lett.*, **21**, 1789–1791.
4. Akahane, Y., Asano, T., Song, B.S., and Noda, S. (2003) High-Q photonic nanocavity in a two-dimensional photonic crystal. *Nature*, **425**, 944–947.
5. Song, Q., Cao, H., Ho, S.T., and Solomon, G.S. (2009) Near-IR subwavelength microdisk lasers. *Appl. Phys. Lett.*, **94**, 061109.
6. Smotrova, E.I., Nosich, A.I., Benson, T.M., and Sewell, P. (2006) Optical coupling of whispering-gallery modes of two identical microdisks and its effect on photonic molecule lasing. *IEEE J. Sel. Top. Quantum Electron.*, **12**, 78–85.
7. Nezhad, M.P., Simic, A., Bondarenko, O., Slutsky, B., Mizrahi, A., Feng, L., Lomakin, V., and Fainman, Y. (2010) Room-temperature subwavelength metallo-dielectric lasers. *Nat. Photonics*, **4**, 395–399.
8. Nezhad, M.P., Tetz, K., and Fainman, Y. (2004) Gain assisted propagation of surface plasmon polaritons on planar metallic waveguides. *Opt. Express*, **12**, 4072–4079.
9. Hill, M.T. *et al.* (2007) Lasing in metallic-coated nanocavities. *Nat. Photonics*, **1**, 589–594.
10. Johnson, P.B. and Christy, R.W. (1972) Optical constants of the noble metals. *Phys. Rev. B*, **6**, 4370–4379.

11. Goebel, E.O., Luz, G., and Schlosser, E. (1979) Optical gain spectra of InGaAsP–InP double heterostructures. *IEEE J. Quantum Electron.*, **15**, 697–700.
12. Korbl, M., Groning, A., Schweizer, H., and Gentner, J.L. (2002) Gain spectra of coupled InGaAsP/InP quantum wells measured with a segmented contact traveling wave device. *J. Appl. Phys.*, **92**, 2942–2944.
13. Hill, M.T. *et al.* (2009) Lasing in metal–insulator–metal sub-wavelength plasmonic waveguides. *Opt. Express*, **17**, 11107–11112.
14. Mizrahi, A., Lomakin, V., Slutsky, B.A., Nezhad, M.P., Feng, L., and Fainman, Y. (2008) Low threshold gain metal coated laser nanoresonators. *Opt. Lett.*, **33**, 1261–1263.
15. Lee, J.H., Khajavikhan, M., Simic, A., Gu, Q., Bondarenko, O., Slutsky, B., Nezhad, M.P., and Fainman, Y. (2011) Electrically pumped sub-wavelength metallo-dielectric pedestal pillar lasers. *Opt. Express*, **19** (22), 21524–21531.
16. Bondarenko, O., Simic, A., Gu, Q., Lee, J.H., Slutsky, B., Nezhad, M.P., and Fainman, Y. (2011) Wafer bonded sub-wavelength metallo-dielectric laser. *IEEE Photonics J.*, **3** (3), 608–616.
17. Ding, Q., Mizrahi, A., Fainman, Y., and Lomakin, V. (2011) Dielectric shielded nanoscale patch resonators. *Opt. Lett.*, **36**, 1812–1814.
18. Khajavikhan, M., Simic, A., Katz, M., Lee, J.H., Slutsky, B., Mizrahi, A., Lomakin, V., and Fainman, Y. (2012) Thresholdless nanoscale coaxial lasers. *Nature*, **482** (7384), 204–207.
19. Yeh, P., Yariv, A., and Marom, E. (1978) Theory of Bragg fiber. *J. Opt. Soc. Am.*, **68**, 1196.
20. Mizrahi, A. and Schächter, L. (2004) Bragg reflection waveguides with a matching layer. *Opt. Express*, **12**, 3156.
21. Miyagi, M., Hongo, A., and Kawakami, S. (1983) Transmission characteristics of dielectric-coated metallic waveguide for infrared transmission: slab waveguide model. *IEEE J. Quantum Electron.*, **19**, 136.
22. Smotrova, A.E.I. and Nosich, A.I. (2004) Mathematical study of the two-dimensional lasing problem for the whispering-gallery modes in a circular dielectric microcavity. *Opt. Quantum Electron.*, **36**, 213.
23. Palik, E.D. (1985) *Handbook of Optical Constants of Solids*, Academic Press.
24. Novotny, L. and Hafner, C. (1994) Light propagation in a cylindrical waveguide with a complex, metallic, dielectric function. *Phys. Rev. E.*, **50**, 4094.
25. Baba, T. (1997) Photonic crystals and microdisk cavities based on GaInAsP-InP system. *IEEE J. Sel. Top. Quantum Electron.*, **3**, 808–830.
26. Baba, T., Fujita, M., Sakai, A., Kihara, M., and Watanabe, R. (1997) Lasing characteristics of GaInAsP-InP strained quantum-well microdisk injection lasers with diameter of 2-10 μm. *IEEE Photonics Technol. Lett.*, **9**, 878–880.
27. Asada, M. and Suematsu, Y. (1985) Density-matrix theory of semiconductor lasers with relaxation broadening model-gain and gain-suppression in semiconductor lasers. *IEEE J. Quantum Electron.*, **21**, 434–442.
28. Liu, Z., Shainline, J.M., Fernandes, G.E., Xu, J., Chen, J., and Gmachl, C.F. (2010) Continuous-wave subwavelength microdisk lasers at $\lambda = 1.53$ μm. *Opt. Express*, **18**, 19242–19248.
29. Van Campenhout, J., Rojo-Romeo, P., Van Thourhout, D., Seassal, C., Regreny, P., Cioccio, L.D., Fedeli, J.-M., and Baets, R. (2007) Thermal characterization of electrically injected thin-film InGaAsP microdisk lasers on Si. *J. Lightwave Technol.*, **25**, 1543–1548.
30. Albert, F., Braun, T., Heindel, T., Schnedier, C., Reitzenstein, S., Höfling, S., Worschech, L., and Forchel, A. (2010) Whispering gallery mode lasing in electrically driven quantum dot micropillars. *Appl. Phys. Lett.*, **97**, 101108.
31. Purcell, E.M. (1946) Spontaneous emission probabilities at radio frequencies. *Phys. Rev.*, **69**, 681.
32. Berman, P. (ed.) (1994) *Cavity Quantum Electrodynamics*, Academic Press, San Diego, CA.
33. Abe, H., Furumoto, T., Narimatsu, M., Kita, S., Nakamura, K., Takemura, Y., and Baba, T. (2012) Direct live cell imaging using large-scale nanolaser array. *IEEE Sens.*, **1565**, 4.

34. Ma, R.-M., Yin, X., Oulton, R.F., Sorger, V.J., and Zhang, X. (2012) Multiplexed and electrically modulated plasmon laser circuit. *Nano Lett.*, **12**, 5396–5402.
35. Noda, S. (2006) Seeking the ultimate nanolasers. *Science*, **314** (5797), 260–261.
36. Burgos, S.P., deWaele, R., Polman, A., and Atwater, H.A. (2010) A single-layer wide-angle negative-index metamaterial at visible frequencies. *Nat. Mater.*, **9**, 407–412.
37. Jacob, Z. and Shalaev, V.M. (2011) Plasmonics goes quantum. *Science*, **28**, 463.
38. Yokoyama, H. (1992) Physics and device applications of optical microcavities. *Science*, **256** (5053), 66–70.
39. Bjork, G. and Yamamoto, Y. (1991) Analysis of semiconductor microcavity lasers using rate equations. *IEEE J. Quantum Electron.*, **27**, 2386–2396.
40. Baida, F.I., Belkhir, A., and Van Labeke, D. (2006) Subwavelength metallic coaxial waveguides in the optical range: role of the plasmonic modes. *Phys. Rev. B*, **74**, 205419.
41. Feigenbaum, E. and Orenstein, M. (2008) Ultrasmall volume plasmons, yet with complete retardation effects. *Phys. Rev. Lett.*, **101** (16), 163902.
42. Benzaquen, R. *et al.* (1994) Alloy broadening in photoluminescence spectra of $Ga_xIn_{1-x}As_yP_{1-y}$ lattice matched to InP. *J. Appl. Phys.*, **75** (5), 2633–2639.
43. Bayer, M. *et al.* (2001) Inhibition and enhancement of the spontaneous emission of quantum dots in structured microresonators. *Phys. Rev. Lett.*, **86** (14), 3168–3171.
44. Vuckovic, J., Painter, O., Xu, Y., Yariv, A., and Scherer, A. (1999) Finite-difference time-domain calculation of the spontaneous emission coupling factor in optical microcavities. *IEEE J. Quantum Electron.*, **35**, 1168.
45. Schawlow, A.L. and Townes, C.H. (1958) Infrared and optical masers. *Phys. Rev.*, **112** (6), 1940.
46. Henry, C. (1982) Theory of the linewidth of semiconductor lasers. *IEEE J. Quantum Electron.*, **18** (2), 259–264.
47. Björk, G., Karlsson, A., and Yamamoto, Y. (1992) On the linewidth of microcavity lasers. *Appl. Phys. Lett.*, **60**, 304.
48. Rice, P.R. and Carmichael, H.J. (1994) Photon statistics of a cavity-QED laser: a comment on the laser-phase-transition analogy. *Phys. Rev. A*, **50** (5), 4318–4329.
49. Pedrotti, L.M., Sokol, M., and Rice, P.R. (1999) Linewidth of four-level microcavity lasers. *Phys. Rev. A*, **59**, 2295.
50. Roy-Choudhury, K. and Levi, A.F.J. (2011) Quantum fluctuations and saturable absorption in mesoscale lasers. *Phys. Rev. A*, **83**, 043827.
51. Strauf, S. *et al.* (2006) Self-tuned quantum dot gain in photonic crystal lasers. *Phys. Rev. Lett.*, **96** (12), 127404/1–127404/4.
52. Khurgin, J.B. and Sun, G. (2012) Practicality of compensating the loss in the plasmonic waveguides using semiconductor gain medium. *Appl. Phys. Lett.*, **100** (1), 011105.

2
Optically Pumped Semiconductor Photonic Crystal Lasers
Fabrice Raineri, Alexandre Bazin, and Rama Raj

2.1
Introduction

The concept of the photonic crystal (PhC) was introduced in 1987 by Yablonovitch [1] and John [2] in the context of the control of the spontaneous emission of a light emitter. Their idea was to design and fabricate a structure enabling the tailoring of the electromagnetic environment of an emitter in order to inhibit [1] or enhance spontaneous emission [2]. More generally, PhC structures presenting a wavelength-scale periodic refractive index modulation in space are the photonic analogs of atomic lattices where the spatially varying dielectric constant plays the role of the atomic potential. The opening of photonic band-gaps (PBGs), frequency ranges within which the propagation of light is forbidden, becomes possible – enabling thereby the tailoring of the properties of these structures. Some examples of fabricated PhCs are represented in Figure 2.1. The periodic refractive index modulation may be 1D, 2D, or 3D, as shown in the figure. The 1D structures are also referred to as *Bragg mirrors*.

During the last 15 years, PhCs have demonstrated their ability to control quasi-perfectly the propagation of light and proved to be of particular use in obtaining photonic resonators of very good quality, which exhibit record high Q/V ratios, where Q is the resonator quality factor and V is the modal volume of the resonant mode. This is, of course, of primary importance for achieving low thresholds in nanolasers.

This chapter is dedicated to optically pumped semiconductor 2D and 3D PhC lasers. Until very recently, optical pumping was the only way to inject carriers into semiconductor based PhCs, in order to obtain laser emission. Electrical injection of PhC lasers was and remains, indeed, a very challenging task as will be shown in the next chapter dedicated to electrically pumped PhC lasers. However, the optical pumping studies have allowed the investigation of particular properties of PhC lasers, illustrating their specificities and showing performances that are likely to be very interesting for future applications, in terms of power consumption, speed, and footprint.

Compact Semiconductor Lasers, First Edition.
Edited by Richard M. De La Rue, Siyuan Yu, and Jean-Michel Lourtioz.
© 2014 Wiley-VCH Verlag GmbH & Co. KGaA. Published 2014 by Wiley-VCH Verlag GmbH & Co. KGaA.

Figure 2.1 Scanning electron microscope (SEM) images of (a) 1D, (b) 2D, and (c) 3D [3] photonic crystals.

The rest of the chapter is divided into three main sections.

> Section 2 deals with the design and the fabrication of semiconductor PhC lasers. We show here that PhC lasers may be classified into two types, one relying on the use of micro/nano cavities obtained by introducing a defect in the periodicity of the structures and the other one based on the use of the slow light regime of propagation, which can be obtained at certain frequencies in either perfectly periodic structures or in PhC waveguides.
> In Section 3, the specific behavior of these nano/micro lasers is presented in detail. We show that these lasers may exhibit light–light power characteristics where no threshold is visible, justifying the qualification of thresholdless laser; they also display ultrafast dynamics.
> Finally, Section 4 is devoted to the main issues that PhC lasers encounter, and which have to be tackled in order to use them in applications. This is followed by a conclusion.

2.2
Photonic Crystal Lasers: Design and Fabrication

PhC optical properties are mostly determined by their photonic band structure, that is, the optical mode frequencies as a function of the wavevector [4]. The band structure depends, obviously, on the opto-geometrical parameters of the PhCs, such as the lattice configuration or the refractive index of the constitutive materials. This diagram is often sketched using numerical tools based on plane-wave expansion [5] or finite-difference time-domain (FDTD) [6] analyses. By choosing the parameters properly, the photonic band diagram can be engineered almost at will. The photonic band structures of a 2D PhC [7] (triangular lattice of air holes drilled into a semiconductor material), and of a 3D PhC [8], are plotted in Figure 2.2a,b.

Looking at the figures, two striking features appear.

- Large PBGs may be obtained when the refractive index contrast between the constitutive materials (about 3 : 1 in this example) of the PhC is sufficiently large and when the lattice parameters are on the order of the optical wavelength in these materials. Within the PBG ranges of frequencies and for every direction of the crystal, the wavevector value has a non-zero imaginary part, attesting to an exponentially decaying electromagnetic field in the structure and to the absence of a propagating mode.
- We can also observe that, at some particular frequencies, the bands flatten when the wavevector value and direction correspond to the high symmetry points of the

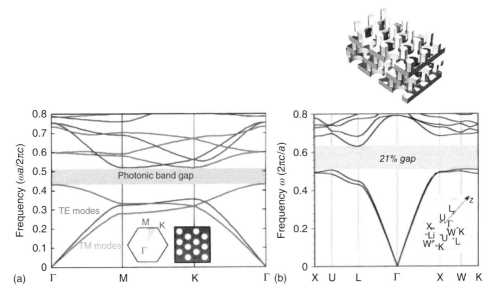

Figure 2.2 (a) Photonic band structure of a 2D PhC made of a triangular lattice of cylindrical air holes drilled into a semiconductor [7]. (b) Photonic band structure of a 3D PhC [8].

first Brillouin zone. At these "band-edge" frequencies, indeed, the group velocity (v_g) goes to zero, implying an increase in the optical density of states [9].

These features allow the realization of the excellent optical resonators on which PhC lasers rely.

In this chapter, we consider PhCs made of III–V semiconductors. These semiconductor compounds combine the very best features required for making PhC lasers, as they provide a large refractive index ($n > 3$) and high gain for a wide range of wavelengths from the visible to the infrared [10], due to the use of quantum confined structures such as quantum wells (QWs) or quantum dots (QDs). They also enable the use of well established technologies such as lithography and dry etching, which facilitate their fabrication and open the way to their eventual use in future applications.

In what follows we present the two types of PhC lasers that have been realized, the micro/nanocavity based lasers, and those based on slow light propagation.

2.2.1
Micro/Nano Cavity Based PhC Lasers

When a point defect is created in the periodic lattice of a PhC presenting a PBG, new states appear within the PBG at certain frequencies. These states correspond to optical modes that are spatially confined within the defect [4]. The proposal of the cavities so-formed has constituted a breakthrough for in-depth studies of light matter interaction – as the Q/V ratios (proportional to intracavity intensity) obtained in these configurations are the highest reported to-date. Indeed, Q-factor values on the order of 10^6 are reachable for a cavity with a modal volume of the order of $(\lambda/n)^3$ [11, 12]. In comparison, the smallest semiconductor microdisk cavities reported [13] exhibit a Q-factor of only 2000 for $V = 0.5(\lambda/n)^3$. These PhC cavities have been used to achieve laser emission at frequencies ranging from the visible to the IR, depending on the active materials used. The cavity properties such as the resonant wavelength, Q and V depend strongly on its design. Lasers based on 2D and 3D PhC cavities are described in what follows.

2.2.1.1 Lasers Based on 2D PhC Cavities

Two-dimensional (2D) semiconductor PhCs are periodic arrangements of cylindrical holes or rods etched into a semiconductor slab surrounded by a low index material such as air or silica, as represented in Figure 2.3, – and these configurations enable the formation of a PBG. Two-dimensional PhC cavities, most of the time, result from the suppression of, or from the modification of the radius of, one or several holes/rods, as well as from a localized shift of the lattice constant along several unit cells of the PhC [14–17]. In these "planar" structures, the propagation of light is ruled not only by the periodic patterning, but also by the stacking of the layers in the direction normal to the plane of periodicity, which usually forms a waveguide. In this manner, 2D PhC cavities may enable a 3D confinement of the light, by combining PBG effects in the plane of periodicity with total internal

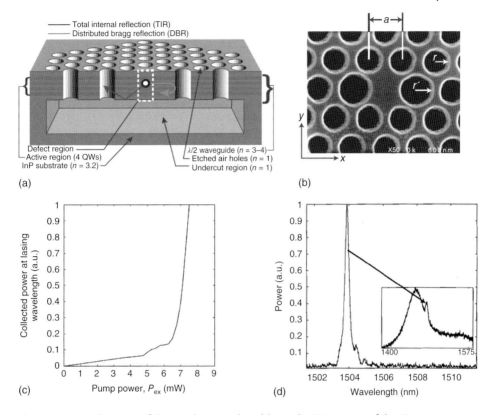

Figure 2.3 (a) Schematics of the 2D PhC cavity based laser. (b) SEM picture of the 2D PhC cavity based laser. (c) Light–light characteristics of the laser. (d) Spectrum of the emitted light just above threshold. (From [18].)

reflection occurring in the third direction, at each semiconductor/low-index layer interface.

The First Demonstration These mechanisms of light confinement were used right from the first demonstration of laser emission with a 2D PhC cavity by Painter *et al.* [18]. In this work, the authors created a cavity by omitting the etching of a hole within a 2D PhC made of a triangular lattice of air holes drilled into a 220 nm thick InGaAsP-based air-bridged membrane, which contained four QWs emitting at telecom wavelength. A schematic representation and a scanning electron microscope (SEM) image of their structure are shown in Figure 2.3a,b. Here the hole radius is about 180 nm and the lattice constant is 515 nm ($\sim\lambda/n$), giving a laser emission around 1500 nm. The triangular lattice is often chosen, as it enables the opening of a large PBG for a transverse-electric (TE) polarized electromagnetic field (electric field parallel to plane of periodicity), which often corresponds to the polarization state of the light emitted by QDs or the strain-compensated QWs used in their work. Two holes are enlarged on the border of the cavity, in order to render the laser emission single-mode. The structure is optically pumped by focusing a pulsed 830 nm laser diode providing 10 ns long pulses at a 4 MHz repetition rate, which are absorbed in the InGaAsP material (barriers and QWs). The emitted light is measured normal to the surface. The light–light characteristics (collected power at the operating wavelength vs pump power) and the emission spectrum just above threshold are given in Figure 2.3c,d. The laser threshold is measured to be 6.75 mW and the emission wavelength was 1504 nm. This first demonstration was achieved on a structure which was far from optimized, in terms of design, as the Q-factor was around 200, which explains, basically, the fairly high threshold values.

Optimization of the 2D PhC Cavity for Laser Emission Increasing the Q-factor of 2D PhC cavities, while keeping their modal volume comparable to a cubic wavelength, has been – for 15 years – one of the important goals of research in this domain. Nowadays, many designs are available to obtain Q factors higher than 10^6 [11, 12, 19, 20] with V comparable to $(\lambda/n)^3$. They all rely on the precise analysis of the losses of the cavities. In this type of structure, losses arise from in-plane coupling and out-of-plane coupling to the outside world. The in-plane losses depend directly on the number of holes etched on the side of the cavity and can in principle be made as low as desired, simply by increasing this number. The out-of-plane losses come from the diffraction of light occurring at the interface of the defect cavity and the periodic mirrors. In a 2D PhC, an in-plane propagating electromagnetic field may be perfectly confined in the third direction, thanks to total internal reflection, as long as the modulus of the in-plane wavevector $k_{//}$ obeys:

$$k_{//} > \frac{2\pi n_s}{\lambda}$$

where λ is the wavelength of light and n_s is the refractive index of the surrounding medium of the PhC. This relation defines what is commonly referred to as the *light cone*. When a defect cavity is considered, the electromagnetic field is confined in a slab volume, the dimensions of which are comparable to the cube of the lattice constant. The distribution of the electromagnetic field amplitude in the wavevector space for the cavity mode can be obtained by Fourier transforming its real-space distribution in the semiconductor slab.

Owing to the small size of such a cavity, the field components may have a rather broad distribution in the reciprocal space with, in general, a significant part lying in the light cone. It can be shown that the Q-factor of 2D PhC cavities depends directly on the amount of the field distribution that lies within the light cone [14]. Several groups [11, 12, 15] proposed and demonstrated that the Q-factors can be significantly improved by diminishing these components.

As an illustration, the real-space and the wavevector-space distributions of an L3 cavity (a cavity formed by three missing holes in a triangular lattice of air holes-type PhC) made in a suspended membrane of silicon are given in Figure 2.4a,b [15]. The air light-cone is denoted by the white circle and delimits the leaky region which corresponds to the part of the field distribution that couples to the continuum of radiative modes. One can see that, in the case of a regular L3 cavity, a significant

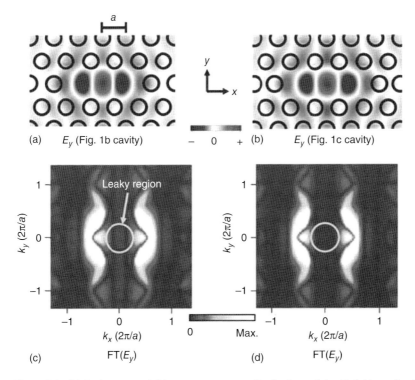

Figure 2.4 (a) Real space and (b) wavevector-space distributions of the E_y field amplitude for the unmodified L3 cavity. (c) Spatial and (d) wavevector-space distributions of the E_y field amplitude for the modified L3 cavity. (From [15].)

part of the distribution of the field lies inside the light cone. The Q factor of such a structure is measured to be around 2600. By shifting the two inner holes of the cavity, the wavevector-space distribution can be modified in such a way that the field components lying within the light cone are strongly reduced. The measurements give a Q factor of 45 000, more than an order of magnitude greater than the regular structure.

General rules on the design of high-Q 2D PhC cavities can be derived from studies carried out in the literature [11, 19, 21]:

- Abrupt changes in the spatial envelope of the field amplitude must be avoided as they generate large field components in the leaky region, especially when the cavity size gets smaller. As stated in [15], "gentle" confinement should be implemented to reduce the amplitude of these components. The best results are obtained when the spatial envelope of the field matches with a Gaussian profile. Indeed, in this way, the envelope of the field in the wavevector space is also a Gaussian that rapidly decays and extinguishes quasi-completely inside the light cone. In order to achieve such a profile, several structures were proposed. Double [22] or multi-heterostructure type PhC cavities [23] are among the possible

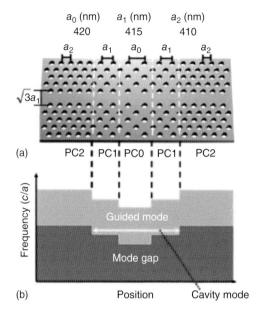

Figure 2.5 (a,b) Schematics of the multi-heterostructure cavity. (From [23].)

designs. Figure 2.5 shows the multi-heterostructure design, which is based on the use of a PhC waveguide (one row of missing holes), where the cavity is created by changing the lattice constant along the direction of propagation in order to increase gradually the mirror strength (imaginary part of the propagation constant) and achieve a Gaussian profile for the field envelope. By increasing the number of sections, Q can reach values above 10^8. Similarly, in [12], the Gaussian profile of the field was obtained by shifting the position of the holes in the direction orthogonal to the PhC waveguide.

- Large losses are associated with modal or impedance mismatch between the evanescent mode within the mirrors and the mode confined in the cavity [24]. Indeed, the transverse mode profiles in the mirrors and the cavity are in general very different, which leads to large scattering at their interfaces. This was pointed out in [25], where the authors examined several different types of PhC cavity and showed that the impedance mismatch can be strongly reduced by creating a taper zone in the lattice, which adapts to the mirror and the cavity modes.
- In order to obtain the smallest leaky region in the wavevector space, the medium surrounding the PhC cavity should possess a refractive index that is as small as possible. This is the reason why most of the studies on PhC cavities concern structures made in an air-bridged membrane. However, recent studies have shown that high Q cavities may be obtained ($Q > 10^7$) by using multi-heterostructure PhC cavities that allow the attainment of a Gaussian field profile.

Improvements in the design of 2D PhC cavities were judiciously implemented in order to obtain low threshold laser emission. For example, the design based on

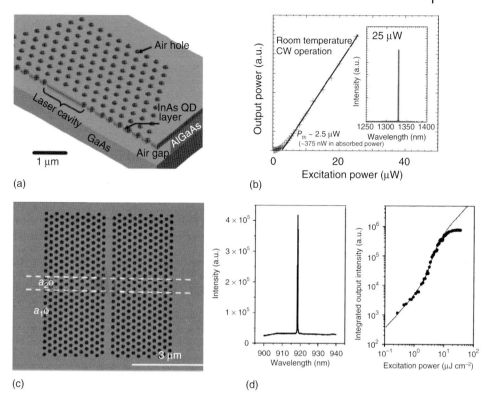

Figure 2.6 Example of demonstration of laser emission using optimized 2D PhC cavities. (a) Schematics of the fabricated L3 PhC cavity based laser. (b) Characteristics of the laser emission. (From [26].) (c) SEM image of double heterostructure PhC cavity based laser. (d) Characteristics of the laser emission. (From [29].)

modified L3-type cavities [15] was used in GaAs-based PhC air-bridged membranes containing QDs [26–28]. Figure 2.6a,b depicts one of the achievements in [26], which is room-temperature laser emission around 1.3 μm, with a threshold corresponding to an absorbed pump power as low as 375 nW. Some PhC lasers were also based on the use of double heterostructure cavities [22] made of GaAs-based [29] (see Figure 2.6c,d) or InP-based [30] materials with embedded QDs. Another interesting design relies on the use of nanobeam or wire cavities [31] and consists of a single row of holes drilled into a single-mode semiconductor ridge-waveguide, as depicted in Figure 2.7a,b. This design gives a very small total footprint $\sim 10\,\mu m^2$, which is 10 times smaller compared to a regular 2D PhC cavity while keeping much the same performance in terms of Q and V. More importantly, this configuration enables high Q factors even in the case where the surrounding medium is (partially) silica, instead of air Md [32], chosen for its mechanical robustness as well as for its improved heat sinking property. The design was implemented to achieve low laser thresholds by taking into account the above mentioned optimization [20, 33–35]. Figure 2.7c,d shows the characteristics and

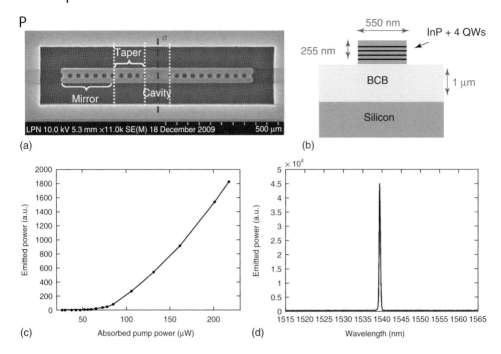

Figure 2.7 Wire cavity laser. (a) SEM image of the top view. (From [35].) (b) Schematics of the transverse profile of the wire cavity. (From [35].) (c) Light–light characteristics of the laser emission. (d) Laser spectrum.

the emission spectrum of such a laser made in an InP-based epitaxial structure containing four QWs bonded onto a silicon wafer by means of a benzocyclobutene (BCB) polymer layer (with the refractive index of BCB being similar to that of silica).

In practice, the performance of these lasers depends strongly on the quality of their fabrication. Triangular lattices of holes are fabricated using a top-down approach. As a typical example, the process flow for the fabrication of the laser investigated in [29] is given in Figure 2.8. The initial step is the growth of the III–V semiconductor layers (generally using either molecular beam epitaxy or metallo-organic chemical vapor deposition), so as to obtain the right stacking. A mask layer patterned with the lattice of holes is firstly fabricated on top of the wafer using high resolution electron-beam lithography followed by reactive ion-etching. The pattern of the mask is then transferred to the semiconductor using plasma etching. The final steps are the removal of the mask and the chemical undercutting of the sacrificial layer, in order to suspend the PhC membrane in air. The main issues in the fabrication are the minimization of the roughness of the etched surfaces and the accurate control of the dimensions of the structure. In particular, the latter has to be down to the nanometer scale to control perfectly the operating wavelength of the laser and to enable the attainment of ultra-high Q-factor values.

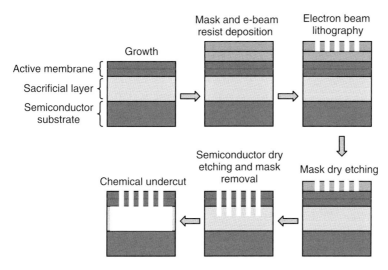

Figure 2.8 Process flow used in the fabrication of a 2D PhC made in an air-bridged membrane of III–V semiconductor. (Please find a color version of this figure on the color plates.)

Laser Based on 2D PhC Cavities Made in a Lattice of Rods Very few demonstrations of a laser operation using a cavity based on 2D lattices of semiconductor rods have been shown so far. This is mainly due to the difficulty in fabricating a PhC completely surrounded by a low index material, which is necessary to ensure vertical confinement of the electromagnetic field. The other inconvenience is that these types of PhCs have PBGs only for transverse-magnetic (TM) polarized waves (magnetic field parallel to the plane of periodicity) [4], so cavity modes exist only for this polarization, which can be a problem when active materials such as strain-compensated QWs or QDs are used, since they usually provide gain only for TE polarized waves.

Nevertheless, this type of cavity has been used in [36] to achieve laser emission at $\lambda \sim 1\,\mu m$. A schematic and an SEM image of the fabricated structure are shown in Figure 2.9a,b. It consists of a hexagonal arrangement of rods encapsulated in polydimethylsiloxane (PDMS) (polymer with a refractive index of about 1.4) and made of GaAs-based materials containing InGaAs QW active material. The cavity ($Q \sim 5000$) is formed by adjusting the position of the rods in the center of the cavity. Here, the PhC is fabricated using a bottom-up approach where the rods are directly obtained by growing the semiconductor material through apertures made in a silicon nitride layer. This approach allows the fabrication of rods with atomically smooth sidewalls, diminishing thereby the optical losses caused by technological processing, compared to the conventional top-down approach. The emission spectrum and the laser light–light characteristics are shown in Figure 2.9c.

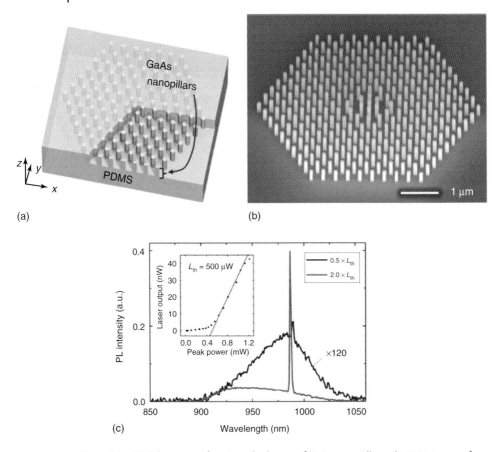

Figure 2.9 (a) Schematics of a triangular lattice of GaAs nanopillars. (b) SEM image of the fabricated structure prior to PDMS encapsulation. (c) Emission spectrum. Inset: laser characteristics. (From [36].)

2.2.1.2 Lasers Based on 3D PhC Cavities

Three-dimensional PhCs provide, in principle, ultimate control on the propagation of light, thanks to their complete PBGs. Several designs of 3D PhC that enable a complete PBG have been proposed in the literature [1, 8], but few of them have been explored experimentally because of the complexity involved in their fabrication. Among the proposed 3D PhCs, woodpile structures have proved to be one the most feasible ones since the standard methods for semiconductor processing can be used in that case. A schematic and SEM images of such a structure are represented in Figure 2.10.

Defect cavities were successfully fabricated using this configuration in GaAs-based material with embedded InAs QDs as the active material (see Figure 2.11a). These structures allowed the demonstration of laser oscillations [38], as indicated by the laser light–light characteristics plotted in Figure 2.11b.

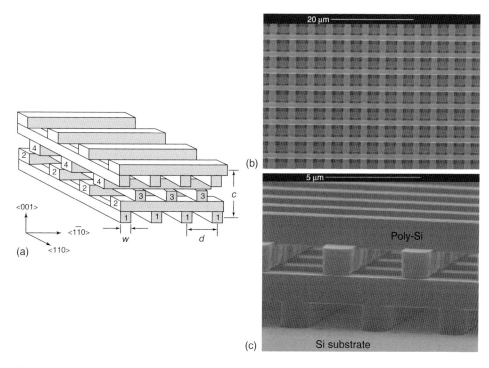

Figure 2.10 (a) Schematics and (b,c) SEM images of a 3D PhC woodpile structure. (From [37].)

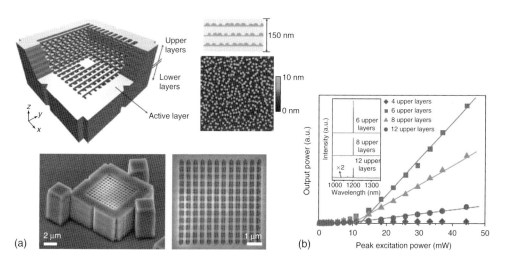

Figure 2.11 (a) Schematics and SEM images of the 3D PhC cavity. (b) Laser light–light characteristics as a function of the upper layer number. (From [38].)

In this work, the researchers demonstrated their prowess in fabricating such a structure by micromanipulating and stacking 2D PhC membranes, a priori obtained by using the standard process described above. They succeeded in assembling 25 layers of 2D PhC and obtained cavities with Q factors as large as 40 000.

2.2.2
Slow-Light Based PhC Lasers: DFB-Like Lasers

The second way to achieve an optical resonator using PhCs is by taking advantage of the low group velocity (v_g) modes that exist at frequencies corresponding to "band-edges" in perfectly periodic structures or to "mode-edges" in PhC waveguides. Usually these modes correspond to wavevectors with the values and directions of the high symmetry points of the first Brillouin zone [9] of the lattice. From the calculated band structure (Figure 2.2), we see that these modes show a null group velocity, which is equivalent to saying that these modes are stationary like in a cavity. This situation can be harnessed to provide a large increase in the optical density of states, which can then be exploited to obtain laser emission, indeed, from the distributed feedback (DFB) induced by the PhC lattice in the entire propagation path of the electromagnetic field [39]. In the following, we shall consider such a type of resonator, realized as a 2D structure, for achieving laser emission. The properties of the resonator depend on the strength of the feedback, as well as on the rate of the light leakage out of the structure due to the coupling to the continuum of radiative modes as shown in [40]. The photon lifetime τ in the resonator can be expressed as:

$$\frac{1}{\tau} = \frac{1}{\tau_a} + \frac{1}{\tau_c}$$

where τ_0 represents the photon lifetime in the absence of coupling to the radiative modes and τ_c is the time constant associated with the losses due to the coupling to the radiative modes. For these resonators, it can be shown in a first approximation that τ_0 is inversely proportional to the curvature (α) of the photonic band and proportional to the surface area of the PhC. The photon lifetime is written as:

$$\tau_0 = \frac{R^2}{2\alpha}$$

where R is the lateral dimension of the PhC. It should be noted that the curvature of the photonic bands is a direct signature of the strength of the DFB, diminishing when the latter increases. α depends to a large extent on the design of the periodic structure. For example, α is greatly reduced in 2D PhCs that are fabricated by fully etching semiconductor air-bridge membranes. Indeed, it takes a much lower value than in the case of shallowly etched gratings like those in regular DFB lasers [41], which is the reason why a large τ_0 can be obtained within rather small structures (~10 μm × 10 μm) using 2D PhC membranes.

As for τ_c, the associated losses can be neglected in the case when the low v_g mode lies below the light line ($1/\tau_c = 0$), if the size of the structure is sufficiently large to avoid any spreading of the field distribution in the wavevector space. The resonator will then be operated in the plane of periodicity. However, as will be shown below, modes above the light line can also be used and are of particular interest for obtaining resonators for laser emission in the direction normal to the plane of periodicity [42].

2.2.2.1 2D PhC DFB-Like Lasers for In-Plane Emission

2D PhC lasers based on the exploitation of the low v_g modes can be designed to obtain light emission perfectly confined in the plane of periodicity of the structure, which is achieved by choosing the parameters of the structure so that a low v_g mode below the light line is at a frequency within the gain bandwidth of the chosen semiconductor active material. These types of laser were demonstrated using perfectly periodic 2D structures made of lattices of cylindrical holes or rods drilled in III–V semiconductor thin slabs or in defect waveguides formed in a 2D lattice of holes drilled in semiconductors. Here, the dimension of the structure as well as the curvature of the band at the particular point of the band structure will fix the Q-factor of the resonator.

Lasers Achieved in Perfectly Periodic Lattices of Holes Drilled in Semiconductor Slabs
Triangular and square lattices of holes etched into III–V semiconductor thin slabs are among the most popular PhC structures used to achieve in-plane emitting band-edge lasers. Figure 2.12 shows an SEM image and the photonic band-diagram of a square lattice of air holes drilled into a 200 nm thick InP-based air-bridged membrane containing seven QWs providing optical gain at wavelengths around 1.55 μm [43]. On the calculated band structure, one can see that a certain number of low v_g modes (denoted by circles on the figure) are situated below the air light-line (border of the black area). These modes are obtained when the in-plane wavevector corresponds to the high symmetry points X and M of the first Brillouin zone of the PhC. The lattice constant (a) and the hole radius (r) were varied on

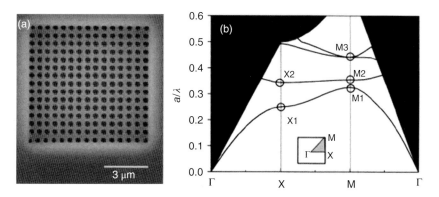

Figure 2.12 (a) SEM image of a band-edge PhC laser made of square lattices of air holes drilled into an InP-based air bridged membrane. (b) Band-structure of the square lattice PhC calculated by plane wave expansion ($a = 550$ nm, $r = 210$ nm). (From [43].)

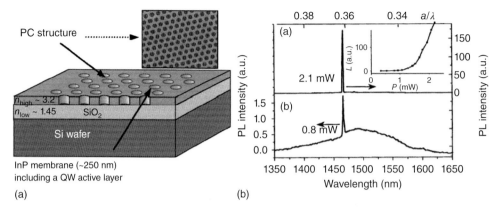

Figure 2.13 (A) Schematics and SEM image of top view of the band edge PhC laser based on a triangular lattice of air holes. (B) Spectrum of the emitted light for two different pump power levels. Inset: laser light–light characteristics. (From [45].)

the fabricated samples, in order to position the X2 and M2 modes within the gain bandwidth of the active material. Laser emission arising from these two modes was demonstrated at room temperature (RT) using structures with a surface area of about 7.5 μm × 7.5 μm (15 holes × 15 holes) using pulsed optical pumping at 980 nm. The measurements were performed by collecting the emitted light from the surface of the samples, which appeared to be rather weak, due to the fact that most of the emitted light escaping from the structure propagates along the unpatterned region of the membrane as guided modes, as indicated by 3D FDTD calculations. The laser thresholds are measured to be quite moderate, around 0.6 mW, which can be explained by the relatively large expected Q factors for these resonators (several thousands) [44].

Similarly, in-plane laser emission was obtained using regular triangular lattices of holes [45] etched into an InP-based membrane transferred onto silicon via SiO_2–SiO_2 wafer bonding (see Figure 2.13). The PhC structure extends over a surface area of 36 μm × 42 μm and is operated at RT at a wavelength of 1450 nm under pulsed optical pumping. The thresholds are measured to be around 1.1 mW when the pump is focused to a 3 μm spot on the sample.

PhC Slow-Light Waveguide Lasers PhC waveguides can be achieved by creating a line defect within the periodicity of a structure that exhibits a PBG (for example, by omitting to drill a row of holes or rods). For frequencies within the PBG range, new states for light propagation are available [46]. They correspond to propagating modes with wavevector parallel to the defect direction or, in other words, to modes that are waveguided along the defect. These PhC waveguides were and still are the subjects of intense research, as they exhibit unique dispersion properties [47] such as zero group velocity modes at certain frequencies. The most studied type of PhC waveguide is the one formed by removing or modifying a row of holes in a 2D triangular lattice of holes that are drilled into a semiconductor

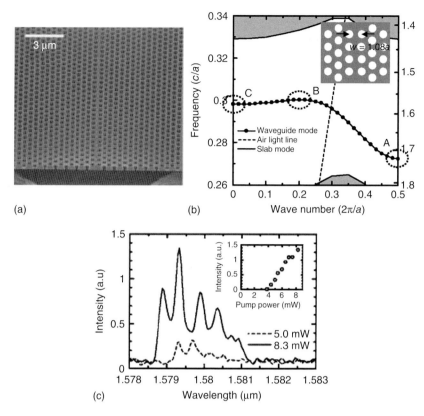

Figure 2.14 (a) SEM image of the fabricated PhC slow light waveguide laser. (b) Dispersion relation of the PhC waveguide. (c) Laser emission spectra for different pump powers corresponding to mode B. Inset: laser light–light curve characteristics (mode B). (From [48].)

membrane (see SEM picture on Figure 2.14a [48]). The waveguides that are so-formed confine TE polarized electromagnetic waves (electric field parallel to the plane of periodicity); the confinement here is total for modes lying under the light-line. The dispersion relation of an example of such a waveguide is plotted on Figure 2.14b. The PhC waveguide is drilled into an InP-based air bridged membrane containing QWs. The ratio of the radius of the holes to the lattice constant is fixed to be 0.29 and the width of the waveguide to $1.08a$. The structure exhibits three different zero group-velocity modes, which are denoted as A, B, and C in Figure 2.16b.

Laser emission originating from the B and C modes was demonstrated by pumping the PhC optically. For B mode operation, the laser light–light characteristics and some spectra of the emitted light are represented in Figure 2.14c. One can see that as the pump excitation is increased, the laser emission becomes multimodal, which results from the combination of the slow light effect with the reflections at

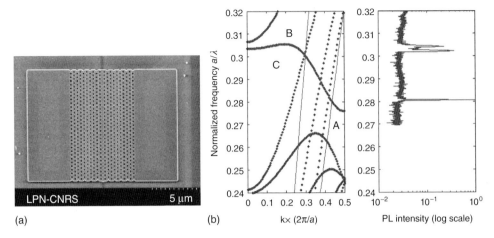

Figure 2.15 (a) SEM image of the PhC slow light waveguide laser. (From [49].) (b) Dispersion relation of the PhC waveguide together with the photoluminescence spectrum obtained under optical pumping.

the interface of the 50 μm long fabricated structure. The laser under investigation was operated using slow modes above the light line, which are intrinsically lossy. In [49], laser emission was also obtained from the mode A below the light-line. Figure 2.15 shows an SEM image of the InP-based fabricated sample, as well as the dispersion relation of the waveguide with the corresponding photoluminescence spectrum, obtained under optical pumping.

The structure was designed to position the theoretically lossless A mode on the maximum of the QW material gain. Low-threshold single-mode laser operation was demonstrated for a 14 μm long structure at the A mode frequency.

2.2.2.2 2D PhC DFB- Like Lasers for Surface Emission

2D PhCs can also be used to obtain laser emission in the direction normal to the plane of periodicity. As can be seen in Figure 2.2a, the band structure of a 2D PhC exhibits flat bands (zero group velocity modes), at certain frequencies, at the Γ-point of the first Brillouin zone (null in-plane wave-vector). Surface normal laser operation can then be achieved by designing the periodic lattice in such a way that the frequency of one of these modes falls within the gain bandwidth of the embedded active material. The mechanism of emission is very similar to that of second order DFB lasers [50]. But, here again, because the 2D lattice is fully etched into the semiconductor membrane waveguide and, as a consequence, the strength of the feedback is very large, the laser thresholds are expected to be relatively low. However, extra care has to be taken in the choice of the mode at the Γ-point – in order to build a resonator with sufficiently small losses to enable laser emission. Indeed, modes at the Γ-point being situated always above the light line, large losses may be expected due to the coupling to the radiative modes. This assertion is not always correct, since the coupling of some of the modes at the Γ-point to the radiative modes is not allowed, owing to the symmetry mismatch of their spatial

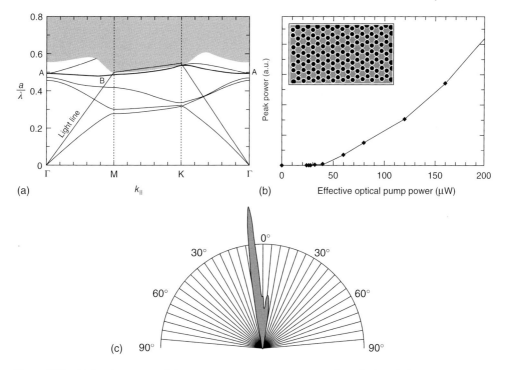

Figure 2.16 (a) Band structure of the graphite lattice of air doles drilled into an InP slab surrounded by SiO$_2$. (From [54].) (b) Measured light–light curve of the laser at room temperature under pulsed optical pumping. (From [54].) Inset: SEM image of the structure. (c) Angle resolved laser emission. (From [55].)

field distributions with those of the radiative modes [51, 52]. Obviously, these are the modes that should be used to build low threshold surface emitting PhC lasers. The demonstration of surface laser emission in [53, 54] relies on the use of these particular modes based, respectively, on a perfectly periodic triangular lattice of air holes drilled into an InP-based air bridged membrane embedding QWs and on a graphite lattice of air holes drilled into an InP based membrane, also with embedded QWs, bonded onto the SiO$_2$ layer.

The photonic band diagram and the light–light curve of the PhC laser explored in [54] are represented in Figure 2.16a,b respectively. Laser emission at RT was obtained at wavelengths around 1.5 µm under pulsed pumping, with a fairly low threshold (40 µW). The remarkable directionality of the emission close to the surface normal was measured and reported in [55] (see Figure 2.16c).

The design of such lasers was further improved by incorporating a mirror placed below the 2D PhC membrane, in order to collect the light emitted in the lower half space, and also to improve the quality factors of the distributed resonators [56–58]. As shown in [56], a large increase in the Q-factor can be obtained by choosing properly the spacing between the mirror and the 2D PhC membrane properly.

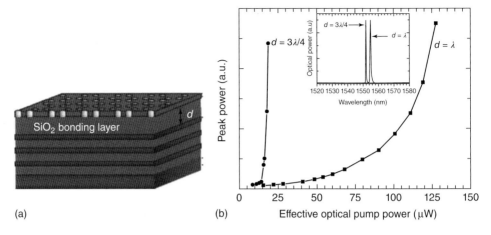

Figure 2.17 (a) Schematics of the investigated structure. (b) Laser light–light curves for structures with $3\lambda/4$-thick and λ-thick SiO_2 bonding layers. (From [56].)

A schematic of the structure used and the light–light curves of the fabricated structures are given in Figure 2.17.

As indicated by the two laser light–light curves, the lowest threshold is obtained when the optical thickness of the separating layer, d, is equal to $3\lambda/4$ – so that the light reflected by the mirror destructively interferes with the light emitted by the PhC laser toward the upper space, decreasing thereby the effective out-of-plane losses of the structure. Conversely, the worse case occurs (no laser emission is observed), when d is equal to λ, as constructive interference occurs in this situation.

2.3
Photonic Crystal Laser Characteristics

This section is dedicated to the description of both the static and the dynamical characteristics of semiconductor PhC lasers. The main features of these lasers are derived by using a standard rate equation model and are illustrated by a few recent demonstrations taken from the literature. This approach will allow us to highlight the peculiarities of such lasers and identify what makes them so interesting for further research and applications.

2.3.1
Rate Equation Model and PhC Laser Parameters

2.3.1.1 Linear Rate Equation Model
As for all semiconductor lasers, the general behavior of most PhC lasers can be approximately described by using a simple rate equation model. For a single mode laser, the evolution of the free carrier density (N) and the photon number (S) in the

cavity mode may be written as [59]:

$$\frac{dN}{dt} = R_{\text{pump}} - R_{\text{sp}}(N) - R_{\text{nr}}(N) - R_{\text{st}}(N, S)$$

$$\frac{dS}{dt} = -\frac{S}{\tau_p} + (\beta R_{\text{sp}} + R_{\text{st}}(N, S))V_a$$

where R_{pump} is the injection rate of carrier density due to pumping, R_{sp} is the spontaneous photon generation rate per unit volume emission, R_{nr} is the non radiative recombination rate per unit volume, and R_{st} is the stimulated photon generation rate per unit volume. Finally, τ_p denotes the photon lifetime, V_a denotes the volume of the active material participating in the laser emission, and β is the spontaneous emission factor.

In order to simplify further discussions, let us consider the functional dependence of R_{sp}, R_{nr}, and R_{st}, as linear in N, such that:

$$R_{\text{sp}}(N) = \frac{N}{\tau_r} \quad \text{and} \quad R_{\text{nr}}(N) = \frac{N}{\tau_{\text{nr}}}$$

where τ_r and τ_{nr} are, respectively, the radiative and non-radiative lifetimes. Note the linear dependence of R_{sp} on N is correct when QDs are used as the active material, since they may be regarded as artificial atoms. (For bulk gain material or QWs, the dependence is usually taken as quadratic.)

The stimulated emission rate term can also be written, in the approximation where the gain dependence on N is linear, as:

$$R_{\text{st}}(N, S) = g(N - N_0)S$$

where N_0 is the carrier density at transparency. The constant g can be expressed using the Einstein's relation for A and B coefficients that links the spontaneous and stimulated emission rates, in the following manner:

$$g = \frac{\beta}{\tau_r}$$

The general framework having being described, we continue with the values of the different parameters evoked in the rate equations model for PhC lasers. Herein, lies the origin of peculiarities of PhC lasers.

2.3.1.2 PhC Laser Parameters

Photon Lifetime and Active Volume As previously described, PhC resonators and especially PhC cavities can exhibit very high Q/V ratios. A typical value for the Q-factor of a 2D PhC cavity fabricated in III–V semiconductor is 10 000, giving a photon a lifetime of about 8 ps for frequencies in the telecom window, while the modal volume is on the order of $(\lambda/n)^3$.

However, the modal volume is not the right parameter to take into account for the rate equations. It is rather V_a that we describe as "the active volume participating to the laser emission." Basically, V_a is the volume of the active material that overlaps spatially with the cavity mode. It may be written as:

$$V_a = \frac{\int \gamma(\vec{r})\epsilon(\vec{r})|E(\vec{r})|^2 dV}{Max(\epsilon(\vec{r})|E(\vec{r})|^2)}$$

where ϵ is the dielectric constant and E is the electric field strength. $\gamma(\vec{r})$ is equal to 1 in the active material and is zero everywhere else. To be more complete, the impact of carrier diffusion should also be taken into account for the calculation of V_a, as some carriers may diffuse into or out of the cavity region during their lifetime [60]. Because of the compactness of PhC lasers, the value of V_a is much smaller than in other types of lasers such as conventional edge-emitters, vertical-cavity surface-emitting lasers (VCSELs), and even microdisk lasers.

Radiative Lifetime and Spontaneous Emission Factor   Right from the initial proposal of the concept of the PhC, the control of spontaneous emission has been the motivation of many studies related to these structures. Indeed, it has been demonstrated that, by placing an emitter like a QD in such a controlled electromagnetic environment, inhibition [61] or acceleration [62–64] of the spontaneous emission may be obtained. In other words, the apparent radiative lifetime of an emitter placed in a PhC can be quite different from that of an emitter placed in a bulk material. When an emitter is placed within a PhC cavity, its radiative lifetime is modified by the Purcell effect [65] in such a way that:

$$\frac{1}{\tau_r} = \frac{F_p}{\tau_{r\_bulk}}$$

where τ_{r_bulk} is the radiative lifetime of the emitter placed in the bulk material. The Purcell factor F_p is given by:

$$F_p = \frac{3}{4\pi^2}\left(\frac{\lambda}{n}\right)^3 \frac{Q_p}{V}$$

where V is the modal volume and n is the refractive index. Q_p is the quality factor, which is linked to the cavity photon lifetime and the dephasing time, T_2, of the emitter by Xu et al. [66]:

$$\frac{1}{Q_p} = \frac{1}{\omega}\left(\frac{1}{\tau_p} + \frac{1}{T_2}\right)$$

These relationships indicate that the Purcell factor can be larger than 1 only when V is on the order of, or smaller than, $(\lambda/n)^3$ (which is, indeed, the case for PhC cavity) and when Q_p is large enough. Obviously, the latter condition depends on both τ_p and T_2. For example, T_2 can range from a few tens to hundreds of picoseconds [62] when QDs are used as the active material and when the structure is operated at cryogenic temperature (around 4 K). In this case, the Purcell factor is often limited by the photon lifetime. F_p values higher than 20 have been reported experimentally for PhC lasers [62]. However, at RT, the Purcell factor is completely determined by the dephasing time of the emitter, which is on the order of a few tens of femtoseconds [67] (i.e., much shorter than τ_p). Hence, a rapid calculation gives F_p as close to unity.

The control of spontaneous emission also has an impact on the value of the spontaneous emission factor β, which is defined as the ratio of the number of photons emitted into the laser mode to the total number of emitted photons. Most of the time, in a laser, a small amount of the spontaneous emission arising from the emitters in the active medium is coupled into the cavity mode that supports the laser emission. This is generally because the emitter spectral linewidth is typically much broader than that of the cavity and only a small fraction of the light is emitted in the right direction, allowing it to couple into the optical mode. β is typically in the range of 10^{-5} to 10^{-4} for edge emitting lasers [59] and in the range of 10^{-5} to 10^{-2} for VCSELs [68, 69].

From the definition of β, we can write the relationship:

$$\beta = \frac{\Gamma_{\text{mode}}}{\Gamma_{\text{all}}} = \frac{F_p}{\gamma + F_p}$$

where Γ_{all} is the rate of overall spontaneous emission and Γ_{mode} is the rate of spontaneous emission into the cavity mode. γ is defined as the ratio of the rate of spontaneous emission in all other modes over the rate of spontaneous emission if the emitters were in the bulk material. We can immediately see that β will tend toward 1 for large values of F_p, but possibly more interestingly, also when γ tends toward 0. This is where PhCs come into play, as they enable the inhibition of spontaneous emission, which is not coupled into the cavity mode thanks to the PBG. As a result, for PhC lasers, β is expected to be orders of magnitude greater than in conventional lasers, and should even approach 1 for optimized structures. Moreover, as the PBG bandwidth can be larger than the emitter linewidth at RT, this assertion holds even for lasers operated at RT.

Non-Radiative Lifetime As this issue is treated in more detail later in this chapter, let us just note here that non-radiative carrier recombination represents one of the major problems encountered while working with semiconductor PhC lasers. The processing of these λ-scale structures induces irremediably the creation of defects at the etched interfaces, which often results in a considerable decrease of τ_{nr}.

2.3.2
The Stationary Regime in PhC Lasers

Taking into account the developments described in Section 2.3.1, the steady-state solutions for the number of photons and the pumping rate may be written as a function of the carrier density:

$$S = \frac{V_a \beta (N/\tau_r)}{(1/\tau_p) - V_a(\beta/\tau_r)(N - N_0)}$$

$$R_{\text{pump}} = \left(\frac{1}{\tau} + \frac{1}{\tau_{\text{nr}}}\right) N + \frac{\beta}{\tau}(N - N_0)S$$

For optically pumped lasers, by assuming that each absorbed pump photon results in an electron–hole pair in the active medium, R_{pump} can be simply expressed as a function of the absorbed pump power P_{abs} as:

$$R_{pump} = \frac{P_{abs}}{h\nu_p V_a}$$

where ν_p is the pump frequency.

The most widely used definition for the laser threshold is that it is reached when the gain compensates for the cavity loss. By considering the stimulated emission to be weak in comparison with spontaneous emission at the laser threshold, the value of the pump power P_{th} necessary to reach it is calculated to be:

$$P_{th} = h\nu_p V_a \left(\frac{1}{\tau} + \frac{1}{\tau_{nr}}\right)\left(N_0 + \frac{\tau_r}{\tau_p V_a \beta}\right)$$

It is then clear that the laser threshold is decreased for small active volumes, long cavity photon lifetimes, and high β values. In consequence, the use of PhC nanocavities may result in low laser thresholds, since they may exhibit high β, small V_a, and long τ_p. Ultimately, using these structures, P_{th} is only limited by the number of carriers that require to be injected to reach material transparency.

The laser light–light curve deduced from the rate equation model is plotted on a log–log scale in Figure 2.18 for different values of β and assuming negligible non-radiative recombination. The other parameters are fixed at the values indicated in the figure and these values are typical for PhC lasers based on nanocavities with embedded QDs as an active medium.

When β is well below 0.1, the transition from the spontaneous emission regime to the stimulated emission regime is marked by a rapid jump in the emission power as a function of the pump power. This is a well known behavior of lasers

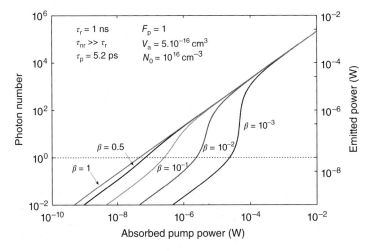

Figure 2.18 Light–light characteristics of PhC lasers for different values of β and negligible non radiative recombination.

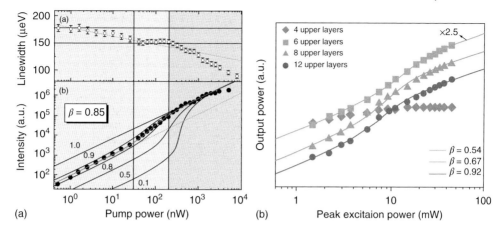

Figure 2.19 (a) Light–light curve measured at 4 K on a 2D PhC cavity patterned in GaAs based suspended membrane embedding one layer of InAs QDs as the active medium. (From [28].) (b) Light–light curve measured at 7 K on a GaAs-based 3D PhC cavity embedding three layers of InAs QDs as the active medium. (From [38].) (Please find a color version of this figure on the color plates.)

where a steep increase in the output power is expected as their threshold is crossed. On the other hand, as β takes larger values and approaches unity, this transition becomes smoother and smoother up to the limit where no transition is observed. These lasers do not show any apparent threshold in their light–light curve and for this reason, are often called *thresholdless lasers* [70, 71].

This peculiarity has been and still is the object of intense research aiming at its experimental demonstration and the understanding of its impact on the properties of the emitted photons.

Some examples of measured light–light curves for high-β PhC lasers with smooth transitions between the spontaneous emission and the stimulated emission regimes are shown in Figure 2.19. In [28], experiments were performed at 4 K on 2D PhC cavities patterned in a GaAs-based air-bridged membrane encompassing one layer of low density InAs QDs emitting at wavelengths around 950 nm. From the fit of the experimental light–light curve, the authors deduced a very high β value of 0.85. Similarly, in [38], the fit of the light–light curve measured at 7 K on a GaAs-based 3D PhC cavity with three layers of embedded InAs QDs resulted in β values ranging from 0.5 to 0.9.

Looking at such laser light–light curves, one may wonder how to determine the laser threshold and whether the classical definition of the laser threshold is appropriate for PhC lasers. In [72, 73], the authors proposed a definition of the laser threshold as the pump power for which the number of emitted photons is equal to 1 (quantum definition of the laser threshold). Indeed, they pointed out in [74] that the system undergoes the transition from the linear operating regime to the non-linear regime when the photon number is unity, whatever the set of parameters taken for the laser. This threshold condition is visualized in Figure 2.18 by the horizontal line. They noted that for lasers with β close to one, the value of the pump power required to reach this condition may be quite different from the one corresponding to the conventional definition. Such a situation occurs when the number of emitted photons at transparency, S_0, is much greater than unity. In these systems, the nonlinear regime occurs before population inversion is reached, that is, for pump powers well below the conventional laser threshold.

Moreover, it can be shown that the emitted photons for pump powers close to just above the quantum threshold do not exhibit the same coherence properties as the photons emitted by what is usually considered to be a laser emission. The quantum fluctuations associated with the small number of photons at threshold cause the output intensity of the laser to fluctuate strongly with chaotic statistics that can persist well above the stimulated emission (quantum) threshold [75, 76]. The result of this situation is that the photons are emitted with non-Poissonian statistics, unlike in conventional lasers.

The emission statistics are usually characterized by measuring the second order or higher order autocorrelation functions, $g^{(n)}(0)$, defined by:

$$g^{(n)}(0) = \frac{\langle E(t)^n E^*(t)^n \rangle}{\langle E(t) E^*(t) \rangle^n}$$

where $\langle \rangle$ stands for the integration over time and $*$ for the complex conjugate. For a chaotic state (thermal light), $g^{(n)}(0) = 1/n!$ and for a coherent state $g^{(n)}(0) = 1$. References [28, 77, 78] are examples of articles where $g^{(2)}(0)$ was measured as a function of the pump power for lasers built from 2D PhC membranes with embedded QDs. In [78], the authors also measured the higher order autocorrelation

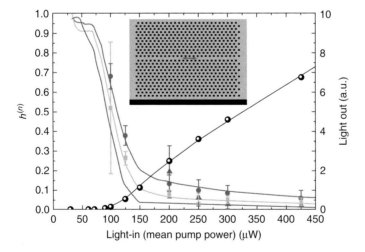

Figure 2.20 Emission statistics at room temperature of a 2D PhC laser made in an InP membrane containing InAsP QDs. Right-hand axis: Experimental (black dots) and calculated (black line) light-in–light-out curve. Left-hand axis: Experimental (red circles, green squares, and blue triangles for $n = 2, 3, 4$, respectively) and calculated (continuous lines of the corresponding colors) values of $h(n)$ as functions of pump power. Inset: SEM image of the sample. (From [78].) (Please find a color version of this figure on the color plates.)

functions. Their results are shown in Figure 2.20, where the light–light curve of the laser is plotted, together with $h^{(n)}$ defined as:

$$h^{(n)} = \frac{g^{(n)}(0) - 1}{n! - 1}$$

such that $h^{(n)}$ equals one for a chaotic state and $h^{(n)}$ is 0 for a coherent state of light. The figure clearly shows that the light emitted below the threshold has the characteristics of thermal light. When the pump power is increased above the threshold, $h^{(n)}$ tends toward 0 but without reaching this value, even for pump powers that are three times the threshold power, which illustrates the presence of chaotic fluctuations, in spite of the predominance of stimulated emission. This type of measurement reveals without ambiguity the value of the laser threshold, and also that these lasers should be operated well above the threshold to reduce their amplitude fluctuations.

It should be noted that most of these measurements were performed at cryogenic temperatures on samples with embedded QDs as the active material. The reason for this choice of temperature is that, at RT, the smooth transition between the two emission regimes is not observed, because of the fairly large amounts of non-radiative recombination. This temperature dependence is even more pronounced when QWs are used instead of QDs as the active material, since the carriers may diffuse more easily toward surface defects. The simulated light–light curves of such lasers are plotted in Figure 2.21 for different values of β. The non-radiative carrier lifetime is set at 200 ps, which is a typical value for InP-based PhCs [79].

As expected, the light–light curve does not exhibit the smooth transition observed in the case where the non-radiative recombinations are negligible, which is, of course, accompanied by an increase of the laser threshold, as some of the injected carriers are not useful for light emission.

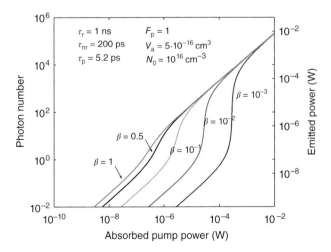

Figure 2.21 Light–light characteristics of PhC lasers with different values of β and $\tau_{nr} = 200$ ps.

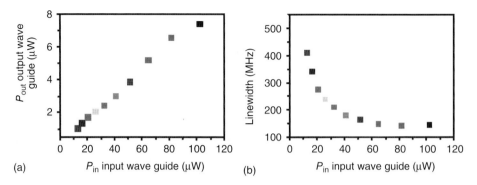

Figure 2.22 (a) Light–light curve far above threshold measured at room temperature in a 2D PhC laser made in an InP membrane embedding a buried heterostructure which contains QWs. (b) Laser width measured as a function of the pump power. (From [81].)

Finally, the last characteristic of laser emission that can be deduced by solving the rate equations in the stationary regime is the emission spectral linewidth. As for all semiconductor lasers, the PhC laser linewidth above the threshold can be obtained from the modified Schawlow–Townes formula [80]:

$$\Delta \nu = \frac{V_a \beta R_{sp}}{4\pi s}(1+\alpha^2) = \frac{V_a \beta h\nu}{4\pi P_{out} \tau_p}\left(N_0 + \frac{\tau_r}{\tau_p V_a \beta}\right)(1+\alpha^2)$$

with α being the linewidth enhancement factor and P_{out} the total output power. Even though $\Delta \nu$ is proportional to V_a and $1/\tau_p$, the linewidth of the emission of PhC lasers is generally of the order of a few gigahertz to hundreds of megahertz, as the intra-cavity photon number is on the order of 100, a rather small quantity when compared to that of conventional lasers.

Linewidth measurements were performed in [81] (Figure 2.22), where the authors obtained a linewidth down to 150 MHz for pump powers 30 times larger than the laser threshold. A comparison with other types of lasers is also provided in that article. They show that, up to now, the measured PhC laser linewidths are at least one order of magnitude greater than that of other lasers. However, when comparing the product $P_{pump} \Delta \nu$, PhC lasers show superior performance, as they require much weaker pump power levels to drive them.

2.3.3
Dynamics of PhC Lasers

After examining PhC lasers in the stationary regime in the previous paragraph, we now shift the focus onto the dynamics of these lasers. Let us first take a look at the dynamics by performing the small-signal analysis of the rate equations. By assuming the dynamic changes in the carrier density and in the photon number away from their steady-state values, induced by a perturbation in the pump power, to be small, one can write the modulation transfer function $H(\omega)$ as a function of the perturbation frequency such that [59]:

$$H(\omega) = \frac{\widetilde{S}(\omega)}{\widetilde{S}(0)} = \frac{\widetilde{N}(\omega)}{\widetilde{N}(0)} = \frac{\omega_R^2}{\omega_R^2 - \omega^2 + j\omega\gamma}$$

where $\widetilde{S}(\omega)$ and $\widetilde{N}(\omega)$ are the frequency-dependent small-signal responses of the number of photons and the carrier density, respectively. ω_R, the relaxation resonance frequency, and γ, the damping factor, can then be expressed as:

$$\omega_R^2 = \gamma_{NP}\gamma_{PN} + \gamma_{NN}\gamma_{PP}$$

$$\gamma = \gamma_{NN} + \gamma_{PP}$$

with

$$\gamma_{NN} = \left(\frac{1}{\tau_{nr}} + \frac{1}{\tau_r}\right) + \frac{\beta}{\tau_r}S \quad \gamma_{NP} = \frac{\beta}{\tau_r}(N - N_0)$$

$$\gamma_{PN} = \frac{\beta}{\tau_r}V_a(1 + S) \quad \gamma_{PP} = \frac{1}{\tau_p} - \frac{\beta}{\tau_r}V_a(N - N_0)$$

As can be seen in these expressions, ω_R and γ depend on the different parameters of the laser but also on the carrier density N and the photon number S obtained in the stationary regime for a chosen pump power. Thus, in order for a comparison of the dynamics to make sense, the different lasers were explored by calculating the small signal responses $|H(\omega)|$ of the lasers for the case where they emit the same steady-state number of photons. Indeed, it is a quite convenient way to compare different systems from a user point of view, as the output power and the necessary pump power to reach it are the features that often determine the potential application. The small signal modulation response is plotted in Figure 2.23 for lasers with different β values, for fixed photon numbers ($S = 1, 10, 100, 1000$).

The 3 dB modulation bandwidth of the lasers can directly be observed on the graphs. As the number of emitted photons is kept small ($S = 1$ or 10), the modulation bandwidth of the emission is limited by the spontaneous emission rate (1 GHz) for the lasers with the smallest β (10^{-3}, 10^{-2}) but is already of the order of 10 GHz when β is greater than 0.1. As S is increased to 100 and 1000, the modulation bandwidth is enhanced for all lasers, but is much bigger for the lasers with the highest β values. In these calculations, we can see that bandwidths larger than 50 GHz may be obtained with reasonable pump powers ($\sim$48 μW) using high β lasers.

Very few papers report on the measurement of the small signal modulation response of PhC lasers. This is mainly due, as indicated later, to the difficulty of operating such lasers in the continuous wave (CW) regime. In [82], the authors examined the small signal modulation response of a PhC laser at RT, emitting at around 1.55 μm (see Figure 2.24a). Their structure is a 2D PhC cavity made in an InP membrane air-bridge. The active material is localized inside the cavity (in a buried heterostructure (BH)) and contains a single InGaAs/InGaAsP QW. The 3 dB-modulation bandwidth was found to be 5.5 GHz when the laser was operated at pump power levels 50 times larger than the laser threshold.

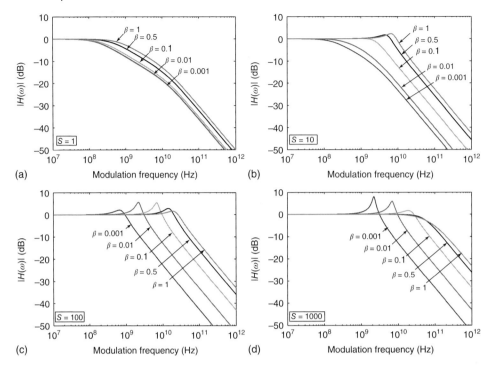

Figure 2.23 Small signal modulation response of PhC lasers with different β values emitting a fixed number of photons ((a) $S = 1$, (b) $S = 10$, (c) $S = 100$, and (d) $S = 1000$). Non radiative recombination is negligible and the parameters of the lasers are fixed identically to the ones of Figure 2.18.

The small signal modulation of 2D PhC lasers was also reported in [83]. In this work, the structures under investigation were 2D PhC cavities made in an InGaAsP membrane bonded onto a sapphire substrate. The membrane contained four QWs to provide gain to the structure. As can be seen in the graph in Figure 2.24b, the authors measured the small signal modulation response for three power levels above the laser threshold and obtained a maximum 3 dB-modulation bandwidth as large as 10 GHz, when the laser was biased at 2.5 times the threshold. Even though these early results are encouraging, they still fall short of the high modulation bandwidth promised by the use of PhC lasers.

The dynamics of PhC lasers were also scrutinized by carrying out gain-switching experiments where the pump is delivered to the structure in the form of short optical pulses (roughly 100 fs long). The dynamic response of the system under pulsed pumping can be calculated by solving numerically the rate equations in the time domain. This response is plotted in Figure 2.25 – again for lasers with different β values and for a fixed output peak power level (here corresponding to a peak number of photons equal to 1000). It is clear that the build-up time of the laser, which is defined as the delay between the pump-pulse arrival and the moment when the maximum of the emitted power is reached, is dramatically decreased as

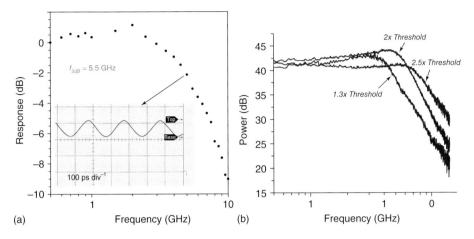

Figure 2.24 Measured small signal modulation response of PhC lasers. (a) From [82]. (b) From [83].

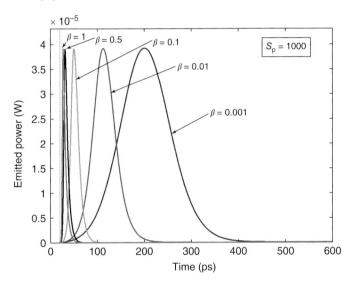

Figure 2.25 Calculated output power under 100 fs long pulsed pumping as a function of time, for lasers with different β factors. Non-radiative recombination is negligible and the parameters of the lasers are fixed identically to those of Figure 2.18. The arrival of the pump pulse is denoted by the dotted line. The energy of the pump pulse is varied from one laser to another by keeping the emitted peak power level constant.

β increases. In this example, the build-up time plummets from 180 ps for $\beta = 10^{-3}$ to less than 7 ps for $\beta = 1$, very close to the limit imposed by the photon lifetime (5.2 ps in the calculation). This fast onset of the emission for lasers with high β can be simply explained by the fact that the spontaneous emission coupled with the cavity mode from which the laser builds up, increases very rapidly in these systems after the arrival of the pump pulse.

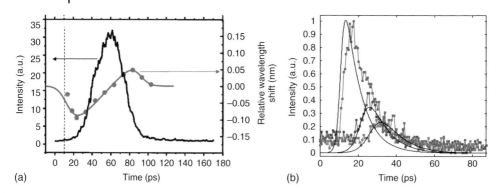

Figure 2.26 Time domain measurements of PhC lasers. (a) Time domain measurement using a streak camera on a GaAs PhC cavity embedding one layer of InAs QDs. (From [29].) (b) Time domain measurements using up-conversion optical gating technique on an InP-based PhC band-edge laser embedding four InGaAs QWs for several pump powers. (From [87].)

In [29, 84–86], the authors measured the ultrafast dynamic response of PhC lasers using a streak camera. An example of such time-domain measurements is represented in Figure 2.26a. In [29], the measurements were performed on a 2D PhC cavity made in a GaAs-based suspended membrane containing a layer of InAs QDs emitting at 920 nm.

The sample was cooled down to 4 K, in order to eliminate the non-radiative recombination of carriers. β is evaluated at 0.67 by fitting the light–light curve of the laser with a standard rate equations model. The build-up and the fall times of the emission are measured to be around 50 and 30 ps for a pump power level five times the laser threshold, showing the possibility of running these lasers at a repetition rate of the order of 10 GHz.

In [87], the dynamics of PhC band edge lasers under ultra short pulsed pumping were measured using an optical gating technique based on the nonlinear mixing (up-conversion) [88] of the laser signal with an intense 100 fs gate optical pulse in a beta barium borate (BBO) crystal. The measured samples were perfectly periodic 2D PhCs made in an InP-based slab with embedded InGaAs QWs emitting at 1.55 μm bonded onto a silicon wafer. The results of these high resolution measurements are shown in Figure 2.26b, where the time-domain evolution of the emission is represented when the laser is pumped with pulses of different energy (260 pJ, 290 pJ, and 500 pJ). For the highest pump level, the build-up and the fall times are found to be 17 and 15 ps. These characteristic times obtained as a function of the pump energy allowed the authors to retrieve a value of β of 3×10^{-2} for this type of laser.

Let us finally state here, once again, that not many researchers have delved deep into the dynamics of PhC lasers. Extensive experimental explorations have yet to be performed. In most of the studies, it is still very difficult to attribute the relatively fast dynamics to enhanced control over the spontaneous emission due to the use of PhCs. Indeed, parameters such as the carrier capture time in QDs or the non-radiative carrier recombination rate impact strongly on the laser dynamics by slowing down or accelerating them. Moreover, a systematic comparison of this type of laser with more conventional structures such as VCSELs is also still missing.

2.4
The Final Assault: Issues That Have Been Partially Solved and Others That Remain to Be Solved Before Photonic Crystal Lasers Become Ready for Application

Reviewing the achievements of PhC nanolasers, it is striking that in the last two decades, a great number of concepts have been brought to fruition. Many of them, like electrical operation, have been few-shot victories (to be precise just three!), which still need to be consolidated but many others, such as low threshold operation, small size, and relative mastery over thermal issues have made a tremendous headway. Most of the research, as is usual in the initial stages, has been conducted on isolated PhC devices. In order to go from the single element PhC device to large scale photonic integrated circuits, an important step needs to be taken: that of optimal coupling – both for acceding the PhC structure in order to activate it and for extraction of the maximum power from the active device. Important too is that, subsequently, this extracted power has to be channeled to other elements of the circuit. Of course, practical application hinges on sufficient output power being produced by these nanolasers, which is essential both for building planar photonic integrated circuits and for using surface emission in applications such as sensors.

Furthermore, the issue of the difficulty of obtaining CW operation at RT in PhC lasers, right from its inception, has occupied a major place in the PhC laser community. This difficulty arises from two very obvious reasons.

Firstly, the confinement of the electromagnetic field, which provides the extraordinary properties of PhCs, is based on the high contrast of refractive indices, both in the horizontal (patterning with holes) direction and the vertical direction – and "What better contrast for a material than with air ?" Thus, air-clad membranes efficiently confine light vertically but at a cost, the poor thermal conductivity of air. In order to overcome this hurdle, several groups have explored the situation through two main channels, heat-sinking substrates and the choice of materials with better thermal conductivity.

The second obvious reason for the difficulty in obtaining RT operation in PhC lasers lies in the very construction of the PhC, etched holes or pillars which expose a huge surface area, rendering it a prey to surface recombination; this has been dealt with, via the use of active materials less sensitive to surface recombination and via optimized processing including surface passivation.

In this section, we review the main achievements in the drive toward obtaining RT CW operation in optically pumped PhC lasers, as well as solving the problem of interfacing these nano-devices with the outside world efficiently.

2.4.1
Room Temperature Continuous Wave Room Temperature Operation of Photonic Crystal Nano-Lasers

2.4.1.1 CW Operation via Nonradiative Recombination Reduction

The nonradiative recombination of carriers has a major impact on the behavior of PhC lasers, as noted in Section 2.3.2. The laser threshold increases with the non-radiative recombination rate $1/\tau_{nr}$, since a substantial part of the pump power is dissipated through this process rather than contributing to the emission. Beyond this issue of increased threshold, non-radiative recombination has an even more detrimental effect, as it leads to the heating of the structure, which then hampers the attainment of sufficiently high gain to obtain laser emission.

Surface recombination is the major source of nonradiative carrier recombination in PhC lasers. Indeed, the processing of these λ-scale structures induces the creation of defects at the etched interfaces, where carriers efficiently recombine nonradiatively. As the pattern dimensions can be well below the diffusion length of the carriers, the nonradiative lifetime τ_{nr} is expected to be significantly shorter than in the bulk material. This characteristic was observed in different types of semiconductor PhCs. In GaAs-based PhCs, the carrier lifetime was measured to be as short as 10 ps, due to the domination of surface recombination of carriers, preventing, thereby, this type of structure from lasing [89]. In InP-based PhCs with embedded InGaAs QWs, the carrier lifetime is shortened to 200 ps, as shown in [79], preventing, in this case also, CW operation.

In order to tackle this issue, two paths can be followed, either bypassing the problem by choosing active materials that are minimally affected by surface recombination (e.g., QDs) or resolving it by acting on the surface recombination through material processing.

2.4.1.1.1 CW Operation in Air Clad PhCs by a Smart Choice of Active Material

PhC Membrane with Embedded Quantum Dots Possibly the first demonstration of CW laser operation in the in-plane direction was made using self-assembled InAs/GaAs QDs emitting at around 1.3 μm [90]. The authors attributed the attainment of the CW regime at RT to the three-dimensional nature of the confinement of carriers in the QDs. Further, the use of QDs reduces the transparency carrier density, leading to low thresholds, which necessarily means less problems related to thermal effects.

The confinement of carriers diminishes the non-radiative recombination, since the diffusion of the captured carriers is greatly suppressed by comparison with other material systems [91]. But, since the modal gain of QDs is typically lower than the gain in QWs, it was necessary to build a high-Q cavity (1.9×10^4), with an increased spontaneous emission coupling factor β. The latter was estimated to be 0.22, as deduced by comparing experimental results with the calculated L–L curves from the coupled rate equations. The Q-factor was optimized by designing a structure (Figure 2.27) for molding the electromagnetic field profile, as in Ref. [15], through a gentle field envelope function.

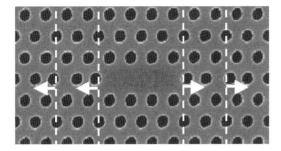

Figure 2.27 SEM image of the L3 defect nanocavity. The first and third nearest air holes at both ends of the cavity are shifted outwards by $0.15a$ as shown by the white arrows [90].

The sample itself consisted of a 250 nm thick GaAs slab layer with five stacked InAs QD layers as the active gain material. A subtle 4 nm-thick $In_{0.16}Ga_{0.84}As$ strain reducing layer was included to provide a handle on the control of wavelength, that is, to shift the emission of the QDs from 1.25 to 1.3 µm. The areal QD density was $\sim 2 \times 10^{10}$ cm^{-2} for each QD layer. With a CW excitation power level of 40 µW, single mode operation was obtained at 1.3 µm. Subsequently the same team used the lasers to explore the spontaneous emission coupling factor β [26].

In a laser cavity formed by a PhC double-heterostructure etched into a 320 nm-thick suspended InP membrane incorporating a single layer of self-assembled InAsP QDs at its vertical center-plane [30], lasing was demonstrated by measuring the second-order autocorrelation function, giving a CW threshold of 115 µW.

PhC Membrane Incorporating Quantum Wires Using a single layer of self-assembled InAs/InP quantum wires (QWires) as active material and a high quality factor of $Q \sim 5.5 \times 10^4$, RT CW emission was obtained in the surface normal direction for an L7 PhC cavity [92]. The QWires had, in fact, the growth morphology of QDs, elongated along the [93] crystal direction, with an average width of 15 nm, height of 3.6 nm, and pitch period 18 nm. The CW operation is here attributed mainly to the reduced dimensionality of the active medium, which entails a reduced threshold for laser emission and thus a reduced temperature sensitivity, as well as a reduction of the linewidth enhancement factor and reduced surface recombination in the QWires.

PhC Membrane Incorporating Large Band-Edge Offset Quantum Wells More recently, RT, continuous-wave lasing with the specific objective of high extraction of energy and enhanced far-field emission directionality was shown in a suspended membrane in coupled-cavity PhC lasers, incorporating InAsP/InP QW material [94]. These surface-emitting lasers had a very low effective threshold power level of 14.6 µW, with a linewidth of 60 pm, and 40% of the surface emitted power concentrated within a small divergence angle. Here the particular properties of the InAsP/InP QWs, with their larger band-edge offset and thus better carrier confinement, as well as the weaker surface recombination as against the commonly used InGaAsP/InP QWs, permit the characteristics obtained for these lasers.

2.4.1.1.2 RT CW Operation with QWs in an Air Cladding Membrane, via "Fine Processing" and Surface Passivation

Fine Processing of InP-Based PhC   To our knowledge, the only study which shows RT, CW PhC laser operation in a suspended membrane using the usual QW active medium (as against one with QWs that have a large band-edge offset) attributes this success to the "fine" processing of the structure [95] and the use of an optimized H0 design, with the high-Q cavity inspired by Akahane *et al.* [15]. The active slab contained five quaternary (InGaAsP) compressively strained QWs.

The structure was etched in a HI/Xe-inductively coupled plasma – which gives very smooth hole sidewalls, as can be seen in Figure 2.28 – and led to a steep decrease in the threshold power level, down to 1.2 µW, as shown in Figure 2.29. This performance was certainly due to the fact that this specific dry etching process induces fewer surface defects and thereby fewer centers for the surface recombination of carriers.

Surface Passivation for Improved Efficiency in PhC Lasers   An essential improvement in the fabrication technique has been under scrutiny for several years: passivation of the surface of the holes that constitute the PhC. More generally, the question of surface recombination has been widely discussed within the "semiconductor physics community" – seeking its reduction – in order to improve the thresholds and augment efficiency in optoelectronic and photonic devices. In the context of PhCs particularly, this issue has been highlighted, as the holes forming the lattice naturally induce an increase in the surface recombination velocity (SRV), while in PhC structures with QW-containing active layers, a large surface area is also exposed,

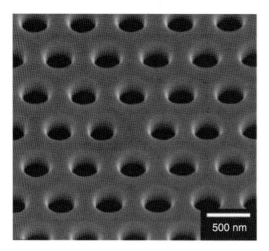

Figure 2.28   SEM image of the fabricated H0 nanolaser. Center two airholes are shifted laterally to form a H0 cavity. (From [95].)

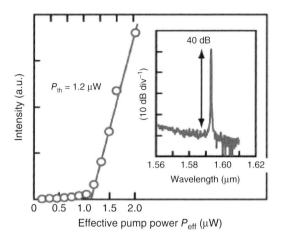

Figure 2.29 CW lasing characteristic of H0 nanolaser with $a = 480\,nm$, $2r/a = 0.62$ [95]: light–light characteristics and lasing spectrum (inset) above threshold.

with damaging repercussions on the efficiency. This nonradiative recombination has been overcome through various passivation methods. An order of magnitude reduction in the nonradiative velocity in GaInAsP/InP microcolumns using CH_4 ECR (electron cyclotron resonance) plasma irradiation was reported in [96]. A more dedicated work, this time exploring different material systems, was reported in [97], where the SRV was determined by absolute photoluminescence efficiency measurements. For instance, in a test sample consisting of a 20 nm thick n-type $In_{0.53}Ga_{0.47}As$ single QW structure with InP cladding layers, the SRV after the etching of the hole was $4.5 \times 10^4\,m\,s^{-1}$. Prior to passivation, the sample was rid of its surface damage by a gentle wet etch in $H_2SO_4:H_2O_2:H_2O$ (1 : 8 : 5000), reducing the SRV threefold. Further improvement was obtained after a 5 min long passivation in a solution of ammonium sulfide $(NH_4)_2S$. The passivation is mainly owing to the removal of surface oxide [98] and its replacement by sulfur atoms, which prevents subsequent adsorption of oxygen.

The ammonium sulfide treatment produces both sulfur-III and sulfur-V bonds, giving a lasting protection to the surface. In PhCs, this passivation method yielded convincing results [60], as shown in Figure 2.30, it lowered the nonradiative recombination rate by more than four times, which also reduced the lasing threshold fourfold (Figure 2.30) in InGaAs/GaAs QD PhC lasers, as well as enabling RT lasing [99].

2.4.1.2 CW Operation via Increased Heat Sinking

2.4.1.2.1 A Comparison of Heat Sinking Between a Membrane and a Bonded PhC Laser

In air clad PhC lasers, other than ones with QDs, the heat evacuation forms a bottleneck, which prevents CW operation. Even in the QD CW lasing case, the

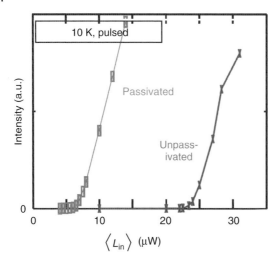

Figure 2.30 Lasing curves for unpassivated and passivated structures: passivation reduces threshold from 24 µW (with SRV of 1.7×10^5 cm s^{-1}) to 6 µW (with SRV of 4×10^4 cm s^{-1}) averaged power [99].

laser operation continues until just a little above the threshold value, which limits its potential application, for instance, in high-speed modulation (see Section 2.3.3).

The other inherent problem is the fact that the output power level is low, which precludes possible applications. Several teams have overcome this hurdle in diverse ways. However, most of these studies concern surface normal operation which has very specific field of deployment. In-plane emission is still the favored feature for PhC lasers, as it is the natural behavior for these structures and it holds promise for large-scale photonic integration; only a few studies have shown CW operation.

Before we embark on a discussion of the achievements world-over in this domain, let us dwell briefly on a systematic examination of the path through which heat is evacuated from an air-suspended membrane, as against a membrane bonded onto a heat-sinking substrate.

In [100] using a pump–probe spatiotemporally resolved thermo-reflectance technique, the heat spreading characteristics of two representative 2D PhCs laser structures (Figure 2.31) were elucidated. One structure (device 1 in Figure 2.31) is a Bloch mode PhC bonded by Au/In onto a substrate [57] with QWs for active medium. The other (device 2 in Figure 2.31) is a suspended membrane incorporating QDs, processed into an L3 cavity [101]. Both the PhCs have a filling factor of around 30%, and both are operated at RT in the CW regime.

The thermal dynamics of two active PhCs yield a thermal dissipation time of 429 ns in device 1, whereas device 2 showed a higher dissipation time of 999 ns. The temporal evolution of the temperature during heating gave a spatially averaged temperature increase of 6.7 K for device 1 and 23.8 K for the membrane device.

The quantitative measurement pin-pointed a striking feature, that is, the localization of the hot-spot in the transferred PhC (device 1) and a spreading of the

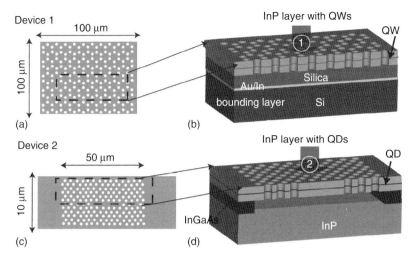

Figure 2.31 (a) Sketch of the top view of the Bloch mode device 1. (b) Cut view inside device 1. (c) Sketch of the top view of the membrane device 2. (d) Cut view inside the membrane device. Arrows correspond to two heating locations studied [100].

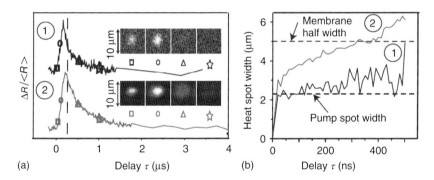

Figure 2.32 (a) Relative reflectivity changes as a function of the delay, when the pump is focused on device 1 (in black) or on the membrane of device 2 (in red). (b) Images of the relative reflectivity change are plotted for 30 (rectangle), 130 (oval), 500 (triangle), and 1000 ns (star). (b) Heat spot width as a function of the delay for the two devices [100].

heated zone in the suspended membrane (device 2), as indicated by Figure 2.32. The investigation showed that, in device 2, the width of the hot-spot increases continuously and spreads over the whole width of the membrane, and finally dissipates through the InP substrate.

The mapping of the diffusivity confirmed that it is the vertical heat evacuation in device 1 that permits CW operation whereas, in device 2, it is the reduction in threshold and a consequently lesser temperature increase due to the active material being QDs that allows CW operation. This study confirmed the importance of having specific heat dissipating paths for these nanolasers.

2.4.1.2.2 CW Operation at RT Obtained by Heat Sinking through a Substrate

Heat Sinking via Al_2O_3 Layer An early [102] demonstration of CW RT surface operation of a 2D PhC laser was made in a structure obtained by the wafer fusion of an InP slab having embedded InGaAsP QWs onto an $Al_{0.98}Ga_{0.02}As$ on GaAs structure. After the patterning of the PhC, the AlGaAs was oxidized to Al_2O_3 by wet oxidation, in order to provide a low index layer below the InP slab. CW lasing operation with surface normal emission from 10 μm diameter cavities was obtained at 1.6 μm, with a pump threshold of 9.2 mW. Surface emission in the CW RT regime was also shown in an InGaAsP slab containing four compressively strained wells and bonded onto a sapphire substrate [83], where, as in the previous case, sapphire serves both as a low index confining layer and an efficient heat sink, allowing CW emission with small cavity sizes, down to 3.2 μm diameter.

Heat Sinking via Au/In or SiO_2 Bonding Layer Surface normal operation near 1.55 μm, in the CW regime, was demonstrated at RT using a low v_g band-edge photonic mode in an InGaAs/InP PhC [57].

This operation was possible because the PhC slab of Figure 2.33 was transferred onto a silicon chip by means of Au/In bonding technology. The choice of a high reflectivity metallic bonding layer provided simultaneously two advantages, an improvement in the photon confinement inside the PhC resonator through an interference phenomenon [56] and efficient thermal dissipation through the substrate. Vecchi and coworkers have also explored several different combinations of material that could simultaneously form a bonding layer and act as a heat sink. By comparing the performance obtained by using both the Au/In bonding and SiO_2 cladding layer on the one hand and a reference device in the form of an InP-based 2D PhC bonded on silicon by means of BCB, the decisive role of the SiO_2 cladding layer in enabling CW laser operation was clearly identified. For a fair comparison, a gold lower mirror layer was also positioned between the BCB and the silicon substrate. Owing to the rear gold mirror and the refractive index of BCB ($n = 1.53$) being very close to that of silica, the optical confinement in the vertical

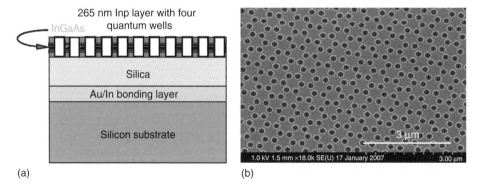

Figure 2.33 (a) Schematic hetero-structure showing the heat sinking silica plus Au/In lower confining layer. (b) SEM image of the top of the PhC structure. (From [57].)

direction in both devices was comparable. The bulk thermal conductivity of BCB, $K_{th} = 0.24$ W (m K)$^{-1}$, is, however, approximately five times lower than that of SiO$_2$, $K_{th} = 1.26$ W (m K)$^{-1}$. The Au/BCB-based PhC structure provided laser emission only under pulsed excitation [58], with a pumping threshold of $\sim$3.4 µJ cm^{-2}. In both structures, the photonic band-edge resonator displayed comparable Q-factor values.

2.4.1.2.3 CW Operation at RT Obtained through the Use of a PhC with Higher Thermal Conductivity

A recent work [103] grappled with many aspects detrimental to the attainment of RT CW operation in PhC lasers. These aspects include the low thermal conductivity of air-cladding layers and the low pumping efficiency due to the discrepancy between the pumping surface area with its associated active volume and the cavity mode volume. The authors have proposed to overcome these problems by introducing an InGaAsP/InP BH active region in an InP air-bridge structure (Figure 2.34).

Owing to the fact that the thermal conductivity is 10 times higher in the InP material that surrounds the quaternary material inside the active region (InGaAsP), the generated heat can escape more easily from the cavity in comparison to the case where the structure is entirely made of InGaAsP. Furthermore, better overlap of the active region with the cavity mode profile provided more efficient pumping, since the BH effectively confines the carrier to the active region. Besides, the BH formed an ultrahigh-Q cavity with small mode volume, since the index of the active region is higher than that of the neighboring InP, giving confinement because of the index modulation effect on the line-defect structure (W1 waveguide) in the BH region. Apart from the better matching between the generated carriers and the optical cavity mode volume provided by the buried active region, the pumping efficiency is further increased by the fact that InP is transparent to the pumping light and therefore the undesired absorption outside the cavity did not contribute to

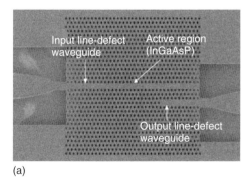
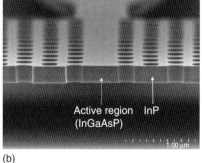

Figure 2.34 (a) SEM image of the fabricated BH-PhC laser with an air-bridge structure. (b) Cross-sectional SEM images of the fabricated BH photonic-crystal nanocavity laser. The air hole diameter is 220 nm. The small active region, 5 µm × 0.3 µm × 0.16 µm in size, is placed within a line-defect PhC waveguide in an InP slab containing three InGaAs QWs between thick InGaAsP barrier layers [103].

heating. Using the optimized configuration of Figure 2.34, Matsuo and coworkers [82] thus obtained a record low threshold input power of 1.5 µW for CW RT operation.

2.4.2
Interfacing and Power Issues

The future of photonic networks has much to gain from the recent advances in PhC structures, both for lasers and for integrated circuits containing several elements. We believe that nanophotonics will play a crucial role by providing solutions to achieve ultra-compact, ultrafast, and power-efficient optical interconnects to be integrated with microelectronic components [104]. One of the main stringent requirements concerns the energy consumption of these devices, which should be less than 10 fJ bit^{-1}. The good news, of course, is that due to their small size and their versatile design, PhCs already provide the functionalities and the high density suited for implementation in a photonic integrated-circuit platform. With respect to the power efficiency, the enhanced light–matter interaction provided by these structures should again, in principle, help to fulfill the requirements. However, for this to become reality, two more essential factors have to be mastered. One is the coupling efficiency, both at the input for the pumping and at the output for the extraction. Related to the latter is the second aspect, that of obtaining sufficient output power to for carry the information. Indeed, the extraction (insertion) of light from (into) PhC cavities has been for a long-time a major drawback to their use in real applications because, in practice, only a small amount of the light stored in these nanocavities is efficiently collected into useful signal. Even though PhC cavities exhibit the most remarkable features for the enhancement of light–matter interaction, devices based on the use of these structures do not hold their promise, in terms of overall power efficiency, because of the problem of interfacing with the outside world.

In the case of planar PhC cavities, the light is generally coupled out from the structure – either by diffraction and coupling to the radiative modes or by using a neighboring waveguide. The choice of the coupling mechanism to interface with the PhC cavities may be dealt with differently, depending on whether we want to interface an isolated device with the external world or we want to interface several devices with each other to form a circuit. Moreover, when lasers are considered, the use of active material for passive functionalities such as waveguiding constitutes an additional problem because of the large absorption of these materials at the lasing wavelength, when they are not pumped.

2.4.2.1 Interfacing an Isolated PhC Cavity-Based Device with the External World
The interfacing of a single PhC cavity-based device such as a laser with the external world does not necessarily require its integration into a photonic waveguide circuit, as the light may be coupled efficiently out of the structure directly through its surface in the vertical direction or by approaching the device through a tapered optical fiber or by fabricating the PhC device close to the edge of the slab.

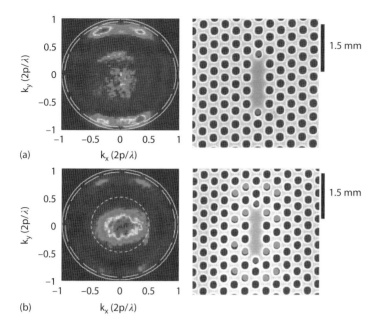

Figure 2.35 Right: SEM images of the unfolded (a) and folded (b) L3 nanocavities. Circles filled in red denote the shifted and shrunk-end holes to boost the Q-factor, those filled in green denote the size modulation at twice the period of the original lattice to achieve the band folding effect. Measured sizes are $a = 437$ nm, $s = 66$ nm, $r_0 = 122$ nm, $r_1 = 97$ nm, and $r_2 = 114$ nm. Left: Measured far-field of the unfolded (a) and folded (b) emission. The white line corresponds to the light line (emission at 90°), while the dashed white line corresponds to N.A. = 0.95, that is, the maximum angle collected in our set-up (~72°). The short dashed line corresponds to 30° (N.A. = 0.5). (From [105].) (Please find a color version of this figure on the color plates.)

Efficient Out-Coupling through the Control of Directionality of Surface Emission Apart from a few exceptions, most of the studies involving planar PhC lasers have been achieved by measuring the light emitted in the direction normal to their surface. Even if the measured light levels are sufficient to characterize the structure properties, most of the emitted light is not collected and cannot be used for further purposes, because these structures usually exhibit complex out-of-plane radiation patterns that are generally not well-suited for surface operations (see for example, Figure 2.35a). Indeed, when looking at the field distribution in the wavevector space of an optimized cavity (see Figure 2.4), it can clearly be seen that most of the field falls close to the first Brillouin zone edge at $k_x = \pi/a$, outside the leaky region, explaining thereby why most of the light cannot be collected through the surface.

Recently, efforts have been made to increase the coupling of PhC cavities to the leaky modes for surface normal operation, while keeping a fairly high-Q factor [105–107]. The problem has been interestingly approached in [106] from the angle of modifying the emission shape itself by introducing a subharmonic periodicity in the lattice of holes surrounding the cavity. The introduction of a subharmonic periodicity equal to twice the lattice constant of the initial PhC by changing the hole size (see Figure 2.35) enables the folding of the main peak of the field distribution in the reciprocal space (at $k_x = \pi/a$) with respect to $k_x = \pi/2a$, giving a replica of this peak at $k_x = 0$ and enhancing the emission of light in the vertical direction. This method was applied in [108] to most of the different designs of high-Q PhC cavities where the authors showed numerically that more than 40% of the emitted light can be collected by using a microscope objective with a numerical aperture of 0.6, positioned above the cavities, while the Q factors are maintained at relatively high values. The concept was also demonstrated experimentally in passive L5 cavities formed by the omission of five holes in a triangular lattice of air holes drilled into semiconductor air-bridge membrane [106] and also for redirection of the laser emission obtained with an InP-based L3 cavity having four embedded InGaAs QWs emitting at 1.55 μm [105] and in an InP-based H0 cavity [109]. Figure 2.35 shows the measurements of the far field emission of these lasers, with and without the subharmonic lattice.

Tapered Fibers to Accede PhC Structures Arguing that the high index of PhC materials required for confinement introduces a spatial mismatch between fiber modes and PhC modes, an evanescent coupling scheme was proposed [110] and implemented [93]. As shown in Figure 2.36, the idea was to bring into the vicinity of the PhC a tapered optical fiber – so that the long evanescent tail of the mode of the tapered fiber penetrates the PhC slab. Initially, this was proposed to couple light into or out of line-defect PhC waveguides. Owing to the large effective index difference between a tapered fiber mode and a normal semiconductor slab waveguide mode, the coupling efficiency is expected to be very weak because the phase-matching condition is not fulfilled (i.e., the requirement for the same wave-vector for the two

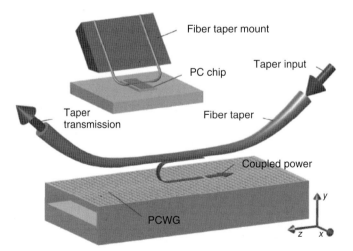

Figure 2.36 Schematic of the coupling scheme [93].

propagating modes or the same effective index). However, by taking advantage of the mode folding at the edge of the first Brillouin-zone, phase-matching becomes possible at a specific wavelength between the two counter-propagating modes belonging to the tapered fiber and the PhC waveguide [110], inducing efficient resonant coupling. Moreover, by varying the position of the tapered fiber, it is possible to tune the coupling strength and the phase-matching wavelength, giving a unique handle for probing of the dispersive and spatial properties of PhC waveguide modes. In fact the key word here is *probe*, as this form of coupling is a very powerful tool for exploring the PhC structure, even though it cannot be used for an extended photonic circuit, as it is limited to probing one isolated element.

In the investigation of laser emission itself, [111] used a sharply curved microfiber positioned above the PhC laser cavity to pump the structure and also to extract the emission. The small curvature radius (∼50 μm) of the microfiber enabled a high degree of localization of the pump, together with minimal interaction with the absorptive region outside of the mode of interest. The 980 nm pump laser light was injected through such a tapered microfiber into a modified L3 cavity by evanescent coupling, in order to be absorbed in the InP/InGaAsP QWs. The emitted photons in the resonant mode close to 1.55 μm were collected by the very same curved microfiber, in both forward and backward directions. A fairly low threshold of 35 μW was shown, with an experimental fiber coupling efficiency η estimated to be about 70% at d_{gap} = 100 nm between the cavity and the fiber. From FDTD simulations, the extracted power at the two fiber ends was calculated to be 84% of the total emitted power, that is, 42% at each end of the fiber.

Edge-Emitting PhC Lasers In order to collect the in-plane emitted power from a wavelength size PhC laser, another possibility consists in fabricating the structures close to the edge of the semiconductor slab to avoid additional losses induced by the material absorption. In [112], 2D PhC edge-emitting lasers were fabricated by following this path and using an InP-based double heterostructure cavity containing QWs (see Figure 2.37). The number of periods left between the edge of the sample and the cavity was reduced to five, permitting redirection of the emission into a single direction and attainment of 120 μW peak output power.

The same team later improved [113] their cavity and obtained an edge-emitted peak collected power level of 540 μW, with 27% quantum differential efficiency.

2.4.2.2 Interfacing Active-PhC Cavity-Based Devices within an Optical Circuit

In the early exploration [114] of PhC coupling, the aim was to fulfill the requirement of combining several elementary blocks with minimal loss, so as to build photonic integrated circuits associating at least two PhC elements and to produce a low-loss system. By fabricating a hexagonal microcavity and a channel waveguide separated by two to five crystal rows and examining their interaction, it was shown that a low-loss coupled system was indeed possible. The use of a waveguide neighboring the PhC cavity to couple light in and out was implemented over the last decade in many different ways. Many studies (for example, [115–117]) have also used

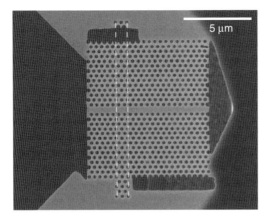

Figure 2.37 Top view SEM image of a fabricated photonic crystal double-heterostructure (PhC DH) cavity, with five PhC cladding periods on the left. The heterostructure region is delineated with white dotted lines [112].

waveguides designed directly inside the PhC slab, to either side- or butt-couple PhC cavities. In some designs, the PhC waveguide can then be coupled to a regular step-index wire waveguide [118] that demonstrates generally better performance, in terms of footprint and losses than W-like PhC waveguides.

In the approximation where the presence of the waveguide weakly perturbs the cavity mode, the evolution of the intra-cavity electromagnetic field can be described by coupled mode theory (CMT) [119]. In this theory, the cavity optical losses are determined by two independent terms one related to the intrinsic losses of the cavity, that is, the losses in the absence of the waveguide, and the other related to the coupling to the waveguide mode. These are described respectively by the quality factors Q_0 and Q_c. The coupling efficiency, η, can then be written as:

$$\eta = \frac{1/Q_c}{(1/Q_0) + (1/Q_c)}$$

It is then clear that, in order to obtain a large efficiency, the coupling losses must be much larger than the intrinsic cavity losses ($Q_c \ll Q_0$).

In what follows, we restrict ourselves to the case of the integration of PhC lasers with optical passive waveguides. In such an optical circuit, as already mentioned, the passive waveguides should not be fabricated in the same material as that of the lasers, in order to avoid absorption losses during propagation at the emission wavelength. To overcome this issue, two approaches were explored, the first one being an all III–V semiconductor based approach [82, 120] just as for telecom devices, making use of epitaxial regrowth to separate the active regions from the passive ones and the second one relying on the heterogeneous integration of III–V active materials on to silicon-on-insulator (SOI) passive optical circuitry [49].

PhCs Lasers Integration with Passive Optical Waveguides through Epitaxial Regrowth To achieve the integration of active devices such as lasers with passive ones such

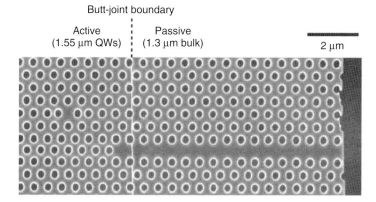

Figure 2.38 SEM image of the fabricated nanolaser and waveguide in the InGaAsP active/passive-integrated PhC slab [122].

as waveguides, one possibility consists in spatially organizing the active material (QWs, QDs, QWires), only where the lasers are to be located. This was achieved in an InP-based system where the active zones were made of InGaAsP materials with embedded QWs emitting at 1.55 μm and where the passive zones were made of transparent bulk InGaAsP [120–122] or bulk InP [82].

In [120–122], PhC lasers were integrated together with a passive line-defect waveguide. For these studies, the authors used, for the active/passive PhC slab, a two-step epitaxial regrowth technique by metal organic chemical vapor deposition. First, the laser wafer including the InGaAsP quantum-well active layer was prepared and patterned into relatively large rectangular mesas (larger than the cavity size), using dry etching. The passive material (1.3 μm InGaAsP + InP) was then selectively grown in the etched areas. The PhC membrane was then processed using the regular process flow described in Figure 2.8. An SEM image of such a fabricated PhC structure is shown in Figure 2.38.

Here [122], the cavity-based laser is positioned close to the boundary between the active and the passive zones, so that the absorption remaining in the unpumped region of the structure does not induce significant additional losses. In [120, 121], the laser emission is based on the use of the low v_g mode of a PhC defect waveguide. In the latter paper, the light was collected from the W1 passive waveguide output using a lensed fiber with 20% efficiency at a fairly high peak power level of 170 μW.

A more sophisticated implementation of the regrowth technique in the framework of PhC lasers was achieved by Matsuo and coworkers [82] by limiting the size of the active region (to $4\,\mu m \times 0.3\,\mu m \times 0.15\,\mu m$) to match the cavity mode profile (see Figure 2.34). The so-called BH region provides gain only where it is needed and enables, simultaneously, increased carrier confinement, which reduces nonradiative recombination within the active material, and the formation of the cavity because of its higher refractive index, together with enhanced thermal heat sinking due to its being embedded within an InP membrane (see Section 2.4.1.2.3). Of course the fabrication of such optimized structures is not simple and necessitates state-of-the-art technology both for the growth and the processing, where the main difficulty resides in the alignment of the PhC with respect to the small active region. In [123], the mode profiles are analyzed in detail to optimize the output coupling of such a PhC laser. The output power levels in these RT operating CW lasers can go up to $\sim$100 µW. The experimental confirmation, with the control over the different coupling aspects and input/output, as well as the coupling between the output line-defect waveguide and a single-mode fiber for connection to the outside world, gave a very convincing 53% external differential quantum efficiency. By showing that direct injection of the pump right into the active region allows the collection of a very large part of the emission, this study illustrates the fact that every increment of improvement takes one a long way in these miniature devices.

III–V PhC Lasers Heterogeneously Integrated onto SOI Waveguides Circuitry Heterogeneous integration of III–V semiconductor structures on to SOI is considered to be one of the key technologies for next generation on-chip optical interconnects. Indeed, this hybrid platform combines the best of both material systems. On one hand, the transparency of silicon at telecom wavelengths and its high-index contrast with silica allows the fabrication of extremely compact low-loss single mode waveguides ($\sim$1 dB cm^{-1}) [124–126] that can be used to route information through a circuit. Critically, the mature complementary metal-oxide semiconductor (CMOS) fabrication processing technology renders possible large-scale integration of functional optical devices, including integration with complex electronic circuitry. On the other hand, III–V semiconductors are of course extremely versatile for making laser sources and active devices, thanks to their direct electronic band gap. Heterogeneous integration of III–V semiconductor structures on SOI waveguides has recently led to the realization of laser sources [127, 128] which is very exciting for further photonic integration with electronics.

In [49, 129, 130], the concept of hybrid III–V semiconductor/SOI photonics was scaled down to nanophotonics by integrating III–V PhC lasers coupled to SOI waveguides. An example of a hybrid structure under investigation is schematically represented in Figure 2.39a and an SEM image is shown in Figure 2.39b. The hybrid structure is a two-optical level structure where the lower level is made of a single mode SOI waveguide and the top level is made of an InP-based active PhC (either PhC waveguide or nanocavity) that contains InGaAsP QWs emitting around 1.55 µm. The two levels are separated by a thin low-index layer. Light channels from one level to the other through evanescent wave coupling. The hybridization relies here on the adhesive bonding of the InP-based heterostructure on the SOI wires, through the use of the planarizing polymer BCB (see Chapter 5 of this book). The success of the fabrication depends, obviously, on the quality of the two parts (SOI and InP structures) and to a very large extent on the accuracy and the repeatability of the alignment of the PhC structures with the subjacent waveguide, which was demonstrated to be better than 30 nm using mixed and matched Deep-UV/electron-beam lithography [129].

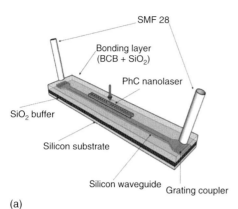 

Figure 2.39 Hybrid III–V semiconductor photonic crystal on SOI waveguide circuitry. (a) Schematic view of the structure. The InP-based photonic-crystal wire cavity nanolaser is positioned on top of silicon on insulator strip waveguide. The two structures are separated by a low-refractive index bonding layer constituted of BCB and SiO_2. Light is extracted from the grating couplers at the extremities via SMF28 fibers. (b) SEM images of the fabricated sample. The SOI waveguides can be seen through the bonding layer aligned with the cavities. Inset: SEM image close-up of a wire cavity [130].

Relatively exhaustive explorations of coupled laser operation from slow-light PhC waveguides [49] and high-Q photonic wire nanocavities [130] to single mode SOI waveguides were carried out. Particularly, evanescent wave coupling was minutely studied in samples where the PhC lasers were based on the use of photonic wire cavities (see Section 2.2.1.2), where the light is trapped between two mirrors formed by a single row of holes drilled into a single mode ridge waveguide. A low threshold is obtained for hybrid lasers in the telecom window within a footprint as small as $5\,\mu m^2$. A thorough investigation of the evanescent wave coupling in the device as a function of the structure parameters, as well as the optical parameters, gives a complete picture of the hybrid system. Optimal parameters may thus be stipulated to obtain tailor-made coupling factors.

In the samples investigated, the photonic wire nanocavities had quality factors as high as 10^5, with modal volumes close to the limit of $(\lambda/2n)^3$ even though the structures were not suspended in air. The SOI level was made of a 220 nm thick Si layer on a $2\,\mu m$ SiO_2 buffer layer. The Si layer was etched down in order to form waveguides of widths, w, varying from 300 to 550 nm, in steps of 50 nm. Grating couplers were etched at a distance from the circuitry to allow coupling with cleaved single-mode optical fibers [131]. The InP wafer was sputter coated with a SiO_2 layer, the thickness of which was chosen to be 200, 300, or 400 nm, before being put in contact with the BCB-coated SOI wafer.

It was shown that the maximum coupling efficiency is obtained for the thinnest low-n layer, as expected, and for $w = 400$ nm – which gives the smallest value of Q_c – the quality factor associated with the evanescent wave coupling to the SOI waveguide. Indeed, for $w = 400$ nm, the SOI waveguide has an effective index of

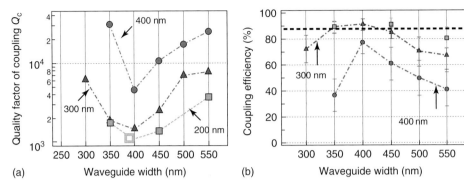

Figure 2.40 Characteristic coupling figures. (a) Measured quality factor associated to the evanescent wave coupling (Q_c) for various SOI waveguide width w and for three low-index layer thicknesses. (b) Coupling efficiency of the emitted light into the SOI wire as a function of the structure parameters. The uncertainty on the coupling efficiency is plotted on the figure with segments. We indicate by the black dotted line the estimated boundary in terms of coupling efficiency between lasing and nonlasing structures [130].

2.25 for wavelengths around 1.55 μm, with a wavevector that corresponds to that of the main peak of the field distribution in the reciprocal space of the InP-based photonic wire cavity mode. This phase-matching enables the maximum transfer of light from the PhC region to the SOI waveguide and thus the minimum Q_c. The coupling efficiency, η, of the emitted light from the PhC into the SOI photonic wires (Figure 2.40a,b) has been plotted for the different separations and shows the possible range of values for η, taking into account the uncertainties in Q_0 (the quality factor of the uncoupled cavity). By adjusting the parameters of the structures, it was demonstrated that it is possible to vary the efficiency, at will, from around 20% to more than 90% by controlling the evanescent wave coupling strength, Q_c. Yet, the cases corresponding to the highest η values do not necessarily correspond to the ideal configuration, as they also give the largest optical losses in the system, which may result in an increase in the laser threshold or, even worse, in a nonlasing situation. The boundary between the lasing and the nonlasing situations is indicated by the dotted line in Figure 2.40b.

2.5
Conclusions

During the last 20 years, remarkable parallel progress in the understanding of PhCs physics and in their fabrication has allowed impressive demonstrations of different key functionalities which could constitute breakthroughs for applications such as sensing or all optical processing in terms of sensitivity, power efficiency, compactness, and bandwidth. Laser emission is one of these functionalities and

PhC laser is the perfect example to show how nanophotonics can bring to the table its specificity to improve the performance of various functionalities.

Indeed PhCs, with their lattice dimensions on the order of the optical wavelength, allow quasi-perfect control of the electromagnetic field propagation. They enable, in particular, the fabrication of possibly the best resonators to date at optical wavelengths, with record Q over V ratios. Myriad designs are now available to build such excellent resonators in different types of systems.

When they incorporate active materials, these resonators render possible the achievement of lasers with ultra small footprints (from few square micrometers to a few hundreds square micrometers) showing very special behavior both in their static characteristic and in their dynamical operation thanks to the enhancement of the light matter interaction and control of the spontaneous emission. Indeed, low threshold or even thresholdless operation has been demonstrated and ultrafast dynamics with modulation bandwidth well beyond 20 GHz has been observed.

Yet, the "glory" of PhCs lasers is still to come. Even though the very fast onset of the research in this field has allowed some outstanding demonstrations, it is only in the last few years that the issues linked to the very nature of PhCs lasers, namely the poor heat sinking, the large nonradiative recombination of carriers and their difficult interfacing, have been partially addressed. These issues have hindered for a long time the possible use of PhCs lasers for applications, since stable RT CW operation could not be realized and the output power levels were limited to few nanowatts. In this chapter, up to this point, the question of electrical injection was omitted, as this subject is treated in the following chapter of this book. Of course, this issue too represents an obstacle to the use of semiconductor PhC nanolasers. So far, electrical injection of PhC lasers operating at optical wavelengths has been successfully achieved by three groups [132–134], which shows the difficulty of realizing such structures. Indeed the whole point is to find a solution to efficient injection of the current into such tiny lasers without spoiling the optical properties of the resonators due to absorptive metallic electrical contacts. Recent progress on this subject consists of either creating a lateral p-i-n junction [133, 134] or smartly designing the electrical contacts [132].

The time is now ripe to test the use of these PhC nanolasers in system experiments and in demonstrators to show their potential in terms of low power consumption and ultrafast operation.

References

1. Yablonovitch, E. (1987) Inhibited spontaneous emission in solid state physics and electronics. *Phys. Rev. Lett.*, **58**, 2059.
2. John, S. (1987) Strong localization of photons in certain disordered dielectric superlattices. *Phys. Rev. Lett.*, **58**, 2486.
3. Chong, H.M.H., De La Rue, R.M., O'Faolain, L., Krauss, T.F., Belabas, N., Levenson, A., Raineri, F., Raj, R., Sagnes, I., Coquillat, D., Astic, M., Delaye, P., Lalanne, P., Frey, R., and Roosen, G. (2006) 3D photonic crystals based on epitaxial III-V semiconductor structures for nonlinear optical interactions. Proceedings of SPIE 6182, Photonic Crystal Materials and Devices III, April 18, 2006, p. 618211.

4. Joannopoulos, J.D., Johnson, S.G., Winn, J.N., and Meade, R.D. (2008) *Photonic Crystals: Molding the Flow of Light*, 2nd edn, Princeton University, Princeton, NJ.
5. Johnson, S.G. and Joannopoulos, J.D. (2001) Bloch-iterative frequency-domain methods for Maxwell's equations in planewave basis. *Opt. Express*, **8** (3), 173–190.
6. Taflove, A. (1995) *Computational Electrodynamics: The Finite-Difference Time Domain Method*, Artech House, Norwood, MA.
7. Joannopoulos, J.D., Villeneuve, P.R., and Fan, S. (1997) Photonic crystals: putting a new twist on light. *Nature*, **386**, 143–149.
8. Johnson, S.G. and Joannopoulos, J.D. (2000) Three-dimensionally periodic dielectric layered structure with omni-directional photonic band gap. *Appl. Phys. Lett.*, **77**, 3490.
9. Sakoda, K. (2005) *Optical Properties of Photonic Crystals*, Springer-Verlag, Berlin, Heidelberg.
10. Saleh, B.E.A. and Teich, M.C. (2007) *Fundamentals of Photonics*, Wiley Series in Pure and Applied Optics, John Wiley & Sons, Inc., New York.
11. Asano, T., Song, B.-S., Akahane, Y., and Noda, S. (2006) Ultrahigh-Q nanocavities in two-dimensional photonic crystal slabs. *IEEE J. Sel. Top. Quant.*, **12**, 1123–1134.
12. Kuramochi, E., Notomi, M., Mitsugi, S., Shinya, A., Tanabe, T., and Watanabe, T. (2006) Ultrahigh-Q photonic crystal nanocavities realized by the local width modulation of a line defect. *Appl. Phys. Lett.*, **88**, 041112.
13. Baba, T. (2003) Low-threshold lasing and purcell effect in microdisk lasers at room temperature. *IEEE J. Sel. Top. Quant.*, **9**, 1340–1346.
14. Vuckovic, J., Loncar, M., Mabuchi, H., and Scherer, A. (2002) Optimization of the Q factor in photonic crystal microcavities. *IEEE J. Quantum Electron.*, **38**, 850–856.
15. Akahane, Y., Asano, T., Song, B.S., and Noda, S. (2003) High-Q photonic nanocavity in a two-dimensional photonic crystal. *Nature*, **425**, 944.
16. Zhang, Z. and Qiu, M. (2004) Small-volume waveguide-section high Q microcavities in 2D photonic crystal slabs. *Opt. Express*, **12**, 3988–3995.
17. Srinivasan, K. and Painter, O. (2002) Momentum space design of high-Q photonic crystal optical cavities. *Opt. Express*, **10**, 670–684.
18. Painter, O., Lee, R.K., Scherer, A., Yariv, A., O'Brien, J.D., Dapkus, P.D., and Kim, I. (1999) Two-dimensional photonic band-gap defect mode laser. *Science*, **284**, 1819–1821.
19. Quan, Q. and Loncar, M. (2011) Deterministic design of wavelength scale, ultra-high Q photonic crystal nanobeam cavities. *Opt. Express*, **19**, 18529–18542.
20. Ahn, B.H., Kang, J.-H., Kim, M.-K., Song, J.-H., Min, B., Kim, K.-S., and Lee, Y.-H. (2010) One-dimensional parabolic-beam photonic crystal laser. *Opt. Express*, **18**, 5654–5660.
21. Velha, P., Rodier, J.C., Lalanne, P., Hugonin, J.P., Peyrade, D., Picard, E., Charvo, T., and Hadji, E. (2006) Ultra-high-reflectivity photonic-bandgap mirrors in a ridge SOI waveguide. *New J. Phys.*, **8** (9), 204.
22. Song, B.-S., Noda, S., Asano, T., and Akahane, Y. (2005) Ultra-high-Q photonic double heterostructure nanocavity. *Nat. Mat.*, **4**, 207–210.
23. Tanaka, Y., Asano, T., and Noda, S. (2008) Design of photonic crystal nanocavity with Q factor of $\sim 10^9$. *J. Lightwave Technol.*, **26**, 1532–1539.
24. Lalanne, P. and Hugonin, J.P. (2003) Bloch-wave engineering for high-Q, small-V microcavities. *IEEE J. Quantum Electron.*, **39**, 1430–1439.
25. Sauvan, C., Lecamp, G., Lalanne, P., and Hugonin, J.P. (2005) Modal-reflectivity enhancement by geometry tuning in photonic crystal microcavities. *Opt. Express*, **13**, 245–255.
26. Nomura, M., Iwamoto, S., Kumagai, N., and Arakawa, Y. (2007) Temporal coherence of a photonic crystal nanocavity laser with high spontaneous emission coupling factor. *Phys. Rev. B*, **75**, 195313.

27. Atlasov, K.A., Calic, M., Karlsson, K.F., Gallo, P., Rudra, A., Dwir, B., and Kapon, E. (2009) Photonic-crystal microcavity laser with site controlled quantum-wire active medium. *Opt. Express*, **17**, 18178–18183.
28. Strauf, S., Hennessy, K., Rakher, M.T., Choi, Y.S., Badolato, A., Andreani, L.C., Hu, E.L., Petroff, P.M., and Bouwmeester, D. (2006) Self-tuned quantum dot gain in photonic crystal lasers. *Phys. Rev. Lett.*, **96**, 127404.
29. Braive, R., Barbay, S., Sagnes, I., Miard, A., Robert-Philip, I., and Beveratos, A. (2009) Transient chirp in high-speed photonic-crystal quantum-dot lasers with controlled spontaneous emission. *Opt. Lett.*, **34**, 554–556.
30. Hostein, R., Braive, R., Le Gratiet, L., Talneau, A., Beaudoin, G., Robert-Philip, I., Sagnes, I., and Beveratos, A. (2010) Demonstration of coherent emission from high-Q photonic crystal nanolasers at room temperature. *Opt. Lett.*, **35**, 1154–1156.
31. Foresi, J.S., Villeneuve, P.R., Ferrera, J., Thoen, E.R., Steinmeyer, G., Fan, S., Joannopoulos, J.D., Kimerling, L.C., Smith, H.I., and Ippen, E.P. (1997) Photonic-bandgap microcavities in optical waveguides. *Nature*, **390**, 143–145.
32. Md Zain, A.R., Johnson, N.P., Sorel, M., and De La Rue, R.M. (2008) Ultra high quality factor one-dimensional photonic crystal/photonic wire microcavities in silicon-on-insulator (SOI). *Opt. Express*, **16**, 12084–12089.
33. Zhang, Y., Khan, M., Huang, Y., Ryou, J., Deotare, P., Dupuis, R., and Lončar, M. (2010) Photonic crystal nanobeam lasers. *Appl. Phys. Lett.*, **97**, 051104.
34. Gong, Y., Ellis, B., Shambat, G., Sarmiento, T., Harris, J.S., and Vuckovic, J. (2010) Nanobeam photonic crystal cavity quantum dot laser. *Opt. Express*, **18**, 8781–8789.
35. Halioua, Y., Bazin, A., Monnier, P., Karle, T.J., Sagnes, I., Roelkens, G., Van Thourhout, D., Raineri, F., and Raj, R. (2010) III-V photonic crystal wire cavity laser on silicon wafer. *J. Opt. Soc. Am. B*, **27**, 2146–2150.
36. Scofield, A.C., Kim, S.-H., Shapiro, J.N., Liang, A.B., Scherer, A., and Huffaker, D.L. (2011) Bottom-up photonic crystal lasers. *Nano Lett.*, **11**, 5387–5390.
37. Lin, S.Y., Fleming, J.G., Hetherington, D.L., Smith, B.K., Biswas, R., Ho, K.M., Sigalas, M.M., Zubrzycki, W., Kurtz, S.R., and Bur, J. (1999) A three-dimensional photonic crystal operating at infrared wavelengths. *Nature*, **394**, 251–253.
38. Tandaechanurat, A., Ishida, S., Guimard, D., Nomura, M., Iwamoto, S., and Arakawa, Y. (2010) Lasing oscillation in a three-dimensional photonic crystal nanocavity with a complete bandgap. *Nat. Photon.*, **5**, 91–94.
39. Sakoda, K. (1999) Enhanced light amplification due to group-velocity anomaly peculiar to two- and three-dimensional photonic crystals. *Opt. Express*, **4**, 167–176.
40. Letartre, X., Monat, C., Seassal, C., and Viktorovitch, P. (2005) Analytical modeling and an experimental investigation of two-dimensional photonic crystal microlasers: defect state (microcavity) versus band-edge state (distributed feedback) structures. *J. Opt. Soc. Am. B*, **22**, 2581–2595.
41. Kogelnik, H. and Shank, C.V. (1971) Stimulated emission in a period structure. *Appl. Phys. Lett.*, **18**, 152–154.
42. Pottage, J.M., Silvestre, E., and Russell, P.St.J. (2001) Vertical-cavity surface-emitting resonances in photonic crystal films. *J. Opt. Soc. Am. A*, **18**, 442–447.
43. Kwon, S.-H., Ryu, H.-Y., Kim, G.-H., Lee, Y.-H., and Kim, S.-B. (2003) Photonic band edge lasers in two-dimensional square-lattice photonic crystal slabs. *Appl. Phys. Lett.*, **83**, 3870–3872.
44. Kwon, S.-H. and Lee, Y.-H. (2004) High index contrast 2D photonic band-edge laser. *IEICE Trans. Electron.*, **E87-C**, 308–315.
45. Monat, C., Seassal, C., Letartre, X., Regreny, P., Rojo-Romeo, P., Viktorovitch, P., Le Vassor d'Yerville, M., Cassagne, D., Albert, J.P., Jalaguier,

E., Pocas, S., and Aspar, B. (2002) InP-based two-dimensional photonic crystal on silicon: in plane Bloch mode laser. *Appl. Phys. Lett.*, **81**, 5102–5104.

46. Johnson, S.G., Villeneuve, P.R., Fan, S., and Joannopoulos, J.D. (2000) Linear waveguides in photonic-crystal slabs. *Phys. Rev. B*, **62**, 8212–8222.

47. Notomi, M., Yamada, K., Shinya, A., Takahashi, J., Takahashi, C., and Yokohama, I. (2001) Extremely large group-velocity dispersion of line-defect waveguides in photonic crystal slabs. *Phys. Rev. Lett.*, **87**, 253902.

48. Kiyota, K., Kise, T., Yokouchi, N., Ide, T., and Baba, T. (2006) Various low group velocity effects in photonic crystal e defect waveguides and their demonstration by laser oscillation. *Appl. Phys. Lett.*, **88**, 201904.

49. Halioua, Y., Karle, T.J., Raineri, F., Monnier, P., Sagnes, I., Roelkens, G., Van Thourhout, D., and Raj, R. (2009) Hybrid InP-based photonic crystal lasers on silicon on insulator wires. *Appl. Phys. Lett.*, **95**, 201119.

50. Kazarinov, R.F. and Henry, C.H. (1985) Second-order distributed feedback lasers with mode selection provided by first-order radiation losses. *IEEE J. Quantum Electron.*, **QE-21**, 144–150.

51. Paddon, P. and Young, J.F. (2000) Two-dimensional vector-coupled-mode theory for textured planar waveguides. *Phys. Rev. B*, **61**, 2090.

52. Ochiai, T. and Sakoda, K. (2001) Dispersion relation and optical transmittance of a hexagonal photonic crystal slab. *Phys. Rev. B*, **63**, 125107.

53. Ryu, H.Y., Kim, S.H., Park, H.G., Hwang, J.K., Lee, Y.H., and Kim, J.S. (2002) Very-low threshold photonic band-edge lasers from free-standing triangular photonic crystal slabs. *Appl. Phys. Lett.*, **80**, 3883.

54. Mouette, J., Seassal, C., Letartre, X., Rojo-Romeo, P., Leclercq, J.-L., Regreny, P., Viktorovitch, P., Jalaguier, E., Perreau, P., and Moriceau, H. (2003) Very low threshold vertical emitting laser operation in InP graphite photonic crystal slab on silicon. *Electron. Lett.*, **39**, 526–528.

55. Seassal, C., Monat, C., Mouette, J., Touraille, E., Ben Bakir, B., Hattori, H.T., Leclercq, J.-L., Letartre, X., Rojo-Romeo, P., and Viktorovitch, P. (2005) InP bonded membrane photonics components and circuits: toward 2.5 dimensional micro-nano-photonics. *IEEE J. Sel. Top. Quant.*, **11**, 395–407.

56. Ben Bakir, B., Seassal, C., Letartre, X., Viktorovitch, P., Zussy, M., Di Cioccio, L., and Fedeli, J.M. (2006) Surface-emitting microlaser combining two-dimensional photonic crystal membrane and vertical Bragg mirror. *Appl. Phys. Lett.*, **88**, 081113.

57. Vecchi, G., Raineri, F., Sagnes, I., Yacomotti, A., Monnier, P., Karle, T.J., Lee, K.-H., Braive, R., Le Gratiet, L., Guilet, S., Beaudoin, G., Talneau, A., Bouchoule, S., Levenson, A., and Raj, R. (2007) Continuous-wave operation of photonic band-edge laser near 1.55μm on silicon wafer. *Opt. Express*, **15**, 7551–7556.

58. Vecchi, G., Raineri, F., Sagnes, I., Lee, K.-H., Guilet, S., Le Gratiet, L., Van Laere, F., Roelkens, G., Van Thourhout, D., Baets, R., Levenson, A., and Raj, R. (2007) Photonic-crystal surface-emitting laser near 1.55μm on gold-coated silicon wafer. *Electron. Lett.*, **43**, 343–345.

59. Coldren, L.A., Corzine, S.W., and Masanovic, M.L. (2012) *Diode Lasers and Photonic Integrated Circuits*, Wiley Series in Microwave and Optical Engineering, 2nd edn, John Wiley & Sons, Inc., Hoboken, NJ.

60. Englund, D., Altug, H., Fushman, I., and Vuckovic, J. (2007) Efficient terahertz room temperature photonic crystal nanocavity laser. *Appl. Phys. Lett.*, **91**, 071126.

61. Fujita, M., Takahashi, S., Tanaka, Y., Asano, T., and Noda, S. (2005) Simultaneous inhibition and redistribution of spontaneous light emission in photonic crystals. *Science*, **308**, 1296–1298.

62. Laurent, S., Varoutsis, S., Le Gratiet, L., Lemaître, A., Sagnes, I., Raineri, F., Levenson, A., Robert-Philip, I., and Abram, I. (2005) Indistinguishable

single photons from a single quantum dot in a two-dimensional photonic crystal cavity. *Appl. Phys. Lett.*, **87**, 163107.
63. Badolato, A., Hennessy, K., Atatüre, M., Dreiser, J., Hu, E., Petroff, P.M., and Imamoglu, A. (2005) Deterministic coupling of single quantum dots to single nanocavity modes. *Science*, **308**, 1158–1161.
64. Englund, D., Fattal, D., Waks, E., Solomon, G., Zhang, B., Nakaoka, T., Arakawa, Y., Yamamoto, Y., and Vuckovic, J. (2005) Controlling the spontaneous emission rate of single quantum dots in a 2D photonic crystal. *Phys. Rev. Lett.*, **95**, 013904.
65. Purcell, E.M., Torrey, H.C., and Pound, R.V. (1946) Resonance absorption by nuclear magnetic moments in a solid. *Phys. Rev.*, **69**, 37.
66. Xu, Y., Lee, R.K., and Yariv, A. (2000) Quantum analysis and the classical analysis of spontaneous emission in a microcavity. *Phys. Rev. A*, **61**, 033807.
67. Bigot, J.-Y., Portella, M.T., Schoenlein, R.W., Shank, C.C., and Cunningham, J.E. (1990) Two-dimensional carrier-carrier screening studied with femtosecond photon echoes, in *Ultrafast Phenomena VII*, C.B. Harris, E.P Ippen, G.A. Mourou and A.H Zewail (eds) Springer Series in Chemical Physics, vol. **53**.
68. Koyama, F., Morito, K., and Iga, K. (1991) Intensity noise and polarization stability of GaAlAs–GaAs surface emitting lasers. *IEEE J. Quantum Electron.*, **27**, 1410–1416.
69. Kuksenkov, D.V., Temkin, H., Lear, K.L., and Hou, H.Q. (1997) Spontaneous emission factor in oxide confined vertical-cavity lasers. *Appl. Phys. Lett.*, **70**, 13–15.
70. Kobayashi, T., Segawa, Y., Morimoto, A., and Sueta, T. (1982) Technical Digest of the 43th Fall Meeting of Japanese Society of Applied Physics, Paper 29a-B-6, September 1982.
71. Yokoyama, H. (1992) Physics and device applications of optical Microcavities. *Science*, **256**, 66–70.
72. Bjork, G. and Yamamoto, Y. (1991) Analysis of semiconductor microcavity lasers using rate equations. *IEEE J. Quantum Electron.*, **27**, 2386–2396.
73. Bjork, G., Karlsson, A., and Yamamoto, Y. (1994) Definition of a laser threshold. *Phys. Rev. A*, **50**, 1675–1680.
74. Yamamoto, Y. and Bjork, G. (1991) Laser without inversion in microcavities. *Jpn. J. Appl. Phys.*, **30**, L2039.
75. Hofmann, H.F. and Hess, O. (2000) Coexistence of thermal noise and squeezing in the intensity fluctuations of small laser diodes. *J. Opt. Soc. Am. B*, **17**, 1926.
76. van Druten, N.J., Lien, Y., Serrat, C., Oemrawsingh, S.S.R., van Exter, M.P., and Woerdman, J.P. (2000) Laser with thresholdless intensity fluctuations. *Phys. Rev. A*, **62**, 053808.
77. Nomura, M., Kumagai, N., Iwamoto, S., Ota, Y., and Arakawa, Y. (2009) Photonic crystal nanocavity laser with a single quantum dot gain. *Opt. Express*, **17**, 15975–15982.
78. Elvira, D., Hachair, X., Verma, V.B., Braive, R., Beaudoin, G., Robert-Philip, I., Sagnes, I., Baek, B., Nam, S.W., Dauler, E.A., Abram, I., Stevens, M.J., and Beveratos, A. (2011) Higher-order photon correlations in pulsed photonic crystal nanolasers. *Phys. Rev. A*, **84**, 061802(R).
79. Raineri, F., Cojocaru, C., Monnier, P., Levenson, A., Raj, R., Seassal, C., Letartre, X., and Viktorovitch, P. (2004) Ultrafast dynamics of the third-order nonlinear response in a two-dimensional InP-based photonic crystal. *Appl. Phys. Lett.*, **85**, 1880–1882.
80. Henry, C.H. (1982) Theory of the linewidth of semiconductor lasers. *IEEE J. Quantum Electron.*, **QE-18**, 259–264.
81. Kim, J., Shinya, A., Nozaki, K., Taniyama, H., Chen, C.-H., Sato, T., Matsuo, S., and Notomi, M. (2012) Narrow linewidth operation of buried-heterostructure photonic crystal nanolaser. *Opt. Express*, **20**, 11643–11651.
82. Matsuo, S., Shinya, A., Kakitsuka, T., Nozaki, K., Segawa, T., Sato, T., Kawaguchi, Y., and Notomi, M. (2010)

High-speed ultracompact buried heterostructure photonic-crystal laser with 13 fJ of energy consumed per bit transmitted. *Nat. Photon.*, **4**, 648–654.

83. Bagheri, M., Shih, M.H., Wei, Z.-J., Choi, S.J., O'Brien, J.D., and Dapkus, D.P. (2006) Linewidth and modulation response of two-dimensional microcavity photonic crystal lattice defect lasers. *IEEE Photon. Technol. Lett.*, **18**, 1161–1163.

84. Altug, H., Englund, D., and Vuckovic, J. (2006) Ultrafast photonic crystal nanocavity laser. *Nat. Photon.*, **2**, 484–488.

85. Ellis, B., Fushman, I., Englund, D., Zhang, B., Yamamoto, Y., and Vuckovic, J. (2007) Dynamics of quantum dot photonic crystal lasers. *Appl. Phys. Lett.*, **90**, 151102.

86. Englund, D., Altug, H., and Vuckovic, J. (2009) Time-resolved lasing action from single and coupled photonic crystal nanocavity array lasers emitting in the telecom band. *Appl. Phys. Lett.*, **105**, 093110.

87. Raineri, F., Yacomotti, A.M., Karle, T.J., Hostein, R., Braive, R., Beveratos, A., Sagnes, I., and Raj, R. (2009) Dynamics of band-edge photonic crystal lasers. *Opt. Express*, **17**, 3165–3172.

88. Bouché, N., Dupuy, C., Meriadec, C., Streubel, K., Landreau, J., Manin, L., and Raj, R. (1998) Dynamics of gain in vertical cavity laser amplifiers at 1.53 µm using femtosecond photoexcitation. *Appl. Phys. Lett.*, **73**, 2718–2720.

89. Yacomotti, A.M., Raineri, F., Vecchi, G., Sagnes, I., Strassner, M., Le Gratiet, L., Raj, R., and Levenson, A. (2005) Ultra-fast nonlinear response around 1.5 µm in 2D AlGaAs/AlOx photonic crystal. *Appl. Phys. B*, **81**, 333–336.

90. Nomura, M., Iwamoto, S., Watanabe, K., Kumagai, N., Nakata, Y., Ishida, S., and Arakawa, Y. (2006) Room temperature continuous-wave lasing in photonic crystal nanocavity. *Opt. Express*, **14**, 6308–6315.

91. Kounoike, K., Yamaguchi, M., Fujita, M., Asano, T., Nakanishi, J., and Noda, S. (2005) Investigation of spontaneous emission from quantum dots embedded in two-dimensional photonic-crystal slab. *Electron. Lett.*, **41**, 1402–1403.

92. Martinez, L.J., Alén, B., Prieto, I., Fuster, D., Gonzalez, L., Gonzalez, Y., Dotor, M.L., and Postigo, P.A. (2009) Room temperature continuous wave operation in a photonic crystal microcavity laser with a single layer of InAs/InP self-assembled quantum wires. *Opt. Express*, **17**, 14993–15000.

93. Barclay, P.E., Srinivasan, K., Borselli, M., and Painter, O. (2004) Efficient input and output fiber coupling to a photonic crystal waveguide. *Opt. Express*, **7**, 697–699.

94. Huang, J., Kim, S.-H., Gardner, J., Regreny, P., Seassal, C., Postigo, P.A., and Scherer, A. (2011) Room temperature, continuous-wave coupled-cavity InAsP/InP photonic crystal laser with enhanced far-field emission directionality. *Appl. Phys. Lett.*, **99**, 091110-1–091110-3.

95. Nozaki, K., Kita, S., and Baba, T. (2007) Room temperature continuous wave operation and controlled spontaneous emission in ultrasmall photonic crystal nanolaser. *Opt. Express*, **15**, 7506–7514.

96. Ichikawa, H., Inoshita, K., and Baba, T. (2001) Reduction in surface recombination of GaInAsP microcolumns by CH_4 plasma irradiation. *Appl. Phys. Lett.*, **78**, 2119–2121.

97. Boroditsky, M., Gontijo, I., Jackson, M., Vrijen, R., Yablonovitch, E., Krauss, T., Cheng, C.-C., Scherer, A., Bhat, R., and Krames, M. (2000) Surface recombination measurements on III-V candidate materials for nanostructure light-emitting diodes. *J. Phys.*, **87**, 3497–3504.

98. Oigawa, H., Fan, J.-F., Nannichi, Y., Sugahara, H., and Oshima, M. (1991) Universal passivation effect of $(NH_4)2Sx$ treatment on the surface of III-V compound semiconductors. *J. Appl. Phys.*, **30**, L322–L325.

99. Englund, D., Vuckovic, J., and Altug, H. (2007) Low-threshold

surface-passivated photonic crystal nanocavity laser. *Appl. Phys. Lett.*, **91**, 071124.

100. Moreau, V., Tessier, G., Raineri, F., Brunstein, M., Yacomotti, A., Raj, R., Sagnes, I., Levenson, A., and De Wilde, Y. (2010) Transient thermoreflectance imaging of active photonic crystals. *Appl. Phys. Lett.*, **96**, 091103.

101. Brunstein, M., Braive, R., Hostein, R., Beveratos, A., Robert-Philip, I., Sagnes, I., Karle, T.J., Yacomotti, A.M., Levenson, J.A., Moreau, V., Tessier, G., and De Wilde, Y. (2009) Thermo-optical dynamics in an optically pumped photonic crystal nano-cavity. *Opt. Express*, **17**, 17118–17129.

102. Hwang, J.K., Ryu, H.Y., Song, D.S., Han, I.Y., Park, H.K., Jang, D.H., and Lee, Y.H. (2000) Continuous room-temperature operation of optically pumped two-dimensional photonic crystal lasers at 1.6 μm. *IEEE Photon. Technol. Lett.*, **12**, 1295–1297.

103. Chen, C., Matsuo, S., Nozaki, K., Shinya, A., Sato, T., Kawaguchi, Y., Sumikura, H., and Notomi, M. (2011) All-optical memory based on injection-locking bistability in photonic crystal lasers. *Opt. Express*, **19**, 3387–3395.

104. Miller, D.A.B. (2009) Device requirements for optical interconnects to silicon chips. *Proc. IEEE*, **97**, 1166–1185.

105. Haddadi, S., Le-Gratiet, L., Sagnes, I., Raineri, F., Bazin, A., Bencheikh, K., Levenson, J.A., and Yacomotti, A.M. (2012) High quality beaming and efficient free-space coupling in L3 photonic crystal active nanocavities. *Opt. Express*, **20**, 18876–18886.

106. Tran, N., Combrié, S., and Rossi, A. (2009) Directive emission from high-Q photonic crystal cavities through band folding. *Phys. Rev. B*, **79**, 041101(R).

107. Portalupi, S.L., Galli, M., Reardon, C., Krauss, T., O'Faolain, L., Andreani, L.C., and Gerace, D. (2010) Planar photonic crystal cavities with far-field optimization for high coupling efficiency and quality factor. *Opt. Express*, **18**, 16064–16073.

108. Tran, N., Combrié, S., Colman, P., Rossi, A., and Mei, T. (2010) Vertical high emission in photonic crystal nanocavities by band-folding design. *Phys. Rev. B.*, **82**, 075120.

109. Narimatsu, M., Kita, S., Abe, H., and Baba, T. (2012) Enhancement of vertical emission in photonic crystal nanolasers. *Appl. Phys. Lett.*, **100**, 121117–121120.

110. Barclay, P.E., Srinivasan, K., and Painter, O. (2003) Design of photonic crystal waveguides for evanescent coupling to optical fiber tapers and integration with high-Q cavities. *J. Opt. Soc. Am. B*, **20**, 2274–2284.

111. Hwang, I., Kim, S., Yang, J., Kim, S., Lee, S., and Lee, Y. (2005) Curved-microfiber photon coupling for photonic crystal light emitter. *Appl. Phys. Lett.*, **87**, 1311107.

112. Lu, L., Mock, A., Yang, T., Shih, M.H., Hwang, E.H., Bagheri, M., Stapleton, A., Farrell, S., O'Brien, J., and Dapkus, P.D. (2009) 120 μW peak output power from edge-emitting photonic crystal double heterostructure nanocavity lasers. *Appl. Phys. Lett.*, **94**, 111101.

113. Lu, L., Mock, A., Hwang, E.H., O'Brien, J., and Dapkus, P.D. (2009) High-peak-power efficient edge-emitting photonic crystal nanocavity lasers. *Opt. Lett.*, **34**, 2646–2648.

114. Smith, C.J.M., De La Rue, R.M., Rattier, M., Olivier, S., Benisty, H., Weisbuch, C., Krauss, T.F., Houdré, R., and Oesterle, U. (2001) Coupled guide and cavity in a two-dimensional photonic crystal. *Appl. Phys. Lett.*, **78**, 1487–1489.

115. Song, B.-S., Noda, S., and Asano, T. (2003) Photonic devices based on in-plane hetero photonic crystals. *Science*, **300**, 1537.

116. Notomi, M., Shinya, A., Mitsugi, S., Kuramochi, E., and Ryu, H. (2004) Waveguides, resonators and their coupled elements in photonic crystal slabs. *Opt. Express*, **12**, 1551–1561.

117. Waks, E. and Vuckovic, J. (2005) Coupled mode theory for photonic crystal

118. Mekis, A. and Joannopoulos, J.D. (2001) Tapered couplers for efficient interfacing between dielectric and photonic crystal waveguides. *J. Lightwave Technol.*, **19**, 861–865.
119. Suh, W., Wang, Z., and Fan, S. (2004) Temporal coupled-mode theory and the presence of non-orthogonal modes in lossless multimode cavities. *IEEE J. Quantum Electron.*, **40**, 1511–1518.
120. Watanabe, H. and Baba, T. (2006) Active/passive-integrated photonic crystal slab microlaser. *Electron. Lett.*, **42**, 695–696.
121. Watanabe, H. and Baba, T. (2008) High-efficiency photonic crystal microlaser integrated with a passive waveguide. *Opt. Express*, **16**, 2694–2698.
122. Nozaki, K., Watanabe, H., and Baba, T. (2008) Photonic crystal nanolaser monolithically integrated with passive waveguide for effective light extraction. *Appl. Phys. Lett.*, **92**, 021108.
123. Matsuo, S., Shinya, A., Chen, C.-H., Nozaki, K., Sato, T., Kawaguchi, Y., Taniyama, H., and Notomi, M. (2011) 20-Gbt/s directly modulated photonic crystal nanocavity laser with ultra-low power consumption. *Opt. Express*, **19**, 2242–2250.
124. Dumon, P., Bogaerts, W., Wiaux, V., Wouters, J., Beckx, S., Van Campenhout, J., Taillaert, D., Luysaert, B., Bienstman, P., Van Thourhout, D., and Baets, R. (2004) Low-loss SOI photonic wires and ring resonators fabricated with deep UV lithography. *IEEE Photon. Technol. Lett.*, **16** (5), 1328–1330.
125. Vlasov, Y. and McNab, S. (2004) Losses in single-mode silicon-on-insulator strip waveguides and bends. *Opt. Express*, **12**, 1622–1631.
126. Gnan, M., Thoms, S., Macintyre, D.S., De La Rue, R.M., and Sorel, M. (2008) Fabrication of low-loss photonic wires in silicon-on-insulator using hydrogen silsesquioxane electron-beam resist. *Electron. Lett*, **44**, 2.
127. Fang, A.W., Park, H., Cohen, O., Jones, R., Paniccia, M.J., and Bowers, J.E. (2006) Electrically pumped hybrid AlGaInAs-silicon evanescent laser. *Opt. Express*, **14**, 9203–9210.
128. Van Campenhout, J., Rojo Romeo, P., Regreny, P., Seassal, C., Van Thourhout, D., Verstuyft, S., Di Cioccio, L., Fedeli, J.-M., Lagahe, C., and Baets, R. (2007) Electrically pumped InP-based microdisk lasers integrated with a nanophotonic silicon-on-insulator waveguide circuit. *Opt. Express*, **15**, 6744–6749.
129. Karle, T.J., Halioua, Y., Raineri, F., Monnier, P., Braive, R., Le Gratiet, L., Beaudoin, G., Sagnes, I., Roelkens, G., van Laere, F., Van Thourhout, D., and Raj, R. (2010) Heterogeneous integration and precise alignment of InP-based photonic crystal lasers to complementary metal-oxide semiconductor fabricated silicon-on-insulator wire waveguides. *J. Appl. Phys.*, **107**, 063103.
130. Halioua, Y., Bazin, A., Monnier, P., Karle, T.J., Roelkens, G., Sagnes, I., Raj, R., and Raineri, F. (2011) Hybrid III-V semiconductor/silicon nanolaser. *Opt. Express*, **19**, 9221–9231.
131. Taillaert, D., Van Laere, F., Ayre, M., Bogaerts, W., Van Thourhout, D., Bienstman, P., and Baets, R. (2006) Grating couplers for coupling between optical fibers and nanophotonic waveguides. *Jpn. J. Appl. Phys.*, **45** (8A), 6071–6077.
132. Park, H.-G., Kim, S.-H., Kwon, S.-H., Ju, Y.-G., Yang, J.-K., Baek, J.-H., Kim, S.-B., and Lee, Y.-H. (2004) Electrically driven single-cell photonic crystal laser. *Science*, **305**, 1444–1447.
133. Ellis, B., Mayer, M.A., Shambat, G., Sarmiento, T., Harris, J., Haller, E.E., and Vuckovic, J. (2011) Ultralow-threshold electrically pumped quantum dot photonic-crystal nanocavity laser. *Nat. Photon.*, **5**, 297–300.
134. Matsuo, S., Takeda, K., Sato, T., Notomi, M., Shinya, A., Nozaki, K., Taniyama, H., Hasebe, K., and Kakitsuka, T. (2012) Room-temperature continuous-wave operation of lateral current injection wavelength-scale embedded active-region photonic-crystal laser. *Opt. Express*, **20**, 3773–3780.

3
Electrically Pumped Photonic Crystal Lasers: Laser Diodes and Quantum Cascade Lasers

Xavier Checoury, Raffaele Colombelli, and Jean-Michel Lourtioz

3.1
Introduction

Molding the flow of light and confining photons in the same way as electrons in quantum structures were the first two challenges for photonic crystals (PhCs) in the early stages of their development [1, 2]. The achievement of high-quality factor microcavities with ultrasmall mode volumes [3, 4] and the demonstration of all-optical devices combining such cavities with tiny waveguides [5] have indeed proven the capabilities of PhCs for the control and manipulation of light during the last decade. Results obtained have also paved the way toward many miniature optical devices such as optical splitters, Mach–Zehnder modulators [6], superprisms [7, 8], resonant-type interferometers [9] and optically pumped lasers, as described in the preceding chapter of this book. However, an external optical source is needed to drive the aforementioned devices, and in many cases an integrated laser diode is desired as the primary source.

From a historical point of view, laser diodes were processed into gratings, now coined as one-dimensional PhCs, in the 1970s in the wake of the seminal article by Kogelnik and Shank [10, 11] with the development of distributed feedback (DFB) laser diodes and distributed Bragg reflector (DBR) laser diodes. More than two decades later, it could then be thought that the extension to two- and even three-dimensional PhC laser diodes would easily follow the progress in the fabrication of two- and three-dimensional photonic structures. Clearly, the evolution has not come up fully to these expectations. The development of electrically pumped 2D and 3D PhC lasers has been real, but slower than was originally imagined. Many reasons can be given for this slow evolution. For all laser diodes in general, the recombination of charge carriers at the walls of the PhC etched holes appears as a serious drawback for an efficient luminescence produced by current injection into the lasing area. For high-Q microcavity lasers realized in suspended membranes in particular, the large density of injected carriers irremediably leads to parasitic thermal heating and free-carrier absorption (through nonradiative recombination, in particular), which in turn severely degrade

Compact Semiconductor Lasers, First Edition.
Edited by Richard M. De La Rue, Siyuan Yu, and Jean-Michel Lourtioz.
© 2014 Wiley-VCH Verlag GmbH & Co. KGaA. Published 2014 by Wiley-VCH Verlag GmbH & Co. KGaA.

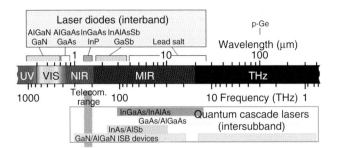

Figure 3.1 Wavelength (frequency) positions of semiconductor laser emissions, from UV to terahertz waves. Electrically pumped photonic-crystal lasers are essentially near-infrared laser diodes emitting in the telecommunication range and quantum-cascade lasers emitting in the mid- to far-infrared domain.

the cavity quality factor and increase the lasing threshold at the same time. Finally, the replacement of one-dimensional DFB or DBR gratings by lateral two-dimensional PhCs in conventional waveguide lasers has to face the remarkable optimization of DFB and DBR laser diodes that has been accomplished for more than three decades in the field of optical telecommunications. The same type of remark applies to vertical cavity surface-emitting lasers (VCSELs), where 2D PhCs have to compete with well optimized, low-loss, ultra-reflecting Bragg layer stacks.

In brief, the achievement of efficient laser emission from electrically pumped 2D PhC lasers at near-infrared and visible wavelengths remains a serious challenge, whatever the laser size. A more favorable situation occurs in the long wavelength or terahertz frequency domain, where solid-state lasers are still in their adolescence (Figure 3.1). Not only are electrically pumped quantum-cascade lasers (QCLs) good candidates for efficient laser emission, but PhC structures can also provide an elegant means for controlling the laser beam. Combining both structures, that is, quantum cascades in multi quantum wells (QWs) for electrons and PhCs for photons, then appears as a promising way for future development in this spectral region. The use of electrical-like or LC-type resonators with metallo-dielectric elements also offers the possibility of designing truly subwavelength resonators and lasers [12–14].

In what follows, we present a synthetic overview of the most recent advances accomplished in the field of electrically pumped photonic crystal lasers. The first part of this chapter focuses on two-dimensional PhC lasers at near-infrared and visible-UV wavelengths including microcavity lasers, waveguide lasers, and VCSELs, respectively. Advantages and drawbacks in the use of 2D PhCs are discussed in detail. The second part of this chapter is dedicated to electrically pumped, quantum cascade, PhC, terahertz, and mid-infrared lasers. Particular attention is paid to the concept of the subwavelength laser in the far infrared. This is followed by concluding remarks and consideration of prospects for further developments.

3.2
Near-Infrared and Visible Laser Diodes

Injecting a few electrons and holes into an ultrasmall laser diode to produce a "thresholdless" laser emission remains the "holy Grail" toward which PhC "micro-nanocavity" experiments are directed. This is indeed a tremendous challenge in terms of fabrication. Meanwhile, PhCs offer many other opportunities to control electromagnetic waves in electrically pumped semiconductor lasers, whether the laser is based on a "micro-nanocavity" in a membrane, a waveguide on a substrate, or a multilayer stack for vertical emission. It is then the purpose of this part to present a general survey of the different PhC geometries that have been explored during the last few decades, in the case of electrically pumped sources at near-infrared and visible wavelengths.

3.2.1
Photonic Crystal Microcavity Lasers

3.2.1.1 Photonic Crystals in Suspended (or Sustained) Membrane

The first optically pumped PhC microcavity laser with emission near $\lambda = 1500$ nm was reported by the Caltech group as long ago as 1999 [15], but 5 years more were needed for the demonstration of a $\lambda \approx 1500$ nm PhC laser with electrical injection, by the KAIST group [16]. Indeed, one of the most crucial problems in implementing an electrically injected free-standing slab PhC laser is the precise fabrication of a p–i–n structure or, in other words, the fabrication of doped semiconductors and metal electrodes in the close vicinity of the microcavity, without introducing too large perturbations of the highly confined optical mode. Only a few emblematic demonstrations have been carried out so far, using two distinct approaches.

The electrical injection scheme used by Park and coworkers [16] was a PhC laser emitting in the vertical direction through a post below a H1 cavity, that is, a cavity with one missing hole in a PhC (Figure 3.2). More precisely, the photonic structure was an optimized H1 cavity where the PhC hole size was chirped around the H1 defect. The resulting resonance quality factor was $Q \approx 2500$, close to the simulated one of 3480, and the mode volume was calculated to be near 0.68 $(\lambda/n)^3 = 0.058 \, \mu m^3$. Electrons, that were injected from the top electrode, and holes, that were injected through the post, recombined in the central layer made of six InGaAsP QWs. One important result obtained with this laser structure was the record-level spontaneous emission coupling factor β of 0.25 estimated from the light–current characteristics. Let us recall that the β-factor is the ratio of the spontaneous emission rate into the cavity mode enhanced by the Purcell effect [17] to the emission rate in all the electromagnetic modes including the cavity one [18, 19]. A β-factor equal to 0.25 means that 25% of the spontaneous emission is coupled to the useful cavity mode. However, the emitted optical power was only a few nanowatts and the threshold current was around 260 μA, under pulsed excitation at room temperature. The relatively large threshold was attributed to several current leakage paths in the fabricated structure, in particular to the

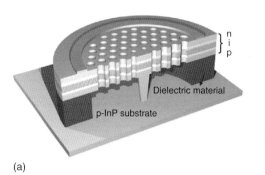 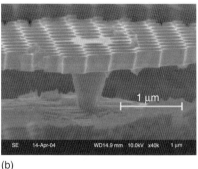

(a) (b)

Figure 3.2 First electrically pumped, photonic crystal, microcavity laser realized at telecom wavelengths [16]. (a) Schematic diagram of current injection. The height of the central InP post is 1.0 μm. The post is diamond-shaped, with 0.64a by 0.51a in diagonal directions, where a is the photonic-crystal period. The diameter of etched mesa is 50 μm, and the inner radius of the AuGe n-electrode is 13 μm. Doping densities of the top n-layer and bottom p-layer are 2.7×10^{19} and 2.5×10^{18} cm^{-3}, respectively. (b) Cross-sectional SEM image. From an intentionally broken sample, the region around the central post is clearly shown. (Reprinted with permission from AAAS).

nonradiative recombination at the semiconductor air hole interfaces. No striking new result in terms of an "electrically pumped microcavity laser" was reported for the seven years that followed this first demonstration. The fabrication of laser devices similar to that represented in Figure 3.2 appeared to be very challenging. Besides, the laser structure itself was fragile, and the use of a post was not compatible with the new cavity designs developed in the period between 2003 and 2008 with Q-factors ranging from 10 000 to 1 000 000, either in silicon [20, 21] or in GaAs [22].

Following a different approach based on the use of a lateral p–i–n junction (see Figure 3.3), Ellis and coworkers from Stanford University [23] demonstrated a new version of an electrically pumped PhC laser in 2011. The lateral junction approach is indeed more robust and better adapted to PhC structuring than the one demonstrated by the KAIST group. It has proved to be successful and well suited to the fabrication of GaAs PhC LEDs (light-emitting diodes) [24] and of Si PhC cavity based modulators [25] and detectors [26, 27]. As seen in Figure 3.3a, the laser cavity used by the Stanford group is an L3 modified cavity similar to the one reported in Ref. [20]. It is fabricated in a suspended GaAs membrane that incorporates three layers of high-density (300 dots μm^{-2}) InAs quantum dots (QDs) with a peak emission wavelength near 1300 nm. The simulated quality factor of the cavity is 115 000 while the experimental one measured at threshold was only 1130, probably because of the free-carrier absorption.

The localized p–i–n junction defined by ion implantation and aligned with a PhC cavity allowed the Stanford group to reach a record low threshold of only 181 nA at a temperature of 50 K and 287 nA at 150 K under continuous electrical injection (Figure 3.3b). This represented an impressive three orders of magnitude improvement compared with the previous demonstration, by KAIST. The beta factor is between 0.29 and 0.41, slightly higher than the one reported by KAIST.

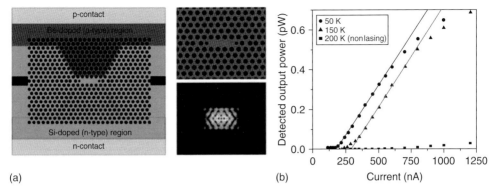

Figure 3.3 (a) Design of the electrically pumped photonic-crystal laser reported in [23]. (Please find a color version of this figure on the color plates.) Left: Schematic of the laser. The p-type doping region is indicated in red, and the n-type region in blue. The width of the intrinsic region is narrow in the cavity region to direct current flow to the active region of the laser. A trench is added to the sides of the cavity to reduce leakage current. Right: modified three-hole defect photonic-crystal cavity design (top) and a FDTD simulation of the E-field of the cavity mode in such a structure (bottom). (b) Experimental output power (detected on the spectrometer) as a function of the current through the laser at 50 K (blue points), 150 K (green points) and 200 K (nonlasing; black points). Red lines are linear fits to the above threshold output power of the lasers, which are used to define the threshold current levels. It is estimated, from the collected output power that the total power radiated above threshold is on the order of a few nanowatts. (Reprinted with permission from Macmillan Publishers Ltd: *Nat. Photonics* **5**, 297–300 copyright 2011).

The total power radiated by the laser is estimated to be on the order of a few nanowatts well above threshold, which represents a radiative efficiency of less than 1%. However, the maximum operating temperature was 150 K. In order to achieve room temperature operation with such nanolasers, future work will have certainly to concentrate on the reduction of the leakage currents including surface carrier recombination in the fabricated structure. An investigation in this direction was the development of a wavelength (lambda)-scale embedded active-region PhC (LEAP) laser by Matsuo and coworkers at NTT [28]. This structure, which consisted of an extremely small, embedded active region in a straight-line defect waveguide in an InP-PhC slab, allowed for a strong reduction of the thermal resistance together with a strong confinement of both photons and carriers in the active region (see Figure 2.34). In a first series of experiments, room-temperature continuous-wave (CW) lasing behavior was obtained with a 390 μA threshold current and a 1.82 μW output power coupled into a waveguide for a 2.0 mA current injection. The NTT group also demonstrated a very narrow linewidth (143.5 MHz) operation from an optically pumped version of this PhC laser [29]. More recently, an ultralow threshold of 7.8 μA was obtained with the same type of structure except for the sacrificial InGaAs layer which was replaced by a wider gap InAlAs layer thus reducing the leakage of current out of the active region [30, 31]. Figure 3.4a shows the light–current characteristic of this laser, which exhibits the lowest threshold

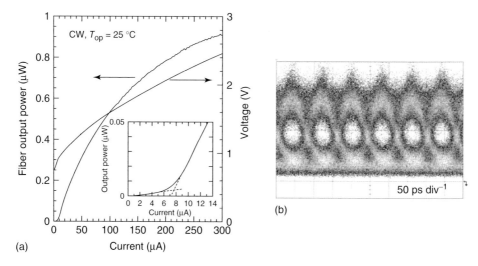

Figure 3.4 (a) Light and voltage versus current curves of LEAP laser under CW operation at room temperature. (b) Eye diagram for a 12.5 Gbps NRZ signal modulation with a PRBS word length of $2^{31} - 1$ © 2013 IEEE. (Reprinted, with permission, from [30]).

current ever reported for any laser at room temperature. For an injection current of 300 μA, almost 1 μW output light could be coupled into an optical fiber. High-speed modulation was also achieved with a 3 dB bandwidth of 16 GHz at 200 μA injection current. Figure 3.4b shows an eye diagram for a 12.5 Gbps nonreturn to zero (NRZ) signal modulation with a pseudo-random binary sequence (PRBS) word length of $2^{31} - 1$. The energy cost per bit was estimated to be 14 fJ b^{-1}.

3.2.1.2 Other Promising Designs

The demonstration of an electroluminescent source is often considered a first step toward the realization of an electrically pumped laser. In this part, we review some LED designs that appear to be promising for the future achievement of PhC microcavity lasers.

In 2008, Kim and coworkers [32] reported the fabrication of a PhC LED structure that presented similarities with the laser structure proposed by Park *et al.* [16]. The main difference stemmed from the use of an active InGaAs QW region that was strictly localized in the PhC microcavity core, but not in its periphery. This arrangement, which was similar to that later proposed by Matsuo *et al.* [28], allowed for an efficient coupling between the optical mode and the gain medium and for a significant reduction of nonradiative carrier recombination at the PhC hole surfaces. Moreover, the use of a buried oxide layer under the PhC membrane also provided a reduction of thermal effects. However, despite these favorable features, lasing at room temperature was not achieved.

Because of its adaptability to many different PhC structures, the lateral p–i–n junction actually appears as the most promising design for electrical injection into micro-nano-lasers. The original design proposed by Ellis *et al.* [23] has been exploited

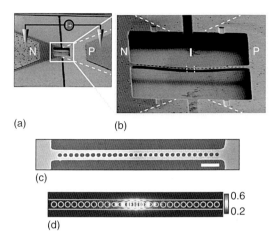

Figure 3.5 (a) Zoomed out and tilted scanning electron microscope (SEM) image of a pair of nanobeam devices [37]. The metal contact pads are observable outside the cavity region and the diagram indicates approximately where electrical probes are placed. The N-type doped region is seen as darker gray and the P-type doped region is outlined via dashed white lines. (b) Zoomed-in SEM of the box region in (a). The nano beam is deflected down by a small amount, probably due to strain at the GaAs/AlGaAs interface. (c) Top view SEM of a nanobeam cavity. The scale bar is 1 μm. (d) FDTD calculated cavity mode electric field magnitude for the designed structure. (After [37], © 2011, AIP Publishing LLC).

in a wide variety of structures by the same team [33, 34]. For instance, the lateral p–i–n junction design has been used for the demonstration of the ultrafast direct modulation of an LED at room temperature. However, perhaps the most promising structure in terms of miniaturization is the electrically injected laser based on a nanobeam cavity. Recently, there has been a renewed interest in nanobeam cavities, which were investigated in the early developments of semiconductor PhCs by Foresi and coworkers in 1997 [35]. Such one-dimensional PhC cavities can indeed exhibit both very small footprints and high-quality factors resulting in optically pumped lasers with a very high β factor of 0.97 [36]. Nanobeam cavities are also suitable for electrical injection as demonstrated very recently by Shambat and coworkers [37], who used a lateral p–i–n junction in the device (Figure 3.5). However, despite a high Q-factor of 2900 and a modal volume around $0.28(\lambda/n)^3$, this structure has not yet been able to yield an electrically pumped emission because of very large nonradiative carrier recombinations at the eched walls. Nevertheless, the emission spectrum was single mode and only 0.5 nm wide (Figure 3.6).

Finally, new solutions can emerge from the hybrid integration of III–V semiconductor PhC lasers with silicon waveguides. This approach should allow the convergence of photonic circuits with CMOS electronic ones. Only optically pumped PhC waveguide hybrid lasers have been demonstrated so far [38]. However, the field is progressing fast, since the first demonstration of an electrically injected hybrid III–V laser on silicon [39], and a breakthrough can reasonably be expected in the near future.

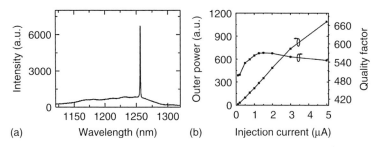

Figure 3.6 (a) Electroluminescence spectrum for a nanobeam device at a forward bias of 5 μA [37]. The cavity fundamental mode is the sharp peak at 1255 nm and the background QD emission is the broad spectrum below. (b) Plot of the cavity output power and Q-factor versus injection current. (After [37], © 2011, AIP Publishing LLC).

As we have now shown, electrically pumped PhC microcavity lasers and LEDs offer impressive results in terms of low threshold, high β-factor, and high speed-modulation. Their development follows and encourages further improvements in fabrication processes, while their characterization and modeling help researchers to gain a better understanding of intricate electromagnetic and transport phenomena at the nanometer scale. In this latter area, microcavity light sources represent invaluable tools. However, from an applications point of view, for example in the telecommunications field, most of them arguably remain laboratory curiosities. Indeed, they typically emit light primarily in a vertical direction and at power levels in the nanowatt range, mainly because microcavities are designed to optimize the resonance quality factor, rather than the emitted power level. Moreover, the majority of ultimately small lasers is so far still unable to operate under continuous excitation at room temperature. The main reasons are the numerous nonradiative electron–hole recombination paths, the high free-carrier optical absorption, and the inefficient thermal management that is inherent in PhC membrane structures. Despite lower Q-factor performance, two dimensional PhCs in the substrate approach therefore remain interesting outsiders, particularly where larger vertical index contrast can be exploited. The following sections will show several completely different PhC geometries, which have been investigated using this approach, during the last decade.

3.2.2
Waveguide Lasers in the Substrate Approach: Weak Vertical Confinement

Edge-emitting ridge-waveguide lasers are key-components for optical telecommunications, since they can deliver high levels of output power in a single mode and in a very stable manner. PhC waveguides in weakly confining vertical structures have been intensively studied at the beginning of the 2000s, as they were believed to exhibit improved properties in terms of threshold current and side-mode suppression ratio (SMSR) compared with standard DFB lasers. The compatibility with classical DFB laser technology, an output power level that can reach 100 mW and better thermal dissipation through the substrate are real advantages of this approach, as compared with the use of a membrane structure. The small index contrast used in this so-called substrate approach obviously leads in turn to much larger mode volumes than in the case of the suspended (or sustained) membrane approach. However, the major limitation of two-dimensional PhC devices with weak vertical confinement stems rather from the leakage of light into out-of-plane directions. Not only must the perspective of small cavities with high confinement be abandoned, but lasers with large β factors have also to be excluded. In a 2D PhC slab based on a classical waveguide slab, the confinement in the vertical direction is determined by the internal reflections of the propagating wave at the lower and upper interfaces of the slab. In practice, one has to design the PhC structure so as to keep the spatial Fourier components of the lasing mode below the light line, that is, far enough from the Brillouin zone center, to avoid optical leakage in the vertical direction. Unfortunately, in the substrate approach, no PhC mode can be simultaneously found in the photonic bandgap and also below the light line defined by $\omega = k \times c/n$, where n is the refractive index of the cladding layers (see for instance Figure 3.7a). In contrast, we may always find cladding modes for which the wavevector component parallel to the guide axis, $k_{//}$, has the same amplitude as the wavevector of the guided mode. In other words, all guided modes in the photonic bandgap can couple to radiative cladding modes, and are leaky modes. Deep etching, by as much as $\sim 5\,\mu m$, of the air holes is generally required to minimize the leakage losses toward the substrate. In some cases, it happens that

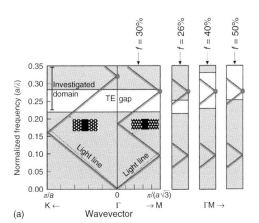
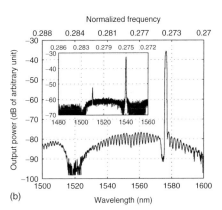

Figure 3.7 (a) Left: schematic representation of the triangular lattice band diagram showing how the fundamental waveguide mode folds, in the two lattice directions ΓK and ΓM. Insets show the W3 and W2-3 waveguides, respectively. Right: Evolution of the transverse electric (TE) band diagram near the M point, as a function of the ratio r/a of the hole radius to the lattice period (After [40], © 2006, AIP Publishing LLC). (b) Emission spectrum of the W3-4 waveguide laser fabricated in Ref. [41], with a lattice period of 432 nm. The electrical injection conditions are: pulse duration ~20 ns, repetition rate 20 kHz, peak voltage 8 V. Inset: emission spectrum of the W3-4 waveguide laser with a lattice period of 424 nm, under optical pumping (pulse duration 15 ns, repetition rate 10 kHz). (After [41], © 2005 The Optical Society).

the strong in-plane confinement of light associated with a 2D photonic band gap is not necessary. A possibility then is to etch holes only into the upper cladding layer region of the structure, thereby making the impact of the PhC structure on optical propagation relatively weak. In what follows, we describe the different types of PhC waveguide lasers according to their size, the waveguide orientation with respect to the PhC lattice orientation, and the kind of interaction that is exploited between the guided mode and the lattice of PhC holes.

3.2.2.1 DFB-Like Photonic Crystal Waveguide Lasers

The standard geometry of a PhC waveguide laser is that of a waveguide formed by one or several rows of missing holes in a periodic lattice of holes. Despite the apparent simplicity of these lasers, different situations can be found depending on the lattice symmetry, on the waveguide width (one, three, or more missing rows) and on the waveguide direction relative to the PhC (ΓK, ΓM, …). Lasing always occurs for modes with low-group velocity since, for these modes, the light–matter interactions are enhanced and a high reflectivity is reached at the waveguide ends because of a strong impedance mismatch. Slow light conditions are mainly obtained either at the center or at the edges of the Brillouin zone, where mode folding occurs. Typically, the lowest thresholds are achieved for the fundamental mode, which is best confined in the waveguide core and exhibits the lowest scattering losses. A schematic view of the fundamental waveguide mode folding is represented in Figure 3.7a for the ΓK and ΓM directions of a triangular lattice PhC in the substrate approach.

For a waveguide oriented along the ΓK direction, the second folding of the dispersion curve occurs at the Γ point for a frequency close to the transverse electric (TE) photonic bandgap. Large airhole radii or, equivalently, large air filling factors (>40%) are required for true photonic bandgap operation. For a waveguide oriented along the ΓM-direction, the third folding of the dispersion curve occurs at the M-point within the gap, even at moderate air filling factors. The periodic modulation of the refractive index is strong in this case, while the 2D photonic bandgap prevents lateral losses. The strong modulation of the refractive index represents the main difference between such PhC lasers and classical DFB lasers. Under optical pumping, it has been shown that single-mode laser emission, with a reduced threshold and increased efficiency, is obtained when the laser mode folds deeply into the bandgap [40]. One possibility for adjusting the frequency range covered by the photonic bandgap consists in modifying the airhole radius, since the photonic bandgap mid-frequency and width strongly depend on the air filling factor. In contrast, the dispersion curve of the fundamental mode is independent of the airhole radius. Lasing in a W3-4 PhC waveguide, the width of which is determined by alternating 3 and 4 missing holes, has been achieved on an indium phosphide (InP) substrate under pulsed electrical injection at room temperature with a SMSR of 40 dB [41]. As expected, lasing was found to occur at the third folding of the fundamental mode in the photonic bandgap. A mini-stopband was detected near the laser line but at a slightly shorter wavelength, as is typical in classical DFB lasers. In other words, gain-loss competition led to the device lasing on the long-wavelength side of this stop-band. The laser spectrum of this electrically pumped

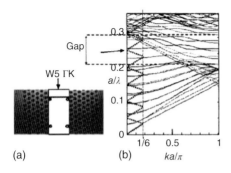
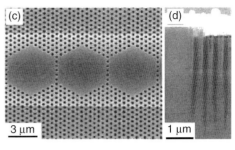

Figure 3.8 Wide photonic-crystal waveguide laser structures. (a) Schematic view of the W5 laser with a periodic constriction of the waveguide width. (b) Dispersion diagram of this W5 waveguide where the zig-zag line shows the fundamental mode folding due to the periodic constriction. The lasing wavelength is indicated by an arrow (After [42], © 2004 AIP Publishing LLC). (c) Scanning electron microscopy (SEM) image detailing three cavities of a coupled-cavity waveguide (CCW) laser (top view) (After [44], © 2002 AIP Publishing LLC). The brighter horizontal rectangle defines the area where the pump current is injected. (d) Cleaved edge SEM image showing the fully processed laser structure with the etched photonic crystal centered around the quantum-well region, polymer insulation layer (dark), and contact metallization (white). The etching depth is about 3.8 μm.

PhC laser is shown in Figure 3.7b. The emission spectrum of the optically pumped version is given in the inset of the same figure.

Another technique to achieve fundamental mode folding into the photonic bandgap is to create a supplementary periodicity in the waveguide instead of adjusting the photonic bandgap frequency itself. This leads to a reduction of the Brillouin zone, while the dispersion curve of the fundamental mode is folded many times. This technique is illustrated in Figure 3.8a,b, where a basic W5 guide (five rows of holes missing in the ΓK direction) is constricted to a W3 geometry every six periods of the PhC matrix [42]. A CW single-mode emission has been obtained under continuous electrical pumping at room temperature with this laser structure. The SMSR was found to be 20 dB. This technique is particularly well suited for large PhC waveguides, for example, with five and more rows of missing holes, since, for these waveguides, the influence of the PhC cladding is too small to ensure single-mode operation with a high SMSR.

The periodically constricted W5 waveguide can also be seen as a system of coupled cavities with low quality factor. Such a system is called a *coupled-cavity waveguide* (CCW) [43]. In fact, the coupling between the modes of identical cavities creates minibands within the photonic bandgap. Each of these minibands corresponds to a particular resonance of the isolated cavities. Because the dispersion curves are nearly horizontal, the guiding structure usually exhibits low-group velocity modes. Even for a situation with weak coupling between cavities, that is, for a large separation between these cavities, the group velocity can indeed be very low. However, a very low-group velocity is accompanied with increased propagation losses in such CCWs, with the consequence that the laser threshold is not necessarily reduced by a large amount. In [44], a CCW with wide hexagonal cavities was fabricated by deep etching of an InGaAsP/InP laser structure (Figure 3.8c,d). A stable single-mode lasing condition was obtained at $\lambda \sim 1.53\,\mu m$. The 150 μm long laser, which was comprised of 40 hexagonal cavities, emitted up to 2.6 mW under CW operation at room temperature. The SMSR was greater than 40 dB. The threshold current was 15 mA. Lasing was found to occur on the short-wavelength edge of the minibands, thereby providing the best spectral match to the material gain.

3.2.2.2 α DFB Lasers

Laser emission under electrical injection has not been reported so far in narrow PhC waveguides such as a W1 channel guide (with one missing row of holes), when using the substrate approach. Even for optically pumped structures, lasing has only been observed for W1 waveguides in square lattice PhCs [45]. Because the active layer is etched, one may suppose that nonradiative recombination at the hole interfaces strongly reduces the available gain in these structures. Even for wide waveguide PhC lasers, the etching of the active layer may reduce the laser efficiency and output power, as compared with standard DFB lasers. One solution to avoid the etching of the active layer and the creation of the nonradiative recombination centers is to etch holes only into the upper cladding layer of the waveguide. Careful design of the PhC has then to be made to ensure single-mode operation with such a weak coupling to the PhC. A broad laser structure emitting at $\lambda = 1000\,nm$ was developed by Hoffmann and coworkers in 2007 [46], who used a rectangular lattice PhC on top of the laser to create a 2D angled-grating DFB mechanism [47, 48] (Figure 3.9a). A similar structure emitting at $\lambda = 1550\,nm$ was proposed in the same year by Zhu and coworkers [49], who emphasized the ability to finely tune the emission wavelength by adjustment of the design parameters (Figure 3.9b). In both cases, the PhC holes are etched only into the upper cladding of the waveguide so as to limit the refractive index modulation. For the wavelength range of interest, resonance conditions in the PhC are satisfied for two pairs of modes that diffract into each other and establish a longitudinal and lateral coherence over the whole PhC structure. Only the modes propagating along the symmetry axis of the semiconductor slab (i.e., normal to the facets) are extracted from the structure, while the other modes are trapped by total internal reflection. The tilt angle ϕ between the cleaved facet and the main axis of the rectangular PhC lattice is, in fact, chosen so as to align the propagation direction of two resonant modes normal to the output laser facets. This laser with a tilted PhC lattice is an extension of the α-DFB laser concept [48].

As a major result in [46], single-mode (longitudinal and transverse) emission was obtained at $\lambda \sim 980\,nm$ with a SMSR of 50 dB and at an injection current of 3 A (Figure 3.9c). The length and width of the laser structure represented in Figure 3.9a were 1 mm and 100 μm respectively. The laser was capable of delivering 100 mW output power in a near-diffraction-limited beam (Figure 3.9d), thereby reflecting the spatial coherence of the electromagnetic field distribution in the whole laser cavity. The maximum laser output power was 140 mW, limited by the available

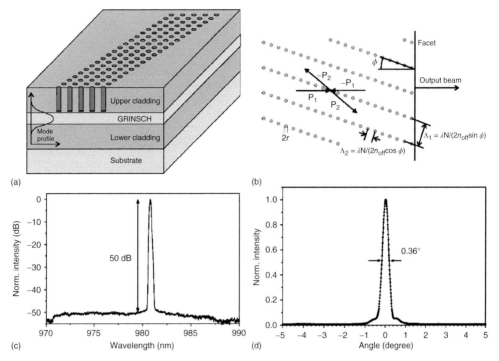

Figure 3.9 (a) Schematic cross section of the DFB laser reported in [46]. The guided mode interacts weakly with an array of holes etched into the top of the laser structure only. The photonic-crystal lattice arrangement is schematized in (b). Λ_1 is the transverse lattice constant, Λ_2 is the longitudinal lattice constant, and ϕ is the facet tilt angle. P_1, P_{-1}, P_2, and P_{-2} are the four modes of the structure. (c) Emission spectrum of a 100 μm wide and 1 mm long-DFB laser at 3 A drive current. (d) Far-field profile of the same laser. The full width at half maximum is 0.36°. (After [46], © 2007 AIP Publishing LLC).

current in the device. Further optimizations of the laser threshold (1.3 A) and of the current–power slope efficiency (0.08 W A^{-1}) appear to have a reasonable probability of being achieved [46].

3.2.2.3 Ridge Waveguide Lasers with Photonic Crystal Mirrors

The simplest way of using photonic bandgap structures in the substrate approach is to exploit PhCs as compact and highly reflective mirrors at the end of ridge waveguides. PhC mirrors also offer the possibility of revisiting earlier single-mode laser configurations, like the so-called C^3 laser, or cleaved coupled-cavity laser [50]. The first PhC version of this kind was investigated by Happ and coworkers [51]. It consisted of two coupled cavities with a ridge waveguide on the top of a QW structure (Figure 3.10a). The longest cavity was ended by a PhC mirror, the shortest one by a cleaved output facet, and their coupling was achieved with an intermediate laterally etched PhC mirror. This device has been further termed a C^2 laser (instead of C^3) for coupled-cavity laser. In this device, each cavity

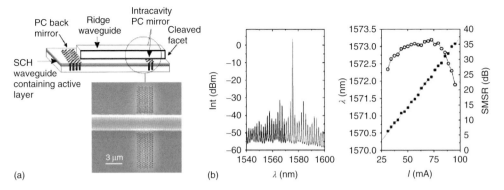

Figure 3.10 (a) Top: schematic structure of the coupled microcavity laser consisting of a high reflectivity back mirror, a lateral intracavity mirror and a cleaved output facet. Bottom: SEM image of the intracavity mirror. The length of the first cavity between the back mirror and the intracavity mirror is ~170 μm, while the length of the second cavity between the intracavity mirror and the cleaved facet is ~35 μm [51]. (b) Emission spectrum of a 200 μm long laser with a superstructure in the side modes that stems from the coupled cavities mode selection. Evolutions of the laser wavelength (square) and of the side-mode suppression ratio (circles) with the bias current for a 400 μm long device. (After [51], © 2001 AIP Publishing LLC).

exhibits a series of Fabry–Perot resonant modes with a given spacing between modes. The larger mode spacing of the short cavity leads to a modulation of the Q-factor of the composite cavity. Single-mode lasing occurs at the highest Qs for wavelengths when both cavities are resonant. Numerous modes are detected below threshold for this cavity configuration, but the SMSR is higher than 30 dB. When the pump current is varied, the refractive index of the laser medium changes, and the wavelength of the lasing mode can be tuned continuously from 1570.5 to 1573 nm (Figure 3.10b). For large variations of the injection current, the laser jumps from one mode to another, thereby providing coarse tuning of the emission wavelength.

In fact, the Vernier effect associated with the coupling of the two cavities can be used for continuous broadband tuning of the emission wavelength, provided that the cavities are independently contacted. In this case, the refractive index variation produced in one segment leads to a spectral shift of the corresponding cavity modes while the mode comb of the second cavity remains unperturbed. This property was used in a more elaborate system developed by Manhkopf and coworkers [52], where ridge waveguides were replaced by wide PhC waveguides. In this system, two independent tunable PhC laser diodes were coupled into a single waveguide using a PhC Y-coupler. Each tunable laser consisted of two independently contacted W6–7 PhC waveguide segments of slightly different lengths, terminated by PhC mirrors and coupled to each other through an intermediate PhC mirror. The laser wavelength was tuned by separately adjusting the injection currents in each of the two laser segments. Quasi-continuous tuning was then achieved in a ~30 nm wavelength window with 36 wavelength division multiplexing channels spaced

0.8 nm apart. The simplicity of this all-PhC fabrication approach makes such a tunable laser design an interesting source for monolithic integration into highly integrated photonic circuits. However, since the PhC is etched through the active layer, nonradiative recombination takes place, as in DFB-like PhC waveguide lasers. This behavior is in contrast with the approach of Figure 3.10, where the PhC structure is only etched laterally or near to the laser extremities.

Tunable, single-mode emitting, coupled-cavity lasers are currently used for gas sensing. For example, Bauer and coworkers have used a C^2 laser similar in design to the one of Figure 3.10 but based on an $In_xGa_{1-x}Sb$ QW structure that emits in the 2 µm spectral range [53]. This laser was single mode and continuously tunable for small variations of the driving current. It also presented mode hops for driving currents equal to 4.4, 7.22, and 8.9 times the laser threshold current. The wavelength ranges where a continuous tuning was obtained corresponded to spectral domains of strong absorption for numerous gases, such as NH_3, N_2O, or CO_2. Using this laser device and recording the light–output characteristics with an InGaAs photodiode, the authors successfully measured the CO_2 concentration in CO_2-enriched atmospheres at room temperature.

More recently, devices based on the C^2 PhC concept have also been used in a methane optical sensor [54]. The active layer consisted of a GaInAsSb/AlGaAsSb QW structure that emitted at 2.35 µm. As seen in Figure 3.11a, the design is again similar to the one of Figure 3.10, except for the ridge waveguide that is interrupted at a few micrometers distance from the intracavity PhC mirror, so as to enhance coupling effects between the two cavities. This modified design allows for better electrical isolation between the 1093 and 453 µm long cavities and a higher output power per facet of up to ~4.5 mW. The emission spectrum shown in Figure 3.11b reveals a SMSR of 14 dB for injection currents of $I_L = 62$ mA and $I_S = 0$ mA in the long cavity and the short cavity, respectively. The latter behaves in this case as a

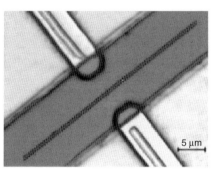
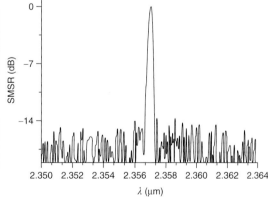

Figure 3.11 Antimonide-based 2.3 µm coupled-cavity lasers reported in [54]. (a) Optical microscope image of the intracavity region with a PhC crossing the ridge waveguide. (b) Emission spectrum at $I_L = 62$, $I_S = 0$ mA, at 24 °C, showing a side-mode suppression ratio of 14 dB. © 2012 IET. (Reprinted, with permission, from [54]).

passive reflector, filtering the lasing mode. On the basis of the C² PhC concept, the laser emission was tuned to coincide with the strong absorption lines of methane in the experiments. Low methane concentrations were measured in a gas cell by using the tunable laser radiation and detecting the acoustic wave produced by the absorption of light, that is, by photoacoustic spectroscopy. Methane concentrations as low as 400 ppb by volume were detected, thereby demonstrating the efficiency of such an optical sensor based on the use of integrated tunable PhC lasers.

3.2.3
Photonic Crystal Surface-Emitting Lasers (PCSELs)

VCSELs are a familiar geometry for semiconductor lasers, in which one or more emitting layers is or are located in a small cavity that is vertically bounded by two semiconductor Bragg reflectors. One advantage of these lasers, as compared with waveguide edge-emitting lasers, is their capability of producing output beams with a small divergence and their relatively easy coupling to optical fibers. Two-dimensional VCSEL arrays are also potentially key elements for application to optical chip-to-chip interconnects. The reasons for interest in the use of two-dimensional PhCs in the VCSEL structure are twofold. Firstly, this is one of the technical approaches for achieving a single-mode operation while maintaining a large emitting area and then a high laser efficiency, as well as a high output power [55]. Two-dimensional PhCs are used for the purpose of laterally modulating the refractive index of the VCSELs rather than for creating a 2D photonic bandgap. In other words, the PhC VCSEL "mimics" a slice of microstructured semiconductor fiber, with all benefits associated with transverse mode control.

A second reason for interest in the use of 2D PhC structures is to replace the upper DBR in long-wavelength VCSELs of the (Al)GaInAs/InP system, where typical thicknesses of Bragg layer stacks lie in the 5–10 µm range for the achievement of high reflectivity (> 99.8%). In 2007, Boutami and coworkers reported vertical Fabry–Perot cavities that were used to realize optically pumped VCSELs at $\lambda = 1.55$ µm, without incorporating a DBR [56]. The DBR was replaced by a two-dimensional PhC membrane, which was operated on a slow Bloch mode at the Γ-point of the transverse electric (TE) band diagram [57]. In the same year, Huang and coworkers [58] reported an electrically pumped VCSEL incorporating a high-index-contrast subwavelength grating. Not only was the epitaxial thickness of the VCSEL mirror reduced, but the fabrication tolerance also increased at the same time. More recently, a very low threshold current density of 667 A cm^{-2} has been achieved at $\lambda = 1.55$ µm by using a hybrid laser structure where the PhC was etched into the middle of a ridge waveguide that formed a horizontal Fabry–Perot cavity [59].

In fact, neither a horizontal nor a vertical cavity is in principle needed to obtain surface light emission from a PhC laser structure provided that the PhC area is large enough to ensure sufficient in-plane feedback. Such a laser structure is referred to as a *photonic crystal surface-emitting laser* (*PCSEL*) instead of a VCSEL. It is also called a *surface-emitting band-edge laser* in the sense that the vertical emission is achieved

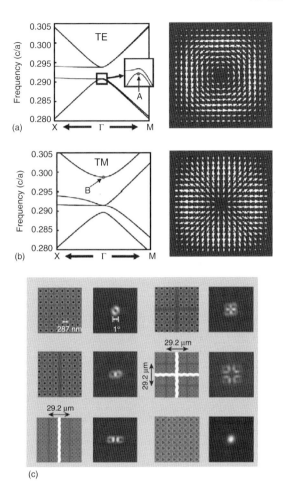

Figure 3.12 (a) Left: band structure of a square lattice photonic crystal for TE-like modes in the vicinity of the folded Γ-point. Four band-edges are shown that enable lasing oscillation. Right: beam pattern and polarization of the cavity mode at band-edge A. (b) Same pictures as in Figure 3.12a for TM-like modes. The beam pattern and polarization are now calculated at band-edge B. (c) A range of beam patterns produced by electrically pumped photonic-crystal surface lasers with engineered lattice points and/or lattice phases [60]. SEM images of crystal structures (left panels) and observed beam patterns (right panels). From top to bottom and left to right: circular lattice points with no lattice shifts, with one shift, with parallel double shifts, with crossed shifts and with double-crossed shifts, triangular lattice points with no shifts (After [60], reprinted by permission from Macmillan Publishers Ltd © 2006).

for low-group velocity modes at the Γ point of the Brillouin zone, in much the same way as DFB-like emissions are obtained at the edge extremities of the Brillouin zone. For illustration, Figure 3.12a,b shows a typical band structure for a square lattice PhC calculated in the vicinity of the folded Γ-point for transverse-electric (TE)-like and transverse magnetic (TM)-like modes, respectively.

In the case of semiconductor laser diodes emitting on band-to-band transitions, waves propagating in the active layers, which form the 2D cavity mode, preferentially have a TE-like polarization. As can be seen in Part 3.3 of this chapter, the reverse situation occurs for QCLs, which emit on TM-polarized intersubband transitions. In both cases, four bands are present at the Γ-point of the square lattice band structure. For TE polarization, detailed analyses reveal that the cavity monopole mode of band-edge A, with the lowest frequency (Figure 3.12a), has the largest quality factor among the four band-edge modes [61, 62]. Lasing occurs at this edge with a tangential polarization and a doughnut-shaped beam spot in the far field. For TM polarization, the largest quality factor is obtained for the monopole mode of band-edge B, which has the highest frequency (Figure 3.2b). Here again, the laser beam is doughnut shaped in the far field, while it now possesses a radial polarization.

Experiments carried out by Miyai and coworkers [60] in the InGaAs/GaAs system at $\lambda \sim 980$ nm have not only confirmed these analyses, but also demonstrated the possibility of producing beam patterns on-demand. This capability is illustrated in Figure 3.12c, which shows beam patterns measured for six slightly different versions of the square lattice PhC. In all cases, the emitted power was larger than 45 mW, and the beam divergence was less than 2°, corresponding to a $50 \times 50\,\mu m^2$ area of coherent oscillations. Single-lobed and various multilobed forms were obtained in the far field, depending on the translational shifts introduced into the crystal lattice. Replacing circular patterns (holes) by triangular ones led to a circular, single-lobed, beam instead of a doughnut-shaped beam. As asymmetry was introduced around the lattice points in the x-direction, destructive interferences could no longer exist in the far field, in this direction.

Of particular interest is the development of surface-emitting optical sources in the visible and UV regions. Lighting applications are indeed a powerful motivation for the development of GaN-based LEDs with high extraction efficiency. Other important applications of UV LEDs and semiconductor lasers include disinfection and sterilization of surfaces and utensils, water disinfection and decontamination, UV curing processes in industry, data storage in video discs, phototherapy, and medical diagnostics. It has been shown that the use of two-dimensional PhCs can significantly enhance light extraction in GaN-based LEDs by coupling guided modes, which would otherwise remain trapped in the higher refractive index InGaN layers [63–65]. One difficulty in the creation of an electrically pumped GaN-based PCSEL lies in the fabrication of a high-quality 2D GaN/air periodic structure close to an unstructured active layer with optical gain. An interesting technique consists in growing the active QW layers and the p-doped cladding over the PhC layer after deposition of silicon dioxide (SiO_2) at the bottom of ~ 100 nm deep airholes [66] (Figure 3.13). The GaN overgrowth proceeds laterally, capping the top of the airholes, whereas the deposited SiO_2 has a growth blocking action. The periodic arrangement of air holes is then preserved during the process (Figure 3.13c).

The device fabricated in [66] had the typical characteristics of a PCSEL. Four luminescence peaks were observed below threshold, corresponding to the four band-edges at the Γ-point, while lasing occurred at the band edge with the

3.2 Near-Infrared and Visible Laser Diodes

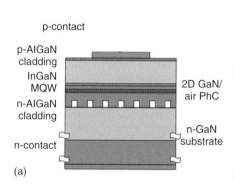
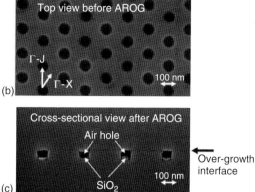

Figure 3.13 (a) Schematic structure of the GaN-based photonic crystal surface-emitting laser (PCSEL) fabricated in [66]. (b) Scanning electron microscope (SEM) image of the triangular lattice of air holes with a period of 186 nm, a diameter of 85 nm, and a depth of 100 nm, formed in an epitaxial GaN/AlGaN layer above the GaN substrate. (c) Cross-sectional SEM image of the PCSEL, showing a well defined GaN/air periodic structure inside the GaN epitaxial layer (After [66], reprinted with permission from AAAS).

lowest frequency (Figure 3.12a). The emitted wavelength was 406.5 nm. For a $100 \times 100\,\mu m^2$ area, the output power was in the milliwatt range and the beam divergence angle was as small as $\sim 1°$. There are obviously many directions for improving the output power and threshold of the current performance, including, for instance, an improved quality of the active layer, a closer distance between this layer and the PhC and the use of a transparent electrode. However, the fabrication of PhC structures with a small period and narrow line patterns ($a = 186$ and $\phi = 85$ nm in Figure 3.13) will remain an obstacle for cost-effective production of short-wavelength lasers as long as a nanometer scale technology such as electron beam lithography (EBL) will be needed to define the pattern.

3.2.4
Nonradiative Carrier Recombination in Photonic Crystal Laser Diodes

Minimizing nonradiative carrier recombination at the walls of PhC etched holes is a crucial issue for the development of all epitaxial semiconductor-based PhC lasers, and for that of electrically pumped laser diodes, in particular. Nonradiative carrier recombination caused by the presence of interface defects depends on many parameters, such as the material system in which the laser is fabricated (GaN, GaAs, and InP, ...), the structures used to confine carriers and to create the laser gain (QWs, quantum wires, or QDs), the etching process itself, related to the approach (membrane or substrate approach), and the parameters (lattice period, air filling factor, and hole shapes). It is usual to analyze nonradiative carrier recombination through the photoluminescence (PL) decay of the semiconductor material. In the etched areas, nonradiative recombination at the sidewalls reduces the PL decay time,

τ, according to the relation: $1/\tau = F/\tau_0 + v_S(S/A)$, where τ_0 is the (radiative) carrier lifetime of the unprocessed material, F, the Purcell factor, that relates the influence of the photonic crystal on radiative recombination, v_S, the surface recombination velocity and S/A, a measure of the ratio of the amount of sidewall area, S, to the total surface area, A, of the emitting region, per unit cell of the lattice [67, 68]. The v_S parameter itself may be expressed as $v_S = cN_d$, where N_d is the surface defect density and c represents the carrier capture rate of defects. As expected, large air filling factors, that is, large values of S/A, tend to decrease the carrier lifetime.

The PL results shown in Figure 3.14 illustrate the dramatic influence of sidewall damage on the carrier lifetime for two material systems and two etching processes. Figure 3.14a shows PL measurements for PhC nanocavities fabricated by using a reactive ion etching (RIE) process in a GaAs slab with a central stack of four InGaAs QWs [69, 70]. The PL signal is found to decay approximately 18 times

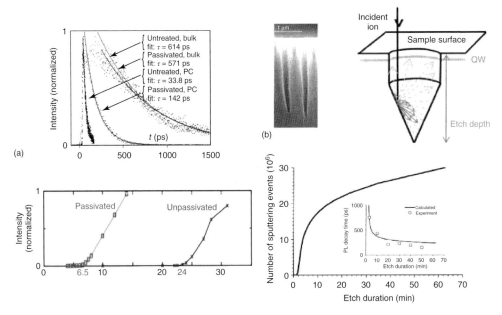

Figure 3.14 (a) Surface passivation of photonic-crystal nanocavity in the InGaAs/GaAs system [70]. Top: Photoluminescence measurements on samples with and without nanocavity. Light gray and dark gray are for passivated and untreated samples, respectively. Bottom: Lasing curves for unpassivated and passivated structures at low temperature (10 K) with pulsed excitation (3.5 ps, 13 ns repetition) (After [70], © 2007 AIP Publishing LLC). (b) Deeply etched photonic crystals in the InGaAsP/InP system [67]. Top: Representative scanning electron microscope cross section of a structure etched for 30 min (left) and schematic sketch of an etched hole showing the incident ion and some of the possible directions for the scattered ion/the ejected etch products (right). Bottom: Calculated total number of accumulated impact events at the position of the quantum well (QW) as a function of the etch duration for $d = 220$ nm hole diameter. The inset shows the measured and calculated carrier lifetimes for $d = 220$ nm. Experimental data are obtained for a period $a = 445$ nm. (After [67], © 2009 AIP Publishing LLC)

faster than in the case of unprocessed GaAs slabs. Clearly, carrier recombination is predominantly nonradiative. The sidewall recombination velocity, v_S, was estimated to be 1.7×10^5 cm s^{-1} at 10 K in [70], not far from values reported, some years ago, for similar InGaAs/GaAs structures at room temperature [71] and for surface recombination in bulk GaAs [72]. This v_S value is also comparable to those measured for other GaAs PhC structures [68]. Fortunately, a significant fraction of the surface defects can be "cured" by surface passivation using, for instance, a (NH$_4$)S treatment [70]. Nonradiative recombination remains dominant, but the carrier lifetime is significantly increased. Using surface passivation, the pump laser threshold was reduced by a factor of four in the experiments by Englund and coworkers [70] (Figure 3.14a, bottom). However, the authors did not comment on the long-term effectiveness of this (NH$_4$)S treatment.

It is known that bulk InP and related InGaAsP/InP heterostructures typically exhibit much smaller surface recombination velocities than GaAs materials [73]. For moderate etch depths like those used in the GaAs experiments mentioned, the v_S value typically decreases to 1.0×10^4 cm s^{-1} or less. In consequence, the approach generally used for circumventing surface recombination in near-infrared suspended-membrane PhC micro(nano)cavity lasers consists in using either the InGaAsP/InP system [16] or an active medium with QD [23]. In the latter case, the carrier diffusion at the interfaces can be dramatically reduced [74], but carrier recombination and current leakage are not totally eliminated in the PhC structure surrounding the QD active medium.

Regarding DFB-like PhC waveguide lasers in the substrate-supported approach, the advantages of the InGaAsP/InP system are compromised by the requirement of large etching depths. Inductively coupled plasma (ICP) etching and chemically assisted ion-beam etching (CAIBE) are generally used for achieving deep etching in semiconductors. Detailed investigations of the CAIBE technique have recently shown that sidewall damage accumulated progressively during the entire etching process time [67].

While the PL decay time of the etched material is found to decrease only by a factor of 2 during the first 10 min of the process (i.e., 0.5–1 µm etch depth), it decreases by a factor of 8 after 50 min etching (i.e., for a 3–4 µm etch depth). This effect is shown in Figure 3.14b, which also shows a PhC hole for a 30 min etching time (left top view) and a schematic sketch of the possible ion trajectories after scattering by the bottom etch surface of the hole (right top view). The active QW used for PL measurements is located 250 nm from the top surface. The cylindrico-conical shape of the holes is typical of deeply etched PhCs in InP [75–78]. In fact, the slow but continuous increase of the number of defects at the position of the QW simply reflects the increase in the total number of accumulated impact events that are calculated from a process model (Figure 3.14b, bottom). This result reveals the dramatic influence of long etch times on the optical quality of PhC structures. Suitable passivation techniques need to be found that eliminate or, at least, palliate the influence of defects generated by etching processes.

One promising solution to the need for strong reduction of the nonradiative surface recombination in PhC laser diodes could be the use of the bottom-up approach, which consists in growing nanopillars or nanowires instead of etching holes. Pillar sidewalls can indeed be perfectly vertical with atomic scale roughness, while a better passivation process is obtained with the growth of lateral shells. Recent experiments on optically pumped bottom-up lasers have demonstrated the feasibility of this approach in the GaAs/InGaAs/GaAs system at $\lambda = 1\,\mu m$ [79]. This approach could be generalized to all semiconductor laser diode systems from the UV-visible to infrared wavelength up to $2\,\mu m$, thanks to the tremendous progress accomplished recently in nanowire growth. At longer wavelengths, the electrically pumped lasers of interest are essentially only QCLs. As will be shown in the following section, these devices are based on intersubband transitions, instead of band-to-band transitions, and do not suffer from nonradiative electron–hole recombinations.

3.3
Mid-Infrared and Terahertz (THz) Quantum Cascade Lasers

QCLs are semiconductor injection lasers which emit in the mid-infrared and terahertz portions of the electromagnetic spectrum. They were first demonstrated at Bell Laboratories (Murray Hill, NJ – USA) in 1994 [80] at mid-IR wavelengths. The extension to the terahertz spectral range was demonstrated in 2002 by an Italian–British collaboration [81]. Unlike interband semiconductor (diode) lasers that emit optical radiation through recombination of electron-hole pairs across the electronic bandgap of the material, QCLs are unipolar devices. The emission is achieved through the use of intersubband transitions in a repeated stack of semiconductor multiple QW heterostructures, an idea first proposed by Kazarinov and Suris in 1971 [82]. Figure 3.15 shows the mid-infrared multi-QW structure developed in the pioneering work by Faist *et al* [80]. It consisted of

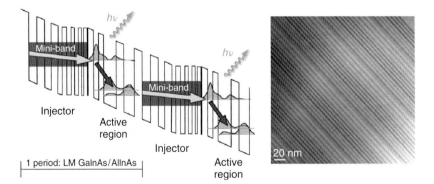

Figure 3.15 (a) Conduction band diagram of a quantum-cascade laser emitting on an intersubband transition at mid-infrared wavelengths [80]. The figure shows the square moduli of the electronic wave functions in the three-quantum-well active regions, which are repeated several times in the multiple quantum-well stack. (b) Transmission electron microscopy (TEM) image of three periods of a QC laser active region (Courtesy of G. Patriarche and I. Sagnes, LPN Marcoussis – France).

three-quantum-well (3QW) active regions periodically spaced by superlattice electron injectors. Each 3QW active region is designed so that the upper laser level is mostly localized in the left-hand well, while the lower laser level wave function is organized so as to reside mostly in the central and right-hand wells. (This configuration produces what is known as a *spatially diagonal transition*). The energy spacing between the lower level of the laser transition and the ground level is designed to be close to the longitudinal optical (LO) phonon energy (~34 meV in GaAs). This leads to resonant LO phonon–electron scattering which can quickly depopulate the lower laser level. Once an electron has undergone a laser transition, it can tunnel into the next period of the structure via the superlattice injector, and then emit another photon. The description provided above covers only the cascade process. Many other QCL designs have been developed since the first successful experiments were carried out. For instance, vertical transition QCLs can be developed, where the upper and lower laser levels are localized in the same wells (vertical transition) [83]. Alternatively, the QCL emission can occur between minibands instead of sub-band levels [84], as in chirped superlattice structures, with the possibility of high current injection [85]. Chirped superlattices can even be engineered so as to operate QCLs without injectors [86]. QCLs can also be based on bound-to-continuum transitions with the possibility of low threshold current densities [87]. A variety of multiple QW structures have thus been proposed to target the terahertz domain, where it is necessary to achieve population inversion between levels whose energy separation is intrinsically small ($h\nu \approx 4-20$ meV) [81, 88]. One common point for all QCLs, due to the cascade process, is the possibility of a high electron-to-photon conversion ratio, and, hence, of high output power levels.

QCLs have become the dominant mid-infrared semiconductor-based laser source. They cover a wide spectral range from $\lambda \approx 2.6$ [89] to $\lambda \approx 24$ μm [90]. Room temperature, CW operation with hundreds of milliwatts of output power has been demonstrated for many wavelengths in the ever-expanding "sweet-spot" region from 3.8 to 10.6 μm [91–95]. QCLs also appear among the most promising solid-state sources in the terahertz, with a spectral coverage extending from ~60 to 240 μm (i.e., 1.25–5.0 THz) [81, 96]. Output power levels up to several tens of, or even one hundred, milliwatts are also attainable at terahertz frequencies [97], but the operating temperature is still limited to ~200 K [98]. While improvements in the design of the QCL active region have somewhat slowed in recent years, more attention has been paid to the photonic laser structure not only for obtaining high power performances but also for achieving (i) a good control of the laser mode and of the laser output beam or (ii) lasers with extremely small modal volumes. Surface-emitting sources have been regarded as attractive alternatives to conventional narrow waveguide lasers that – especially in the long-infrared and THz regions – emit strongly divergent beams with far-field patterns of poor quality. Historically, two main laser geometries – microdisks and 2D PhCs – have been explored. In what follows, results obtained with these two geometries will be presented for the mid-infrared and terahertz, respectively. Considering the ultimate laser dimension achievable in these wavelength ranges, the appropriate question to ask is: "how much smaller than the emitted wavelength can a QCL be"? While optical resonators are constrained to a minimum dimension set by the wavelength (essentially $\lambda/2$), it is known that microwave and radio-frequency oscillators are based on localized circuits (or – equivalently – antennas); their dimensions can be smaller than the wavelength in all the three spatial directions. For terahertz QCLs in particular, it is therefore of great interest to explore the possibility of using microwave- or electronic-inspired resonators with subwavelength sizes. This possibility will be discussed in the last paragraph of this section.

3.3.1
Microdisk QCLs

3.3.1.1 Microdisk QC Lasers at Mid-Infrared Wavelengths

Microdisk and cylinder resonators exhibit ultralow loss "whispering-gallery modes" where light circulates around the curved inner boundary of the resonator, reflecting from the walls with an angle of incidence that is always greater than the critical angle for total internal reflection. This principle allowed the fabrication of the world's smallest lasers at near-infrared wavelengths in the mid-1990s [99–102]. Its extension to mid-infrared QC lasers was appropriate since their inherent TM polarization was expected to inhibit radiation loss from the disk surface. Moreover, the unipolar nature of carrier transport prevented surface recombination effects and, finally, the longer wavelength was more favorable to single-mode emission. The first electrically pumped QC microdisk laser was demonstrated in 1996 using the InGaAs/InAlAs system on InP with a laser emission at $\lambda \sim 5\,\mu m$ [103]. Figure 3.16a,c shows the device structure in the case of a $\sim 2\,\mu m$ thick and $25\,\mu m$ diameter disk on a $\sim 13.5\,\mu m$ InP pedestal. The disk volume is about $8\lambda^3$, while the whispering-gallery mode is contained in only a small percent of this volume. For a still smaller disk with a $17\,\mu m$ diameter, the threshold current was measured to be 2.85 mA at 15 K. However, despite this performance, the in-plane laser emission was hard to collect and the output power was low. A dramatic increase in the output power level was achieved two years later using a "deformed" micro-cylinder geometry [104].

The device side- and top-views are shown in Figure 3.16b. As in the previous case, the InGaAs/InAlAs waveguide core is sandwiched between two semiconductor cladding regions and the laser active region (Figure 3.16c) is designed to emit light at a wavelength of $\approx 5.1\,\mu m$. A tapered InP pedestal is used to support the flat disk. The strong departure from the circular shape is seen from the top view, which shows the quadrupole-like or "stadium" shape of the microdisk. Not only can whispering-gallery modes escape radially from the deformed disk, but within a specific range of deformations a new class of modes, the so-called bow-tie modes [104], appears providing more efficient coupling of the stored energy to the external medium. Representations of the intensity patterns are shown for the two types of modes in Figure 3.16d,e, respectively. Unlike the circularly shaped device, the laser radiation is emitted in specific directions. In the most favorable directions of the radiated far field, the collected power is increased by three orders of magnitude, as

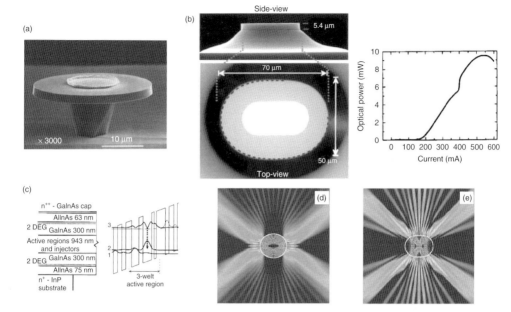

Figure 3.16 (a) Scanning electron microscope (SEM) image of a 25 μm diameter QC disk laser with total height 15.5 μm (After [103], © 1996 AIP Publishing LLC). (b) Left: SEM image of the side and top view of a flattened quadrupolar shaped cylinder laser (After [104], reprinted with permission from AAAS). Right: Light output versus current characteristics measured at 100 K. (c) Left: schematic layer sequence used in the two QC disk lasers (a) and (b). Right: three-quantum-well active region and adjoining quantum-well injector, which are repeated 20 times in the layer sequence. (d) Numerical simulation of the radiation intensity pattern of a chaotic whispering-gallery mode of the cylinder laser. (e) Numerical simulation of the radiation intensity pattern of a bow-tie-mode of the cylinder laser.

compared with conventional circularly symmetric lasers. With a long axis of 70 μm (Figure 3.16b), the optical output power of the laser device then approached 10 mW at $T = 100$ K [104].

3.3.1.2 THz Waves: Several Closely Spaced Demonstrations of Small Volume Micro-Lasers Exploiting Double-Sided Metallic Cavities

Investigations of terahertz microdisk lasers have followed those on their mid-infrared counterparts, as soon as the technology for terahertz QC heterostructures appeared to be sufficiently mature to cover a wide emission spectrum in the long-wavelength range from 60 to 240 μm [81, 105]. The spectral coverage reaches $\lambda \approx 400$ μm if a magnetic field is applied [106]. The major difference with mid-infrared devices stemmed from the use of metal layers to confine the laser modes in the vertical direction. Two types of waveguides are used for terahertz QCLs: the semi-insulating surface-plasmon (SI-SP), and the metal–metal (MM) waveguides. While SI-SP waveguides provide higher output powers and better output beam patterns (edge emission), MM waveguides enable the best high-temperature performance to date and, in general, lower lasing thresholds. MM waveguides indeed offer a near unity optical confinement-factor Γ, that is, an almost perfect overlap of the laser mode with the laser active region. For this reason, they have been preferred in all recent microdisk experiments.

The first results on QC microdisk lasers obtained at terahertz frequencies were reported in 2005 using a circular geometry with 177 GaAs/AlGaAs cascaded active regions and double plasmon confinement in the vertical direction [109]. A threshold current density of 900 A cm^{-2} was measured in pulsed-mode operation at 5 K, for a cylinder radius of 200 µm. The laser emission occurred in the terahertz region between 3.0 and 3.8 THz (i.e., for wavelengths between 79 and 100 µm). Improved devices were reported two years later in two separate experiments by using optimized semiconductor heterostructures and laser geometries [107, 108]. The corresponding devices are shown in Figure 3.17a,b, respectively. Using the device shown in Figure 3.17a, the threshold current density at 6 K was reduced to 120 A cm^{-2} in CW-mode operation, thereby leading to a threshold current of $\sim$4 mA for a 32 µm disk radius (Figure 3.17c) [107]. The 2.9 THz ($\lambda \approx 114$ µm) laser radiation was extracted thanks to the inevitable light scattering at the microdisk outer boundary. The lasing mode was identified as the whispering-gallery mode with radial and azimuthal numbers $N = 1$ and $M = 4$, respectively and the estimated modal volume ($\sim 0.7 \times (\lambda_{eff})^3$) was found to be less than one cubic wavelength. A more detailed analysis of the lasing mode behavior as a function of the disk radius was performed in [108] (Figure 3.17d). Owing to the narrow laser gain bandwidth and the broad intermodal spacing in the case of small micropillars, the laser output was found to be either single mode or bi-modal for micropillar radii varying from 25 to 50 µm, and a larger range of azimuthal numbers was explored.

3.3.2
Photonic Crystal QCLs: Surface Emission and Small Modal Volumes

3.3.2.1 Mid-Infrared Quantum Cascade Lasers with Deeply Etched Photonic Crystal Structures

The typical thickness of semiconductor layer stacks in conventional mid-infrared QCL waveguides ranges from $\sim$5 to 10 µm, depending on the lasing wavelength. This point is illustrated in the SEM (scanning electron microscope) images of the microdisk lasers in Figure 3.16. Therefore the fabrication of PhC structures can be very challenging, due to the required etch depths. An additional difficulty stems from the fact that QC waveguide lasers emit in TM-polarized modes. The use of triangular arrays of etched air holes does not provide a complete photonic bandgap, as can be achieved in near-infrared waveguide lasers instead. Two-dimensional arrays of fully separated semiconductor pillars represent a possible alternative, but disconnected structures need a planarization step during fabrication in order to apply a metallic contact covering the entire structure. The planarization step is difficult, and care must be taken to avoid an inhomogeneous distribution of the injection current in such QCLs with deeply etched PhC structures. On the other hand, QCLs are indeed an ideal laser system: optical scattering from the etched

3.3 Mid-Infrared and Terahertz (THz) Quantum Cascade Lasers

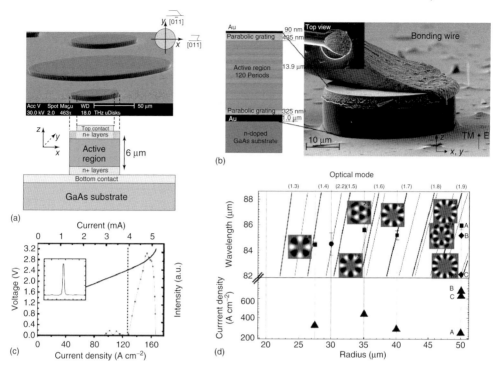

Figure 3.17 (a) Top: scanning electron microscope (SEM) image of the terahertz microdisks fabricated in [107]. Bottom: schematic device geometry. The radii of microdisks are: 32/45/95 μm. Those of the deposited metal are: 25/37.5/87.5 μm. Hatched regions are the portions of $n+$ layer that are removed to increase the resonator Q-factor (After [107], © 2007 AIP Publishing LLC). (b) Vertical structure (left) and SEM image (right) of the bonded 20 μm radius pillar fabricated in [108]. (c) Light–voltage–current characteristics measured at 6 K and in CW mode for microdisks shown in (a) with $r = 25$ μm radius. Inset: emission spectrum for an injection current of 4.75 mA. (d) Electric field patterns and dispersions (black lines) of the whispering-gallery modes calculated versus radius for the micropillar shown in (b). The dashed vertical lines correspond to the radii of the measured pillar (After [108], © 2007 AIP Publishing LLC).

patterns is considerably reduced, owing to the small ratio between the scale of the etching-induced surface roughness and the operating wavelength. Furthermore, the leakage currents due to surface recombination, which are usually substantially increased by the etching of the semiconductor material, are not present because QC lasers are unipolar devices.

The first PhC mid-infrared QC laser was demonstrated by R. Colombelli et al. in 2003 [110]. Figure 3.18a shows the device structure. A high refractive index contrast two-dimensional PhC structure was used to form an in-plane micro-resonator that simultaneously provided feedback for laser action and diffracted light vertically from the semiconductor surface. A key element of the design was the use of a surface-plasmon waveguide for vertical confinement, thanks to the metal contact layer

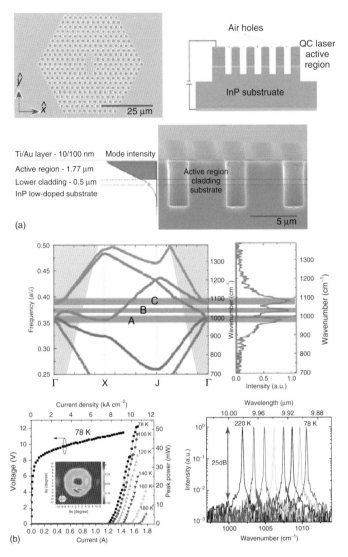

Figure 3.18 (a) From left to right: top-view scanning electron microscope (SEM) image of the first photonic crystal (PhC) mid-infrared quantum-cascade laser [110], a schematic cross section of the device, the layer structure, with the intensity profile of the surface-plasmon mode, and a cross-section SEM image of three air holes (After [110], reprinted with permission from AAAS). (b) From top to bottom: the band diagram with the luminescence spectrum measured in [111], the light–current–voltage characteristics of the laser reported in [111] (left), and the corresponding laser spectra for temperatures from 78 to 220 K (right). The inset in the left figure at the bottom shows the far-field distribution measured at 78 K (After [111], © 2010 The Optical Society).

on the top surface. A relatively thin epitaxial InGaAs/InAlAs material could then be used (≈2.4 μm compared with ≈5.2 μm for a standard waveguide at $\lambda = 8$ μm), thereby leading to a moderate depth for the etched holes. Some structures were fabricated with the central cavity (defect mode) as shown in the left-most image of Figure 3.18a. In all cases, the electroluminescence spectrum measured below threshold was found to exhibit three main peaks, corresponding to the three flat-band regions A, B, and C in the calculated band diagram (Figure 3.18b), respectively. Laser action took place near the A-point of the band diagram, whether the cavity was present or not. However, a higher output power level was obtained when the cavity was present. It is worthwhile to note that the optical frequency at the A-point was close to the second-order Bragg frequency of the hole array.

These pioneering investigations of two-dimensional QC lasers clearly demonstrated the possibility of realizing miniaturized devices at mid-infrared wavelengths with the useful properties of surface emission and single-mode operation. The laser size $\sim (5.4 \times \lambda)^2$ was comparable to, and even smaller than, those of the QCL microdisks shown in Figure 3.16b. The devices could therefore be unambiguously classified as micro-lasers. However the emitted power was low, the threshold current density was relatively high, and the lasing regime was restricted to cryogenic temperatures.

It was only 7 years later that a better understanding of this mid-infrared laser system was achieved in a second series of experiments [111]. Basically, the laser structure fabricated in these experiments was similar to that shown in Figure 3.18a, except for the microcavity defect, which was no longer introduced into the slab. The laser active material, the period ($a = 3.6$ μm) and the radius of the air holes ($r/a = 0.3$) were also close to those used in 2003 experiments [110], but with the targeted emission wavelength being $\lambda \sim 10$ μm instead of $\lambda \sim 8$ μm. The major difference stemmed from the use of an optimized technique for the deposition of the top metal layer. A two-mask process was developed to prevent the metal covering the sidewalls and the bottom of the ~6.2 μm deep air holes during the metal deposition. The optical losses in the PhC were then considerably reduced, thereby leading to better laser performance and easier device characterization. As in the earlier experiments, a triple-peaked luminescence spectrum was observed below threshold, while the laser emission took place near the A-point of the band diagram. This confirmed the band-edge nature of the laser emission.

The light–current characteristics and output spectra of the fabricated lasers are shown in Figure 3.18b (bottom pictures). The spectral and spatial properties of the devices are very good. Single-mode emission with ≥25 dB SMSR is obtained at all temperatures, and a doughnut-shaped beam spot is emitted with an angular divergence close to the diffraction limit. In contrast, despite the 50 mW peak output power level produced in pulsed regime, the temperature and threshold current performance still remain lower than those of standard Fabry–Perot waveguide lasers when using the same material. The metallic absorption losses at the top metal surface, on the order of 30–40 cm^{-1}, are likely to be the cause of this result, which would suggest that surface-plasmon waveguides have to be abandoned. However, it must be remembered that such waveguides were intentionally used to reduce the thickness of the multilayer stack to be etched, while simultaneously avoiding excessive radiation diffraction into the substrate. If surface-plasmon waveguides are abandoned, their replacement by purely dielectric PhC waveguides requires, in turn, the development of new mid-infrared structures with much reduced thickness and, simultaneously, low propagation loss. As will be discussed later, a different solution is to reduce the plasmonic losses themselves via metal patterning. PhC structures can then be etched only through the top metal contact layer, while the rest of the laser is preserved. This type of laser structure will hereafter be referred to as *surface-emitting QC lasers* with thin metallic PhC structures.

If one now considers edge-emitting lasers instead of surface-emitting lasers, deeply etched one- and two-dimensional PhCs can be used as the end-mirrors of a Fabry–Perot cavity without modification of the active region of the waveguide. This solution allows more freedom in the design of the PhC itself. Highly reflective cavity mirrors ($R \geq 0.8$) can be obtained, thereby providing the opportunity for ultrashort Fabry–Perot cavity semiconductor lasers. Decreasing the cavity length in turn favors single-mode operation, because of the increasing mode spacing of the resonator. Ultrashort Fabry–Perot lasers were demonstrated at near-infrared wavelengths by several authors [112–116]. The same kind of demonstration was repeated for mid-infrared QC lasers using either one-dimensional or two-dimensional crystals [117–120]. Figure 3.19 shows the laser devices reported in [119, 120]. The Bragg microlaser was fabricated in the InGaAs/InAlAs system on InP with 35 periods of QC active regions and a total layer thickness of 6.7 μm. The emitted wavelength was $\lambda \sim 8.5$ μm, and the Bragg condition for the mirrors at the laser ends was achieved using semiconductor and air-gap thicknesses of 3.0 and 1.9 μm, respectively. Air gaps were etched with an 8 μm depth using the ICP etching technique. The threshold current density of the Bragg laser was found to be ~35% smaller than that of conventional Fabry–Perot lasers with the same length. This performance was consistent with an estimated Bragg mirror reflectivity of 0.7. Single-mode emission was obtained with ~20 dB side-mode suppression at 80 K, for device lengths near 200 μm (Figure 3.19c). The output power level was in the milliwatt range, while the maximum operating temperature was ~290 K.

Nearly the same performance was obtained with two-dimensional mirror lasers (Figure 3.19b), but at a longer emission wavelength ($\lambda \sim 10.4$ μm). The laser structure, of which the total thickness was ~12.5 μm, was fabricated in the GaAs/Al$_{0.45}$Ga$_{0.55}$As material system. The laser reflectors consisted of ~13 μm high, 1.5 μm diameter semiconductor micropillars arranged in a triangular array with a 3.5 μm period, in such a way that broad stop-bands were obtained for the TM-polarized QC laser radiation in the wavelength range of interest [18]. The mirror reflectivity was estimated to be larger than 0.85 for three periods regardless of the orientation (ΓM or ΓK relative to the laser facet) of the PhC. For a 275 μm long device, the single-mode laser output power level was in the milliwatt range (Figure 3.19d), with a maximum operation temperature of 280 K. It is worthwhile to note that both surface- and edge-emitting lasers actually emit in the same range of wavelengths and possess approximately the same active volume. One drawback

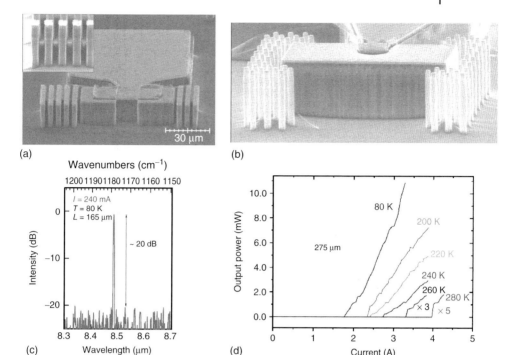

Figure 3.19 (a) Scanning electron microscope (SEM) image of a 50 μm long QC microlaser with distributed Bragg reflectors at both ends of the laser resonator (After [119], © 2007 AIP Publishing LLC). (b) SEM image of a 100 μm long QC microlaser with 2D PhCs at both ends of the laser resonator (After [120], © 2007 AIP Publishing LLC). (c) High resolution single-mode spectrum of a 165 μm long Bragg reflector microlaser measured at 80 K in 100 ns pulse operation. (d) Light–current characteristics for a 275 μm long 2D device emitting at $\lambda = 10.3$ μm. The curves at 260 and 280 K are rescaled for better clarity.

of the surface-emitting laser is the lower value of the maximum operating temperature. In contrast, one of its real advantages is the smaller angular divergence of the emitted beam.

3.3.2.2 Mid-Infrared Quantum Cascade Lasers with Thin Metallic Photonic Crystal Layer

Deeply etched semiconductor structures with high refractive index contrast allow one to obtain strong optical feedback, which is crucial for the implementation of compact QC lasers or, in perspective, microcavity devices that operate on a defect mode of the PhC. However, high-index contrasts are not necessarily required for surface-emitting band-edge QC lasers, where the emission surface has to be reasonably reduced to keep the laser size small, but sufficiently large to obtain a weakly divergent output beam. Moreover, as previously noted for the surface-plasmon based QCLs (Figure 3.18a), the modal field intensity in the *entire* structure presents its maximum at the top metallic surface, which in turn leads to

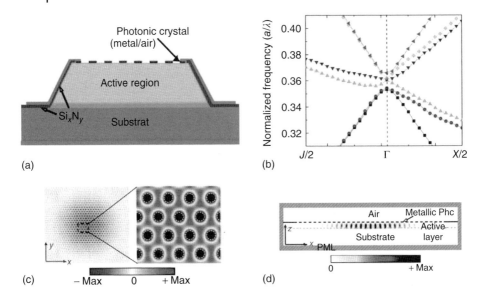

Figure 3.20 (a) Schematic cross section of the QC laser with thin metallic photonic crystal on top [121]. (b) Band diagram around the $a/\lambda = 0.36$ normalized frequency. (c) Electric field distribution (E_z component orthogonal to the semiconductor layers) of the monopole mode with an x–y section taken at the center of the laser active region. (d) Light intensity distribution for the monopole mode in an x–z section across the center, with the simulation domain indicated in gray (After [121], © 2010 SPIE). (Please find a color version of this figure on the color plates.)

relatively high optical losses. A different situation occurs if the pattern of periodic holes is etched only into the device top metal layer, without etching the underlying semiconductor material. This situation is illustrated in Figure 3.20, which represents such a mid-infrared QC laser structure through the field distribution of the laser mode at the normalized frequency of interest ($a/\lambda \sim 0.36$). A close-up of the hexagonal band diagram around this frequency is also represented.

Numerical calculations using a three-dimensional finite-element solver show that the mode with the lowest loss at the Γ-point is a monopolar mode [121], whose electric field distribution (z-component) in the horizontal xy plane is depicted in Figure 3.20c. In the regions covered by the metal, the optical electric field is bound to the metal/semiconductor interface, while it decays almost exponentially into the semiconductor material. On the contrary, in the regions without metalization (below the air holes), the optical electric field is at a maximum at the center of the active layers, while it decays toward the air claddings and the substrate, respectively. As shown in Figure 3.20d, the light intensity indeed reaches its maximum in the active layer below the air holes, which explains why the material loss for the monopole mode can be much lower than for the standard surface-plasmon mode in Figure 3.18a, while still achieving elevated optical gains. In principle, monopolar Bloch modes of infinite structures do not couple to the continuum of radiative modes, since the surface TE and TM fields, which are responsible for the surface emission, are antisymmetric. In reality surface emission can occur, because of the finite size of the structure. However, compared with deeply etched lasers, a larger number of PhC periods, and therefore a somewhat broader structure, are required for band-edge operation, since the effective index modulation along the laser waveguide is lower in the present case.

Figure 3.21 shows the results obtained for typical devices, which were fabricated in the InGaAs/InAlAs system and with the laser emission at $\lambda = 7.5\,\mu m$ [121]. The Ti/Au metallic layer on the device top was ~80 nm thick while the PhC period was 2.7 μm. For devices with 38 periods and an emission area of $\sim(14 \times \lambda)^2$, a single-mode laser emission was obtained at *room temperature* (300 K) in pulsed operation with output power levels in the 20–30 mW range (Figure 3.21b). The SMSR was higher than 20 dB, and the threshold current density (2.7 kA cm^{-2} at 78 K) was three times lower than that of a Fabry–Perot waveguide laser with approximately the same modal volume. These device performances were found to be rather independent of the overall laser shape, which was designed to be either square or hexagonal, or even irregular. In fact, absorbing boundary conditions were present at the sloped sidewalls of the structure (Figure 3.20a), which prevented parasitic feedback to the laser active region [122]. The sidewall slope was produced by the wet etching of the semiconductor mesa.

A doughnut-shaped laser spot was measured (Figure 3.21c) as in previous experiments reported for band-edge lasers. This is a general result: the doughnut shape results from the slowly varying field envelope function associated with the lasing mode at the Γ-point (here the monopole mode). This function is almost constant at the device center. The unpolarized nature of the emitted mode was confirmed by measuring its two orthogonally polarized components with a linear polarizer placed in front of the detector (Figure 3.21c) [121]. Linearly polarized emission was achieved in another series of experiments by modifying the PhC geometry (Figure 3.21d). Elliptic holes were used instead of circular ones, and a π shift phase-delay was introduced to obtain single-lobed emission [123]. The laser beam was emitted normally to the device surface, with a divergence angle as small as $(2.4° \times 1.8°)$.

3.3.2.3 Terahertz (THz) 2D Photonic Crystal Quantum-Cascade (QC) Lasers

The first investigations of terahertz two-dimensional PhC QC lasers started soon after those on their mid-IR counterparts. Because of the much easier fabrication of PhC structures at these long wavelengths, terahertz QCLs are ideally suited for use in PhC structures. Double-sided MM waveguides of moderate, subwavelength height offer the best confinement for the TM-polarized light in the active region, between two metallic layers [124, 125]. This point and the larger PhC period (i.e., the longer distance between motifs) enable, for instance, the fabrication of pillar structures possessing a complete photonic bandgap, with no out-of-plane losses. This opportunity was first exploited to realize selective mirrors for either Fabry–Perot cavity lasers [126] or hexagonal cavity lasers [127]. The pillar structure was further used in the active region of the QC waveguide laser itself, where the

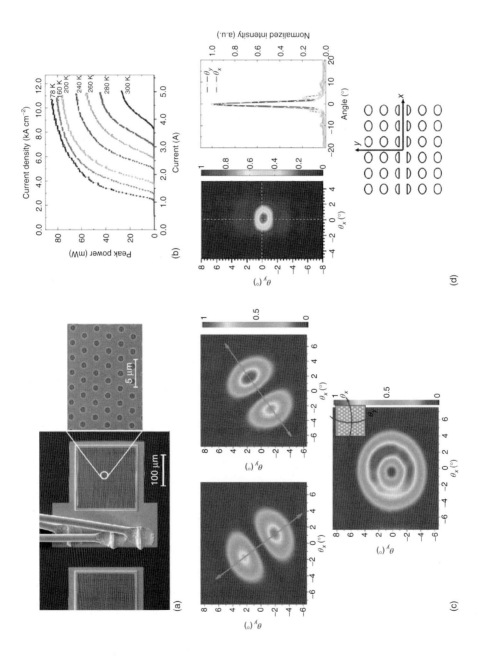

metallic layers of the surface-plasmon waveguide provided the contacts for the electrical injection [128, 129].

Figure 3.22a,b shows SEM images of a terahertz QC laser realized with a triangular lattice of pillars [128]. The latter were created from the high refractive semiconductor QCL gain material embedded and planarized in benzocyclobutene (BCB), which is transparent at terahertz frequencies. At the band edge, the structures act as pure 2D DFB resonators and thus provide the necessary feedback for lasing. The PhC structure has a complete photonic bandgap for TM-polarized light, with flat bands over a large domain in k-space and low-loss band-edge states (Figure 3.22c). Laser emission was indeed obtained in this "slow light" domain of the band diagram. One noticeable result was the reduction of the threshold current compared with regular Fabry–Perot and microdisk lasers with the same gain material and approximately the same sizes. For a 0.1 mm^2 laser footprint, the current density was moderately reduced from 155 to 145 A cm^{-2}, but the absolute threshold current was decreased by a factor of ~4.4 (from 183 to 42 mA).

This result is explained by the fact that for the lasing mode, the antinodes of the in-plane electric field component occur in the low-index waveguide BCB material, but not in the high-index active material. Figure 3.22d shows a typical evolution of the voltage–light–current characteristics with temperature for such terahertz lasers. It is also typical for all terahertz QC lasers. At a given temperature, the output power abruptly increases above threshold, then reaches a maximum and finally decreases to zero at higher current densities. Such a behavior is caused in part by the energy mismatch between the material gain and the optical mode, which limits the dynamics. Sub-band misalignment at high voltage also dramatically lowers the carrier injection efficiency. The minimum threshold current density is ~0.128 kA cm^{-2} in Figure 3.22d.

Despite improvements in threshold performance, the collection efficiency of terahertz, MM, edge-emitting waveguide lasers such as those represented in Figure 3.22 is low, because of the divergent far field emission pattern. Regarding this aspect, surface-emitting lasers appear to be far more advantageous. Investigations of such lasers in the terahertz domain have been very much parallel to those performed in the mid-IR domain. Here again, direct patterning of the PhC into the metallic top waveguide layer has been found to be the most appropriate technique for reduction of the optical losses at the metallic contact while concentrating the laser field in the active region. In contrast, more attention has been paid to the boundary conditions

Figure 3.21 (a) SEM image of a square device fabricated in [121], with a close-up of the metallic photonic crystal on top of the structure. (b) Light-current characteristics measured in pulsed regime (50-ns-long pulses) at different heat-sink temperatures. Note the operation at room temperature (in pulsed regime). (c) Top: far-field patterns measured at 78 K with a linear polarizer placed in front of the detector. Bottom: doughnut-shaped beam spot measured without polarizer (After [121], © 2010 SPIE). (d) Single lobe beam spot (top left) measured for a device with a modified photonic-crystal geometry (bottom) [123]. Unpolarized scans of the far-field patterns along the θ_x and θ_y axes are shown in the top right figure (After [123], © 2010 The Optical Society).

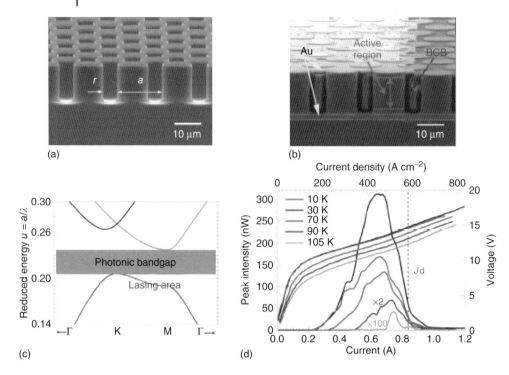

Figure 3.22 Scanning Electron Microscope (SEM) images and performances of the $f \approx 3.5$ THz lasers with pillar arrays described in [128]. (a) Cross-section SEM image of the pillars after dry etching, showing their verticality and sidewall smoothness (~15.5 μm). (b) Side view SEM image of double metal configuration bounding the pillars, which are surrounded by benzocyclobutene (BCB). (c) Photonic-crystal band diagram with the lasing area near the lower bandgap edge. (d) Light–current–voltage characteristics of the laser for various temperatures under pulsed conditions (200 ns pulses). The lattice period is $a = 17$ μm (After [128], © 2007 The Optical Society).

at the laser periphery, since constraints with epitaxial techniques impose device thicknesses of 10/14 μm maximum, for operating wavelengths in the 60–200 μm range. Such extremely subwavelength modes get highly reflected at the device boundaries if precaution is not taken, hence whispering-gallery-like modes instead of true PhC band-edge states can be alternatively excited, depending on whether reflecting or absorbing boundaries are implemented at the periphery of the laser device [130].

In fact, the nature of the lasing mode is in direct relation with the evolution of the laser threshold current, which depends on the PhC lattice parameters. Using mirror boundary conditions, whispering-gallery-like modes have the highest Q-factor and reach lasing before the PhC modes. The threshold current is then almost independent of the r/a parameter, where r is the airhole radius and a is the lattice period. Furthermore, no control on the wavelength is possible. The reverse situation occurs when absorbing boundary conditions are used. The

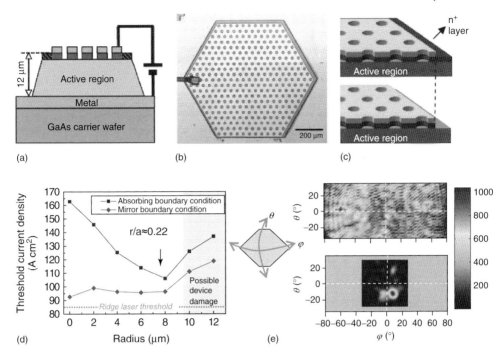

Figure 3.23 (a) Schematic cross section of surface-emitting photonic-crystal terahertz quantum-cascade lasers fabricated in [130]. (b) Optical microscope image of the surface of a typical device. (c) Detailed scheme of the different boundary conditions with or without the top n + contact layer between the metal edge and the mesa periphery (top and bottom panel, respectively). (d) Threshold current density ($T = 10\,\mathrm{K}$) measured as a function of the photonic-crystal hole radius r for the absorbing (top curve) and mirror (bottom curve) boundary conditions, respectively. A constant lattice period ($a = 36.5\,\mu\mathrm{m}$) was used. (e) Far-field patterns measured for devices with absorbing or mirror boundary conditions (top and bottom panel, respectively). The lattice parameters are: $a = 35.4\,\mu\mathrm{m}$ and $r/a = 0.22$ (top panel); $a = 36.1\,\mu\mathrm{m}$ and $r/a = 0.22$ (bottom panel). The (0,0) angular direction corresponds to an emission exactly orthogonal to the device surface (After [130] © 2009 Nature Publishing group).

Q-factor of whispering-gallery-like modes is strongly decreased, and PhC modes have the lowest threshold. A variation of the PhC parameters induces in turn a dramatic variation of the threshold current, while gentle variations lead to tuning of the single-mode spectral emission. All these trends were verified in experiments carried out on GaAs/AlGaAs THz QC lasers emitting in the 2.55–2.9 THz range [130].

Device images are shown in Figure 3.23a–c. A 12 µm thick laser active region was sandwiched between two metal Ti/Au contact layers. The top metal was patterned with a triangular PhC design. Hexagonal mesas were etched down to the bottom metal level to avoid lateral current dispersion, as shown in Figure 3.23b. The wire bonding was applied directly on the top metallization. The top n + GaAs contact layer, approximately 200 nm thick, was removed in the PhC holes, in order to reduce optical losses. In contrast, at the device edges, the n + GaAs layer was either removed or left in place, depending on whether mirror or absorbing boundary conditions were to be used (Figure 3.23c). When the top n + contact was left in place between the metal edge and the mesa periphery, the absorption was high and the boundary behaved like an absorber. When instead it was removed, the mode mismatch at the metal edge led to a high reflectivity and mirror boundary conditions.

As shown in Figure 3.23d, the laser threshold was found to be almost completely independent of parameters for mirror boundary conditions, while it strongly varied with the air hole radius for absorbing boundary conditions. The lowest threshold value (≈ 0.1 kA cm^{-2} at $T = 10$ K) was close to that measured for MM ridge waveguides fabricated from the same epitaxial material. Interest in lasing on PhC modes instead of whispering-gallery-like modes is illustrated in Figure 3.23e, which shows laser far-field patterns measured for the two boundary conditions. When the emission occurs on delocalized PhC modes, the emitted far-field pattern localizes in the angular domain. The output beam exhibits a small number of lobes and even a single one for certain arrangements of the bonding wires [131]. On the contrary, when the whispering-gallery-like mode is excited, the emission is highly nondirectional, with several, unpredictable hot spots distributed over a very wide angular range.

For a laser emission at $\lambda \sim 110$ μm ($f = 2.7$ THz), the emitting surface of the laser represented in Figure 3.23 is already small, of the order of $\sim (6.7 \lambda)^2$. Further reduction of the device area, with the same geometry, unavoidably lowers the device Q-factor and then increases the threshold current density with a simultaneous reduction of the maximum operating temperature. The use of a smaller device area also reinforces the influence of the device edges including the metallic pads for the electrical contacts. This in turn may dramatically degrade the output beam quality. Complementary strategies must then be implemented to avoid the two types of difficulties. Regarding the device Q-factor, a well-established strategy for maintaining a high-Q value in the case of small microcavity lasers is to use photonic heterostructures by using grading of the hole size across the PhC [132]. Although a DFB mechanism – and not the presence of a photonic gap – is responsible for light confinement in band-edge lasers, the same strategy can also be successfully applied in the present case, as shown in [133]. Regarding the beam quality, a single-lobed beam pattern is achieved by introducing a π-phase-shift in the center of the crystal (Figure 3.21d). The π-shift method was in fact successfully applied to obtain single-lobed emission from large-area 2D PhC lasers at near- and mid-infrared wavelengths [123, 134]. Finally, precise modeling of the laser mode field distribution across the whole PhC area can help to find the most appropriate arrangement for the metallic contacts at the top surface [135].

The optimization techniques described above were applied, in an incremental manner, in a series of experiments on 2D surface-emitting terahertz QC lasers with reduced surface dimensions [133, 135]. The basic laser ingredients (active material, MM waveguide, two-dimensional metal patterning, etc.) were the same as those used for the devices in Figure 3.23. The laser emission also occurred in the 2.7–2.9 THz frequency range. Results obtained with the optimization of the

PhC geometry [133] are shown in Figure 3.24a,c. The use of a graded, instead of a uniformly regular PhC, is shown to decrease the threshold current density, while it simultaneously increases the maximum operating temperature by 25 K (from 124 to 149 K). The emitted power level is multiplied by a factor of 5 at its maximum value. This performance is only slightly modified in the presence of a π-shift at the center of the region, while the laser far-field pattern exhibits, in that case, a single lobe, instead of a doughnut shape [133].

Figure 3.24b,d shows results obtained on smaller devices that have the optimal arrangement for the electrical contact [135]. Only 10 periods of a graded PhC are used, with a π-shift at the center of the region, covering a device area of about $(3\lambda)^2$. The threshold current at 78 K is measured to be 1.35 A, instead of 2.4 A, for the devices of Figure 3.24a. The major optimization is that a single-lobed far-field pattern, with a nearly circular shape, is obtained when the bonding pad is placed at the center of the PhC region (Figure 3.24d). The counterintuitive result that the central bonding pad does not perturb the emission pattern is explained by the field distribution of the lasing mode, which is the monopolar Bloch mode depicted in Figure 3.20. This antisymmetric mode couples to the continuum of radiative modes because of the finite size of the structure. It is modulated by a spatially slowly varying envelope function that is almost constant at the device center. As a consequence, the central region does not contribute to surface emission, as has been verified from calculations of the near-field map after removing the contribution of the evanescent components [135]. The absence of central holes in the PhC region and/or the application of a central wire bonding therefore do not affect the far-field distribution.

Using a resonator design similar to that shown in Figure 3.24b (right), but with a broader emitting surface area (600 µm^2) and a QC medium with an optimized current density ($\sim$100 A cm^{-2}), CW laser action has been achieved at 2.7 THz in a single-lobed beam with an output power level of $\sim$300 µW [136].

The output power issue in QC lasers operating on band-edge states is related to the radiative or nonradiative character of the lasing modes. It is easier to visualize it in one-dimensional structures. In such (THz) second-order DFB QC lasers [105, 137], radiative and nonradiative modes exist at the center of the photonic band structure, separated by an energy band gap. The radiative mode exhibits a significantly larger radiation loss because of its field symmetry. Therefore, lasing usually occurs on the nonradiative mode, but with a low power extraction efficiency, which is essentially determined by the finite size of the device. One can easily conclude that much higher power extraction/slope efficiencies would be obtained [138] if it were possible to operate such lasers on intrinsically radiative modes instead.

This possibility has recently been demonstrated using a graded PhC laser system that can be considered as forming a Type-II *graded photonic heterostructure* (GPH) with a position-dependent photonic band gap [139]. In this structure, the nonradiative mode is spatially delocalized, while the radiative modes are confined and their frequencies quantized, very much like holes and electrons in a Type-II QW. The GPH resonator *reverses* the mode competition and activates the mode with

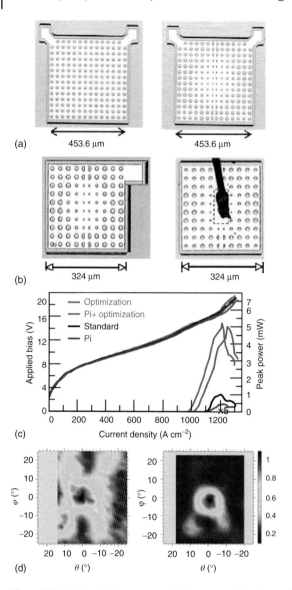

Figure 3.24 Optimized structures of 2D band-edge terahertz QC lasers with small device areas. (a) Microscope images of devices fabricated in [133]. Left: uniform PhC with 8.1 μm hole radius. Right: graded with π-shift and hole radii in the 5.4–10.8 μm range. For both devices, the lattice period is 32.4 μm. Small regions in the two corners of each device are used for wire bonding. (b) Microscope images of devices fabricated in [135]. Left: 10-period graded with π-shift and one bonding pad on the device side. Right: 10 period graded with π-shift and central bonding pad. (c) Voltage–light–current density characteristics of devices fabricated in [133] with and without Q-factor optimization, with and without π-shift ($T = 78$ K) (After [133], © 2010 AIP Publishing LLC). (d) Far-field emission patterns of the devices shown in (b) (After [135], © 2010 AIP Publishing LLC). (Please find a color version of this figure on the color plates.)

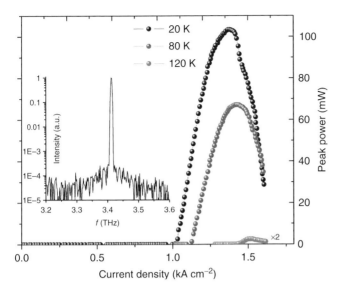

Figure 3.25 Peak output power (surface emission) of a wide ridge graded photonic heterostructure laser [139]. The ridge width is 214 μm, and the central period of the photonic heterostructure is 27.4 μm. Light–current curves are shown at different operating temperatures. The inset shows a typical emission spectrum measured at 20 K. It is spectrally single mode with a side-mode suppression ratio of more than 30 dB (After [139], © 2012 Nature Publishing group).

higher radiation loss, which is never the natural lasing mode in traditional 1D/2D PhC lasers [140]. Record single-mode peak output powers of 103 mW have been obtained at 20 K with such a resonator design in a 3.4 THz QC laser (Figure 3.25) [139]. The external power slope efficiency was measured to be 230 mW A^{-1}, which was more than 10 times higher than that of their DFB counterparts at the same frequency. For the future, one could imagine powerful terahertz sources based on a matrix of such surface-emitting QC lasers provided that all matrix elements could be appropriately controlled in phase.

It is worth mentioning an alternative, very elegant approach for power extraction in terahertz QC lasers, which is based on the concept of third order DFB gratings [141]. Its operating principle is closer to antenna concepts than to PhC ones. One of the main advantages of this architecture is that the far-field divergence is controlled by the device length, not by the device surface [142, 143].

3.3.3
Toward THz QC Lasers with Truly Subwavelength Dimensions

Returning to the main issue of this book, one may wonder how small the terahertz 2D PhC heterostructure lasers of Figure 3.24 could be made. Further size reduction of the device area below ∼$(3\lambda)^2$ should certainly be feasible with improved active region performance and fabrication technology. Similarly, PhC heterostructure *microcavity* terahertz lasers based on small micropillar arrays (cf. Figure 3.22) could be imagined with ultrasmall footprints. However, a minimum number of PhC motifs will be always necessary to provide either an efficient DFB or a strong cavity confinement for laser emission. All elementary microcavities confined by dielectrics only, including microdisks and micropillars, are also unable to provide an effective mode volume V_{eff} smaller than $\sim 2(\lambda/2n_{eff})^3$, where λ is the wavelength, and n_{eff} is the effective refractive index [144–146].

In fact, only the use of metallic or metallo-dielectric microcavities can provide mode volumes below this limit. This point has been illustrated, in the near-infrared to visible domain, through the use of plasmonic cavities in a waveguide configuration [147], with plasmon resonances of nanowires [148], with resonances of nanoparticles [149, 150], or with MM patch-cavities [151, 152]. Terahertz waves offer a much greater opportunity for demonstration of strongly sub-λ metallo-dielectric microcavities, because of the much smaller absorption losses of metals [153]. Their proximity with microwaves and electronics also allows one to borrow device concepts that have been established for many years. The simplest electronic oscillator is the LC circuit (L for inductance, C for capacitance) for which the dimensions are not constrained by the wavelength.

First investigations of an ultrasmall terahertz QC laser mimicking a resonant LC circuit have been reported in 2010 by Walther and coworkers [12]. The laser schematic and a SEM image of the fabricated device are shown in Figure 3.26a,b, respectively.

The resonator consists of two half-circular-shaped "capacitors", connected by a short line that acts as an "inductor". An 8 μm thick QC active region with its gain peaked at 1.5 THz ($\lambda = 200$ μm) [96] is inserted between the "capacitor plates" in a MM waveguide configuration. The central pad is used as the ground connection. The entire device is immersed in a strong magnetic field perpendicular to the QW planes. For a magnetic field strength of 2.3 T and a temperature of 10 K, the emitted laser power was in the 15 nW range. The effective cavity mode volume was estimated by the authors to be $\sim 0.12(\lambda/2n_{eff})^3$, which is largely less than 1/10th of the ultimate volume possible for dielectric cavities. Figure 3.26c shows the emitted laser spectrum as a function of the inductor length of the resonator. This evolution would indicate that the resonator frequency still depends on the cavity length, l, rather than on the inductance L. Indeed, a square-root dependence with both L and l would have been expected for a resonant LC circuit where the inductance varies almost linearly with the device length. Despite the impressively small mode volume and laser footprint, the resonator behavior is more optical-like than electrical-like in this case. In any case, this work certainly reopened the quest for a unification of optics and electronics concepts.

In fact, what matters in electronic circuits, as also in microwave antennas, is the electric current distribution. Unlike the case of optical resonators, where the periodic energy exchange in an oscillator is mostly mediated by the displacement current, $\partial \mathbf{D}/\partial t$, the presence of a real electric current $\mathbf{J}$ is a key ingredient for strongly subwavelength oscillators. Here, it is worthwhile recalling that the responses of metallic PhCs and periodically structured metallic high-impedance

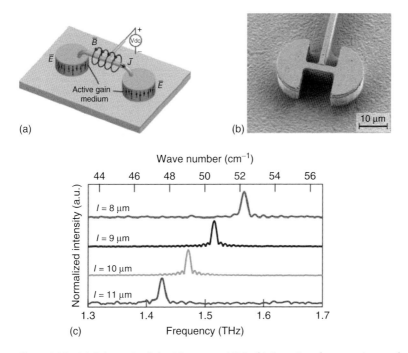

Figure 3.26 (a) Schematic of the LC microresonator-based Laser fabricated in [12]. $\vec{J}$ is the alternating current in the resonator, $\vec{B}$ is the induced magnetic field, and $\vec{E}$ is the electric field. The active gain medium is biased by the voltage source VDC. (b) Scanning electron micrograph image of the LC laser device. (c) Lithographic tuning of the laser frequency by changing the inductor length of the resonator (After [12], reprinted with permission from AAAS).

surfaces in the microwave regime are usually described in terms of LC circuits [18]. Such a description is also familiar in the context of metamaterials [154, 155]. A proof-of-concept of a true electrical-like terahertz resonator has been demonstrated in [13], with subwavelength dimensions in all three spatial dimensions. This resonator is topologically equivalent to a split-ring resonator (SRR), and consists of a semiconductor slab sandwiched between two metallic plates, which are connected to each other by a lateral metallic pad (Figure 3.27a,b).

Because of the lateral connection, charge exchange between the opposing plates is allowed. This activates a resonance with a monopolar electric field distribution, where the plates have electric charges with opposite signs. Figure 3.27c shows the electric field distribution of this mode, which is in fact the fundamental magnetic dipolar mode of a SRR. This is the resonator mode that yields the highest subwavelength confinement within the capacitive gap. It is usually inactive in purely optical resonators, and it is forbidden by the charge conservation law in standard patch-antenna resonators [156]. The monopolar mode has a net magnetic moment along the y-direction, so it can couple, even at normal incidence, with electromagnetic waves with a nonzero magnetic field component along the y-axis.

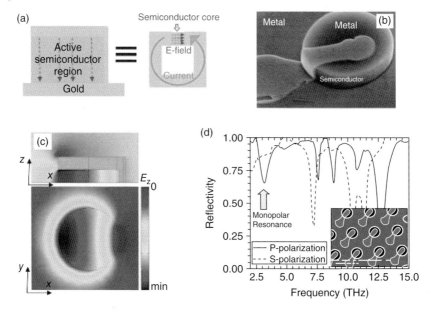

Figure 3.27 (a) Schematic of the split-ring resonator (SRR)-like structure proposed in [13]. (b) Scanning electron microscope image of a typical SRR-like terahertz resonator. (c) Side/top view of the field distribution (electric field component orthogonal to the metallic contacts, E_z, for the monopolar mode of the SRR-like resonator. (Please find a color version of this figure on the color plates.) (d) Typical reflectivity measurement for SRR-like resonators with 6 μm diameter. (The inset shows an SEM image of the resonator array). The solid line corresponds to the polarization which couples to the monopolar resonance, marked by a gray arrow. The dashed line corresponds to the 90°-rotated polarization, which does not couple to the monopolar mode (After [13], © 2012 AIP Publishing LLC).

Figure 3.27d shows the reflection spectrum measured at normal incidence for an array of 6 μm diameter SRR-like resonators with a 1 μm thick semi-insulating (SI)-GaAs slab. A monopolar resonance is well detected at ~3.4 THz (i.e., $\lambda \sim 88$ μm) for the appropriate polarization. This corresponds to a $\lambda_{\text{eff}}/4$ confinement of the terahertz light, where λ_{eff} is the wavelength inside the material (or $\lambda/13$, if the free space wavelength is considered). The maximum confinement achieved in the experiments [13] was for resonators with a diameter of 13 μm, which were operated at a wavelength of 272 μm. A $\lambda_{\text{eff}}/6$ (or $\lambda/20$) confinement was then reached, corresponding to a record low effective mode volume of $V_{\text{eff}} = 0.0025$ $(\lambda/2n)^3$.

More recently, the demonstration of hybrid electronic–photonic subwavelength cavities was reported [14]. The capacitance and inductance of these devices are not interdependent, as in purely photonic cavities, but they can be adjusted almost independently, as in an electronic circuit. The capacitance corresponds to the semiconductor region (emitter, detector, for instance), sandwiched between two metallic plates. The inductance (or, equivalently, the antenna) is in this case a closed metallic loop integrated on the cavity (a closed-loop antenna). As a result, these devices are characterized by dimensions as small as $\lambda_{eff}/9$ in all directions of space (or $\lambda/30$ with the vacuum wavelength) [14].

3.4
Concluding Remarks and Prospects

More than two decades of wavelengths from ∼0.4 to 110 μm have now been covered by electrically pumped two-dimensional PhC lasers. Even if actual applications of PhCs to lasers were to remain limited in scope in the future, their influence on concepts and their role in the drive toward miniaturization of laser systems can no longer be ignored. Ultrasmall high-Q cavities, ultralow threshold currents, narrow beam emission, beam shaping, and high spectral purity are among the most striking features of these lasers, depending on the laser design that is selected. Almost thresholdless lasers with high β-factors above 0.9 have also been achieved in optically pumped *nanobeam* and 3D PhCs [36, 157], with a possible extension to electrical pumping in the future. However, all the remarkable properties of cavities are at the cost of a reduced efficiency in the electrical injection, while a high electron-to-photon conversion is required for practical applications. One must indeed recognize that the transition from an optically pumped laser to a laser diode has taken more time than expected, because of severe technological difficulties arising from electrical injection. The situation is different for QCLs at mid-infrared and far-infrared wavelengths, since they are unipolar lasers that are normally operated using current injection. Fortunately, for all wavelength domains, regular advances in device performance during the last decade have maintained interest in electrically driven PhC lasers.

In the near infrared and for the membrane approach, the threshold current of 7.8 μA recently obtained with a LEAP laser is the lowest threshold ever reported from an electrically pumped laser at room temperature [30, 31]. Other state-of-the-art features of this device are an ultrasmall ∼0.11 μm^3 laser mode volume, an overall ∼5 × 12 μm^2 PhC footprint, output powers in the microwatt range under CW operation, a 3 dB modulation bandwidth of 16.2 GHz, and an energy cost per bit of 14.0 fJ b^{-1} at 12.5 GHz modulation frequency. All these features are of great promise for future applications of PhC lasers to optical interconnects and computercoms. The fabricated device actually combines several decisive improvements achieved in the laser design during the past several years. One decisive improvement is the use of a lateral current injection (lateral p–i–n junction) as in the work by Ellis and coworkers [23]. Another major improvement stems from the embedding of the InGaAlAs-based QW active region limited to within a line defect in the InP membrane layer, thus taking advantage of the thermal conductivity of InP with an increased confinement of photons and electrical carriers in the active region. Further improvements in terms of emitted power and wall-plug efficiency can be expected in the next few years with a further reduction of leakage current through the substrate.

In the conventional substrate approach, SMSRs up to 50 dB have been achieved with near-infrared DFB-like PhC waveguide laser diodes. Near-infrared beams with ultralow (<1°) divergence and far-field patterns controlled on-demand have been produced from electrically pumped PhC surface-emitting lasers. All these results confirm the high potential of PhC lasers in the substrate approach, which also offers good heat sinking and high power generation, even if it is at the cost of lower cavity Q-factors and of larger effective laser mode volumes. Here, the major difficulty stems from the nonradiative carrier recombination at the deeply etched hole sidewalls, which penalizes the laser efficiency, as compared with more conventional structures. Interest in such lasers, in the near infrared could be renewed with the development of appropriate passivation techniques. The so-called bottom-up approach to the realization of PhC structures [79], based on nanopillar (or nanowire) arrays, represents an interesting opportunity, because of the easier passivation of nanopillar (or nanowire) structures. However, the competition will remain difficult in the optical communication wavelength domain, where existing devices have already attained a high level of sophistication.

From this point of view, short-wavelength *UV-visible* PhC lasers certainly represent a more open field for exploration, even though cost-effectiveness will also be a key-point for applications, in the end. The quantum efficiency of III-nitride LEDs has recently been improved up to ~73% through the use of PhCs as light extractors [158]. This progress will encourage similar improvements in laser devices. Besides, interest in III-nitride heterostructures stems from the fact that not only can they produce efficient emission in the UV-visible domain, but they can also exhibit strong intersubband transitions from the near-infrared to the far infrared [159] (see also Figure 3.1). Theoretical investigations even predict that terahertz GaN-based QCLs could be serious competitors with previously developed terahertz QC lasers in the GaAs system [160, 161].

In the mid-infrared, substantial progress has been accomplished in the design of surface-emitting PhC QCLs, since they were first reported in 2003 [110]. Single-mode emission is now obtained at room temperature, in pulsed operating conditions, with a SMSR larger than 20 dB, a peak power level in the range of several tens of milliwatts and a single-lobed output beam with an angular divergence smaller than 10°. The decisive technological ingredient has been the use of a surface-plasmon waveguide with a two-dimensional PhC structure of holes directly patterned into the thin metal layer on top of the waveguide. The performances of these surface-emitting 2D lasers are better than those of standard surface-plasmon Fabry–Perot lasers fabricated from the same wafer and with the same modal volume [121]. However, surface-plasmon mid-IR QC lasers exhibit performances intrinsically largely inferior with respect to state-of-art mid-IR QC lasers based on purely dielectric waveguides.

Clearly, the optimization strategy for mid-infrared lasers has been different from that used at near-infrared wavelengths. While a real effort has been directed toward microcavity lasers in the near-infrared domain, more attention has been paid to surface-emitting band-edge lasers in the mid-infrared. In fact, near-infrared microcavities with an extremely small mode volume and an ultimately high Q-factor can be fabricated using two-dimensional PhCs with a triangular lattice of airholes vertically etched into a suspended waveguide slab. Because mid-infrared QC laser waveguides are operated in the TM mode and not in the TE one, full photonic bandgap, and therefore high-performance, microcavities cannot be implemented using the same technique. As proposed in [162], a triangular lattice of airholes should be etched horizontally into a thin vertical slab that includes the QC active region, in order to mimic the near-infrared situation. However, optimizing such a structure would lead to severe difficulties in terms of fabrication. Alternatively, a honeycomb lattice of vertical airholes could be used, instead of the triangular lattice, to give a full in-plane TM bandgap [163]. Active devices based on this design have not been reported so far, but planar PhC waveguides with such a lattice have been experimentally demonstrated at $\lambda \approx 1.5\,\mu m$ [164]. Nevertheless, despite the absence of a full bandgap, vertically etched PhCs in a QC waveguide can provide efficient optical feedback for band-edge laser action. This result is obtained at the price of a broader emission surface area, of a larger mode volume, and of reduced performances. However, surface emission allows, in turn, a substantial reduction of the output beam divergence.

Potential applications of these lasers and of QC lasers in general are in the domains of infrared spectroscopy, chemical sensing, and gas detection. In most cases, a small laser size is not a requirement for these applications, and the use of the now commercially available powerful QC lasers that are capable of delivering single-mode CW power levels in the hundreds of milliwatts range [91–95, 165–167] appears to be more convenient. Further development of surface-emitting QC lasers will therefore take place in a strongly competitive context.

A much more favorable context occurs for *terahertz* PhC QCLs, because of the much easier fabrication of PhC structures at these long wavelengths, and of the much lower loss of metallo-dielectric structures. Moreover, leakage currents due to surface recombination are not present, since QC lasers are unipolar devices. The so-called Type-II photonic heterostructure recently developed at 3.4 THz ($\lambda = 88\,\mu m$), with graded PhC structures, has produced the largest peak output power (103 mW) ever reported at 20 K from a single-mode semiconductor laser in this wavelength region [139]. The peak power remains as high as 67 mW at 80 K, and the maximum operating temperature has reached 120 K. Using a PhC *surface-emitting* structure, single-mode emission has been achieved at ~ 2.4 THz ($\lambda = 125\,\mu m$), with a lasing surface area as small as $(3\lambda)^2$ [135]. A nearly diffraction-limited laser beam was emitted, with a single-lobed far-field pattern. It is worth noticing that even smaller device footprints can be obtained by exploiting the concept of third order DFB laser, at almost no cost in terms of beam divergence [142]. Furthermore, strongly sub-λ metallo-dielectric terahertz microcavities with a mode volume of 0.0025 $(\lambda/2n)^3$ have been demonstrated [13, 14], and an ultrasmall electrically pumped terahertz QC laser that mimics a resonant LC circuit [12] has been reported. These results open the field for a new generation of electronics-inspired terahertz lasers. Although these lasers are more metamaterial-like, the original PhC concepts will certainly play an important role in the future development of terahertz QC lasers.

References

1. Yablonovitch, E. (1987) Inhibited spontaneous emission in solid state physics and electronics. *Phys. Rev. Lett.*, **58**, 2059.
2. Joannopoulos, J.D., Meade, R.D., and Winn, J.N. (1995) *Photonic Crystals: Molding the Flow of Light*, Princeton University.
3. Asano, T., Song, B.-S., and Noda, S. (2006) Analysis of the experimental Q factors (>1 million) of photonic crystal nanocavities. *Opt. Express*, **14**, 1996–2002.
4. Nomura, M., Tanabe, K., Iwamoto, S., and Arakawa, Y. (2010) High-Q design of semiconductor-based ultrasmall photonic crystal nanocavity. *Opt. Express*, **18**, 8144–8150.
5. Husko, C., De Rossi, A., Combrié, S., Vy Tran, Q., Raineri, F., and Wong, C. W., (2009) Ultrafast all-optical modulation in GaAs photonic crystal cavities. *Appl. Phys. Lett.*, **94**, 021111.
6. Asakawa, K., Sugimoto, M., Watanabe, K., Ozaki, N., Mizutani, A., Takata, Y., Kitagawa, Y., Ishikawa, H., Ikeda, N., Awazu, K., Watanabe, A., Nakamura, S., Ohkouchi, S., Inoue, K., Kristensen, M., Sigmund, O., Borel, P.I., and Baets, R. (2006) Photonic crystal and quantum dot technologies for all-optical switch and logic device. *New J. Phys.*, **8**, 208–220.
7. Jugessur, A., Wu, L., Bakhtazad, A., Kirk, A., Krauss, T., and De La Rue, R. (2006) Compact and integrated 2-D photonic crystal super-prism filter-device for wavelength demultiplexing applications. *Opt. Express*, **14**, 1632.
8. Bernier, D., Le Roux, X., Lupu, A., Marris-Morini, D., Vivien, L., and Cassan, E. (2008) Compact, low crosstalk CWDM demultiplexer using photonic crystal superprism. *Opt. Express*, **16**, 17209–17214.
9. Martinelli, L., Benisty, H., Drisse, O., Derouin, E., Pommereau, F., Legouézigou, O., and Duan, G.H. (2007) Fine impact of lithographic grid irregularities assessed on photonic crystal device selectivity. *IEEE Photonics Technol. Lett.*, **19**, 282–284.
10. Kogelnik, H. and Shank, C.V. (1971) Stimulated emission in a periodic structure. *Appl. Phys. Lett.*, **18**, 152–154.
11. Kogelnik, H. and Shank, C.V. (1972) Coupled-wave theory of distributed feedback lasers. *J. Appl. Phys.*, **43**, 2327–2335.
12. Walther, C., Scalari, G., Amanti, M., Beck, M., and Faist, J. (2010) Microcavity laser oscillating in a circuit-based resonator. *Science*, **327**, 1495.
13. Strupiechonski, E., Xu, G., Brekenfeld, M., Todorov, Y., Isac, N., Andrews, A.M., Klang, P., Sirtori, C., Strasser, G., Degiron, A., and Colombelli, R. (2012) Sub-diffraction-limit semiconductor resonators operating on the fundamental magnetic resonance. *Appl. Phys. Lett.*, **100**, 131113.
14. Strupiechonski, E., Xu, G., Cavalié, P., Isac, N., Dhillon, S., Tignon, J., Beaudoin, G., Sagnes, I., Degiron, A., and Colombelli, R. (2013) Hybrid electronic-photonic sub-wavelength cavities operating at THz frequencies. *Rapid Commun. - Phys. Rev.*, **B 87**, 041408(R).
15. Painter, O., Lee, R.K., Scherer, A., Yariv, A., O'Brien, J.D., Dapkus, P.D., and Kim, I. (1999) Two-dimensional photonic band-gap defect mode laser. *Science*, **284**, 1819–1821.
16. Park, H.-G., Kim, S.-H., Kwon, S.-H., Ju, Y.-G., Yang, J.-K., Baek, J.-H., Kim, S.-B., and Lee, Y.-H. (2004) Electrically driven single-cell photonic crystal laser. *Science*, **305**, 1444–1446.
17. Purcell, E.M. (1946) Spontaneous emission probabilities at radiofrequencies. *Phys. Rev.*, **69**, 681.
18. Lourtioz, J.-M., Benisty, H., Berger, V., Gérard, J.-M., Maystre, D., Chelnokov, A., and Pagnoux, D. (2008) *Photonic*

Crystals, Towards Nanoscale Photonic Devices, 2nd edn, Springer, Berlin, Heidelberg, New York, ISBN: 978-3-540-78346-6.
19. Yamamoto, Y. (1991) *Coherence, Amplification and Quantum Effects in Semiconductor Lasers*, John Wiley & Sons, Inc., New York.
20. Akahane, Y., Asano, T., Song, B.S., and Noda, S. (2003) High-Q photonic nanocavity in a two-dimensional photonic crystal. *Nature*, **425** (6961), 944–947.
21. Kuramochi, E., Notomi, M., Mitsugi, S., Shinya, A., Tanabe, T., and Watanabe, T. (2006) Ultrahigh-Q photonic crystal nanocavities realized by the local width modulation of a line defect. *Appl. Phys. Lett.*, **88** (4), 041112.
22. Combrié, S., De Rossi, A., Tran, Q.V., and Benisty, H. (2008) GaAs photonic crystal cavity with ultrahigh Q: microwatt nonlinearity at 1.55 μm. *Opt. Lett.*, **33** (16), 1908–1910.
23. Ellis, B., Mayer, M.A., Shambat, G., Sarmiento, T., Harris, J.S., Haller, E.E., and Vuckovic, J. (2011) Ultralow-threshold electrically pumped quantum dot photonic-crystal nanocavity laser. *Nat. Photonics*, **5**, 297–300.
24. Long, C.M., Giannopoulos, A.V., and Choquette, K.D. (2009) Modified spontaneous emission from laterally injected photonic crystal emitter. *Electron. Lett.*, **45** (4), 227–228.
25. Tanabe, T., Nishiguchi, K., Kuramochi, E., and Notomi, M. (2009) Low power and fast electro-optic silicon modulator with lateral p-i-n embedded photonic crystal nanocavity. *Opt. Express*, **17** (25), 22505–22513.
26. Tanabe, T., Sumikura, H., Taniyama, H., Shinya, A., and Notomi, M. (2010) All-silicon sub-Gb/s telecom detector with low dark current and high quantum efficiency on chip. *Appl. Phys. Lett.*, **96** (10), 101103.
27. Haret, L.-D., Checoury, X., Han, Z., Boucaud, P., Combrié, S., and De Rossi, A. (2010) All-silicon photonic crystal photoconductor on silicon-on-insulator at telecom wavelength. *Opt. Express*, **18** (23), 23965–23972.
28. Matsuo, S., Takeda, K., Sato, T., Notomi, M., Shinya, A., Nozaki, K., Taniyama, H., Hasebe, K., and Kakitsuka, T. (2012) Room-temperature continuous-wave operation of lateral current injection wavelength-scale embedded active-region photonic-crystal laser. *Opt. Express*, **20**, 3773–3780.
29. Kim, J., Shinya, A., Nozaki, K., Taniyama, H., Chen, C.-H., Sato, T., Matsuo, S., and Notomi, M. (2011) Narrow linewidth operation of buried-heterostructure photonic crystal nanolaser. *Opt. Express*, **20**, 11643–11651.
30. Sato, T., Takeda, K., Shinaya, A., Nozaki, K., Taniyama, H., Hasebe, K., Kakitsuka, T., Notomi, M., and Matsuo, S. (2013) Ultralow-threshold electrically driven photonic crystal nanocavity laser. Conference on Lasers and Electro-Optics, CLEO Digest, Paper ThK3-1.
31. Matsuo, S., Sato, T., Takeda, K., Shinaya, A., Nosaki, K., Taniyama, H., Notomi, M., Hasebe, K., and Kakitsuka, T. (2013) Ultralow operating energy electrically driven photonic crystal lasers. *IEEE J. Sel. Topics Quantum Electron.*, **19** (4), 4900311.
32. Kim, Y.K., Elarde, V.C., Long, C.M., Coleman, J.J., and Choquette, K.D. (2008) Electrically injected InGaAs/GaAs photonic crystal membrane light emitting microcavity with spatially localized gain. *J. Appl. Phys.*, **104**, 123103.
33. Shambat, G., Ellis, B., Majumdar, A., Petykiewicz, J., Mayer, M.A., Sarmiento, T., Harris, J., Haller, E.E., and Vuckovic, J. (2011) Ultrafast direct modulation of a single-mode photonic crystal nanocavity light-emitting diode. *Nat. Commun.*, **2**, 539.
34. Shambat, G., Ellis, B., Petykiewicz, J., Mayer, M.A., Majumdar, A., Sarmiento, T., Harris, J., Haller, E.E., and Vuckovic, J. (2012) Electrically Driven Photonic Crystal Nanocavity Devices, http://arxiv.org/abs/1201.0964 (accessed 30 September 2013).
35. Foresi, J.S., Villeneuve, P.R., Ferrera, J., Thoen, E.R., Steinmeyer, G., Fan, S., Joannopoulos, J.D., Kimerling,

L.C., Smith, H.I., and Ippen, E.P. (1997) Photonic-bandgap microcavities in opticalwaveguide. *Nature*, **390**, 143–145.

36. Zhang, Y., Khan, M., Huang, Y., Ryou, J.H., Deotare, P.B., Dupuis, R., and Loncar, M. (2010) Photonic crystal nanobeam lasers. *Appl. Phys. Lett.*, **97**, 051104.

37. Shambat, G., Ellis, B., Petykiewicz, J., Mayer, M.A., Sarmiento, T., Harris, J., Haller, E.E., and Vuckovic, J. (2011) Nanobeam photonic crystal cavity light-emitting diodes. *Appl. Phys. Lett.*, **99**, 071105.

38. Halioua, Y., Bazin, A., Monnier, P., Karle, T.J., Roelkens, G., Sagnes, I., Raj, R., and Raineri, F. (2011) Hybrid III-V semiconductor/silicon nanolaser. *Opt. Express*, **19** (10), 9221–9231.

39. Fang, A.W., Park, H., Bowers, J.E., Jones, R., Cohen, O., and Paniccia, M.J. (2006) Electrically pumped hybrid AlGaInAs-silicon evanescent laser. *Opt. Express*, **14**, 9203–9210.

40. Checoury, X., Boucaud, P., Li, X., Lourtioz, J.-M., Derouin, E., Drisse, O., Poigt, F., Legouezigou, L., Legouezigou, O., Pommereau, P., and Duan, G.-H. (2006) Tailoring holes for improving the efficiency of single-mode photonic crystal waveguide lasers on InP substrate. *Appl. Phys. Lett.*, **89**, 071108.

41. Checoury, X., Crozat, P., Lourtioz, J.-M., Pommereau, F., Cuisin, C., Derouin, E., Drisse, O., Legouezigou, L., Legouezigou, O.L., Lelarge, F., Poingt, F., Duan, G.H., Bonnefont, S., Mulin, D., Valentin, J., Gauthier-Lafaye, O., Lozes-Dupuy, F., and Talneau, A. (2005) Single-mode in-gap emission of medium-width photonic crystal waveguides on InP substrate. *Opt. Express*, **13**, 6947–6955.

42. Talneau, A., LeGratiet, L., Gentner, J.-L., Berrier, A., Mulot, M., Anand, S., and Olivier, S. (2004) High external efficiency in a monomode full-photonic-crystal laser under continuous wave electrical injection. *Appl. Phys. Lett.*, **85**, 1913–1915.

43. Yariv, A., Xu, Y., Lee, R.K., and Scherer, A. (1999) Coupled-resonator optical waveguide a proposal and analysis. *Opt. Lett.*, **24**, 711–713.

44. Happ, T., Kamp, M., Forchel, A., Gentner, J.L., and Goldstein, L. (2003) Two-dimensional photonic crystal coupled-defect laser diode. *Appl. Phys. Lett.*, **82**, 4–6.

45. Checoury, X., Boucaud, P., Lourtioz, J.-M., Gauthier-Lafaye, O., Bonnefont, S., Mulin, D., Valentin, J., Lozes-Dupuy, F., Pommereau, F., Cuisin, C., Derouin, E., Drisse, O., Legouezigou, L., Lelarge, F., Poingt, F., Duan, G.H., and Talneau, A. (2005) 1.5 micron room-temperature emission of square-lattice photonic-crystal waveguide lasers with a single line defect. *Appl. Phys. Lett.*, **86**, 151111.

46. Hofmann, H., Scherer, H., Deubert, S., Kamp, M., and Forchel, A. (2007) Spectral and spatial single mode emission from a photonic crystal distributed feedback laser. *Appl. Phys. Lett.*, **90**, 121135.

47. Lang, R. J., Dzurko, K., Hardy, A. A., Demars, S., Schoenfelder, A., and Welch, D. F. (1998) Theory of grating-confined broad-area lasers. *IEEE Journal of Quantum Electronics*, **34**, 2196–2210.

48. Paschke, K., Bogatov, A., Drakin, A.E., Güther, R., Stratonnikov, A.A., Wenzel, H., Ebert, G., and Tränkle, G. (2003) Modeling and measurement of the radiative characteristics of high-power α-DFB lasers. *IEEE J. Sel. Top. Quantum Electron.*, **9**, 835.

49. Zhu, L., DeRose, G.A., Scherer, A., and Yariv, A. (2007) Electrically pumped edge-emitting photonic crystal lasers with angled facets. *Opt. Lett.*, **32** (10), 1256–1258.

50. Tsang, T., Olsson, N.A., Linke, R.A., and Logan, R.A. (1983) High-speed direct single-frequency modulation with large tuning note and frequency emission in cleaved-coupled-cavity semiconductor lasers. *Appl. Phys. Lett.*, **42**, 650.

51. Happ, T., Markard, A., Kamp, M., Forchel, A., and Srinivasan, A. (2001) Single-mode operation of coupled-cavity lasers based on two-dimensional

photonic crystals. *Appl. Phys. Lett.*, **79**, 4091–4093.
52. Manhkopf, S., Arlt, M., Kamp, M., Colson, V., Duan, G.-H., and Forchel, A. (2004) Two-channel tunable laser diode based on photonic crystals. *IEEE Photonics Technol. Lett.*, **16**, 353–355.
53. Bauer, A., Muller, M., Lehnhardt, T., Roßner, K., Hummer, M., Hofmann, H., Kamp, M., Hofling, S., and Forchel, A. (2008) Discretely tunable single-mode lasers on GaSb using two-dimensional photonic crystal intracavity mirrors. *Nanotechnology*, **19**, 235202.
54. Jahjah, M., Moumdji, S., Gauthier-Lafaye, O., Bonnefont, S., Rouillard, Y., and Vicet, A. (2012) Antimonide-based 2.3 µm photonic crystal coupled-cavity lasers for CH_4 QEPAS. *Electron. Lett.*, **48**, 277–278.
55. Choquette, K.D., Siriani, D.F., Kasten, A.M., Tan, M.P., Sulkin, J.D., Leisher, P.O., Raftery, J.J. Jr., and Danner, A.J. (2012) *Single Mode Photonic Crystal Vertical Cavity Surface Emitting Lasers*, Advances in Optical Technologies, vol. 2012, Hindawi Publishing Corporation, Article ID 280920.
56. Boutami, S., Benbakir, B., Regreny, P., Leclercq, J.L., and Viktorovitch, P. (2007) Compact 1.55 µm room-temperature optically pumped VCSEL using photonic crystal mirror. *Electron. Lett.*, **43**, 282–283.
57. Boutami, S., Benbakir, B., Letartre, X., Leclercq, J.L., Regreny, P., and Viktorovitch, P. (2007) Ultimate vertical Fabry-Perot cavity based on single-layer photonic crystal mirrors. *Opt. Express*, **15**, 12443–12449.
58. Huang, M.C.Y., Zhou, Y., and Chang-Hasnain, C.J. (2007) A surface-emitting laser incorporating a high-index contrast subwavelength grating. *Nat. Photonics*, **1**, 119.
59. Zheng, W., Zhou, W., Wang, Y., Liu, A., Chen, W., Wang, H., Fu, F., and Qi, A. (2011) Lateral cavity photonic crystal surface-emitting laser with ultralow threshold. *Opt. Lett.*, **36**, 4140–4142.
60. Miyai, E., Sakai, K., Okano, T., Kunishi, W., Ohnishi, D., and Noda, S. (2006) Photonics: lasers producing tailored beams. *Nature*, **441**, 946.
61. Yokoyama, M. and Noda, S. (2004) Finite-difference time-domain simulation of twodimensional photonic crystal surface-emitting laser having a square-lattice slab structure. *IEICE Trans. Electron.*, **E87-C**, 386–392.
62. Sakai, K., Miyai, E., Szkaguchi, T., Ohnishi, D., Okano, T., and Noda, S. (2005) Lasing band-edge identification for a surface-emitting photonic crystal laser. *IEEE J. Sel. Areas Commun.*, **23**, 1335–1340.
63. David, A., Fujii, T., Sharma, R., McGroddy, K., Nakamura, S., DenBaars, S.P., Hu, E.L., and Weisbuch, C. (2006) Photonic-crystal GaN light-emitting diodes with tailored guided modes distribution. *Appl. Phys. Lett.*, **88**, 061124.
64. David, A., Benisty, H., and Weisbuch, C. (2007) Optimization of light-diffracting photonic-crystals for high extraction efficiency LEDs. *J. Dis. Technol.*, **3**, 133–148.
65. Liu, H.-W., Kan, Q., Wang, C.-X., Hu, H.-Y., Xu, X.-S., and Chen, H.-D. (2011) Light extraction enhancement of GaN LED with a two-dimensional photonic crystal slab. *Chin. Phys. Lett.*, **28**, 054216.
66. Matsubara, H., Yoshimoto, S., Saito, H., Jianglin, Y., Tanaka, Y., and Noda, S. (2008) GaN photonic-crystal surface-emitting laser at blue-violet wavelengths. *Science*, **319** (5862), 445–447.
67. Berrier, A., Shi, Y., Siegert, J., Marcinkevicius, S., He, S., and Anand, S. (2009) Accumulated sidewall damage in dry etched photonic crystals. *J. Vac. Sci. Technol. B*, **27**, 1969–1975.
68. Xu, X., Yamada, T., and Otomo, A. (2008) Surface recombination in GaAs thin films with two-dimensional photonic crystals. *Appl. Phys. Lett.*, **92**, 091911.
69. H.Altug, D. Englund, and J. Vuckovic, Ultrafast photonic crystal nanocavity laser, *Nat. Phys.*, **2**, 484-488 (2006).

70. Englund, D., Altug, H., and Vučković, J. (2007) Low-threshold surface-passivated photonic crystal nanocavity laser. *Appl. Phys. Lett.*, **91**, 071124.
71. Hu, S.Y., Corzine, S.W., Law, K.H., Young, D.B., Gossard, A.C., Coldren, L.A., and Merz, J.L. (1994) Lateral carrier diffusion and surface recombination in InGaAs/AlGaAs quantum-well ridge-waveguide lasers. *J. Appl. Phys.*, **76**, 4479.
72. Nelson, R.J., Williams, J.S., Leamy, H.J., Miller, B., Casey, H.C., Parkinson, B.A., and Heller, A., (1980) Reduction of GaAs surface recombination velocity by chemical treatment. *Appl. Phys. Lett.*, **36**, 76.
73. Casey, H.C. and Buehler, E. (1977) Evidence for low surface recombination velocity on n-type InP. *Appl. Phys. Lett.*, **30**, 247.
74. Gérard, J.-M., Cabrol, O., and Sermage, B. (1996) InAs quantum boxes: highly efficient radiative traps for light emitting devices on Si. *Appl. Phys. Lett.*, **68**, 3123–3125.
75. Mulot, M., Qiu, M., Swillo, M., Jaskorzynska, B., Anand, S., and Talneau, A. (2003) In-plane resonant cavities with photonic crystal boundaries etched in InP-based heterostructure. *Appl. Phys. Lett.*, **83**, 1095.
76. Kotlyar, M.V., O'Faolain, L., Wilson, R., and Krauss, T.F. (2004) High-aspect-ratio chemically assisted ion-beam etching for photonic crystals using a high beam voltage-current ratio. *J. Vac. Sci. Technol. B*, **22**, 1788–1791.
77. Berrier, A., Mulot, M., Talneau, A., Ferrini, R., Houdré, R., and Anand, S. (2007) Characterization of the feature-size dependence in Ar/Cl^2 chemically assisted ion beam etching of InP-based photonic crystal devices. *J. Vac. Sci. Technol. B*, **25**, 1.
78. Holzman, J.F., Strasser, P., Wüest, R., Robin, F., Erni, D., and Jäckel, H. 2005) Ultrafast carrier dynamics in InP photonic crystals. *Nanotechnology*, **16**, 949.
79. Scofield, A.C., Kim, S.-H., Shapiro, J.N., Lin, A., Liang, B., Scherer, A., and Huffaker, D.L. (2011) Bottom-up photonic crystal lasers. *Nano Lett.*, **11**, 5387–5390.
80. Faist, J., Capasso, F., Sivco, D., Sirtori, C., Hutchinson, A.L., Chu, S.-N.G., and Cho, A.Y. (1994) Quantum cascade laser. *Science*, **264**, 553.
81. Köhler, R., Tredicucci, A., Beltram, F., Beere, T.H.E., Linfield, E.H., Davies, A.G., Ritchie, D.A., Iotti, R.C., and Rossi, F. (2002) Terahertz semiconductor-heterostructure lasers. *Nature*, **417**, 156–159.
82. Kazarinov, R.F. and Suris, R.A. (1971) Possibility of amplification of electromagnetic waves in a semiconductor with a superlattice. *Fiz. Tekh. Poluprovod.* translated in *Sov. Phys. Semicond.*, **5**, 797.
83. Faist, J., Capasso, F., Sivco, D., Sirtori, C., Hutchinson, A.L., and Cho, A.Y. (1995) Vertical transition quantum cascade laser with Bragg confined excited states. *Appl. Phys. Lett.*, **66**, 538.
84. Scamarcio, G., Capasso, F., Sirtori, C., Faist, J., Hutchinson, A.L., Sivco, D.L., and Cho, A.Y. (1997) High-power infrared (8-micrometer wavelength) superlattice lasers. *Science*, **276**, 773–776.
85. Tredicucci, A., Capasso, F., Gmachl, C., Tredicucci, A., Sivco, D.L., Hutchinson, A.L., and Cho, A.Y. (1998) High performance interminiband quantum cascade lasers with graded superlattices. *Appl. Phys. Lett.*, **73**, 2101.
86. Wancke, M.C., Capasso, F., Gmachl, C., Sivco, D.L., Hutchinson, A.L., Chu, S.-N.G., and Cho, A.Y. (2001) Injectorless quantum cascade lasers with graded superlattices. *Appl. Phys. Lett.*, **78**, 3950.
87. Faist, J., Beck, M., Aellen, T., and Gini, A. (2001) Quantum cascade lasers based on a bound-to-continuum transition. *Appl. Phys. Lett.*, **78**, 3950.
88. Williams, B.S. (2007) Terahertz quantum-cascade lasers. *Nat. Photon.*, Vol. 1, 517.
89. Cathabard, O., Teissier, R., Devenson, J., Moreno, J.C., and Baranov, A.N. (2010) Quantum cascade lasers emitting near 2.6 µm. *Appl. Phys. Lett.*, **96**, 141110.

90. Colombelli, R., Capasso, F., Gmachl, C., Hutchinson, A.L., Sivco, D.L., Tredicucci, A., Wanke, M.C., Sergent, A.M., and Cho, A.Y. (2001) Far-infrared surface-plasmon quantum-cascade lasers at 21.5 μm and 24 μm wavelengths. *Appl. Phys. Lett.*, **78**, 2620–2622.
91. Slivken, S., Evans, A., Zhang, W., and Razeghi, M. (2007) High-power, continuous-operation intersubband laser for wavelengths greater than 10 μm. *Appl. Phys. Lett.*, **90**, 151115.
92. Bai, Y., Darvish, S.R., Slivken, S., Zhang, W., Evans, A., Nguyen, J., and Razeghi, M. (2008) Room temperature continuous wave operation of quantum cascade lasers with watt-level optical power. *Appl. Phys. Lett.*, **92**, 101105.
93. Katz, S., Vizbaras, A., Boehm, G., and Amann, M.-C. (2009) High-performance injectorless quantum cascade lasers emitting below 6μm. *Appl. Phys. Lett.*, **94**, 151106.
94. Bai, Y., Bandyopadhyay, N., Tsao, S., Slivken, S., and Razeghi, M. (2011) Room temperature quantum cascade lasers with 27% wall plug efficiency. *Appl. Phys. Lett.*, **98**, 181102.
95. Yao, Y., Hoffman, A.J., and Gmachl, C.F. (2012) Mid-infrared quantum cascade lasers. *Nat. Photonics*, **6**, 432, and references therein.
96. Walther, C., Fischer, M., Scalari, G., Terazzi, R., Hoyler, N., and Faist, J. (2007) Quantum cascade lasers operating from 1.2 to 1.6 THz. *Appl. Phys. Lett.*, **91**, 131122.
97. Williams, B.S., Kumar, S., Hu, Q., and Reno, J.L. (2006) High-power terahertz quantum cascade lasers. *Electron. Lett.*, **42**, 89–91.
98. Fathololoumi, S., Dupont, E., Chan, C.W.I., Wasilewski, Z.R., Laframboise, S.R., Ban, D., Mátyás, A., Jirauschek, C., Hu, Q., and Liu, H.C. (2012) Terahertz quantum cascade lasers operating up to ~200 K with optimized oscillator strength and improved injection tunneling. *Opt. Express*, **20**, 3866–3876.
99. McCall, S.L., Levi, A.F.J., Slusher, R.E., Pearton, S.J., and Logan, R.A. (1992) Whispering gallery mode microdisk lasers. *Appl. Phys. Lett.*, **60**, 289.
100. Levi, A.F., Slusher, R.E., McCall, S.L., Pearton, S.J., and Hobson, W.S. (1993) Room-temperature lasing action in $In_{0.51}Ga_{0.49}P/In_{0.2}Ga_{0.8}As$ microcylinder laser diodes. *Appl. Phys. Lett.*, **62**, 2021.
101. Deng, H., Deng, Q., and Deppe, D.G. (1996) Native-oxide laterally confined whispering-gallery mode laser with vertical emission. *Appl. Phys. Lett.*, **69**, 3120.
102. Baba, T. (1997) Photonic crystals and microdisk cavities based on GaInAsP-InP system. *IEEE Select. Top. Quantum Electron.*, **3**, 808.
103. Faist, J., Gmachl, C., Striccoli, M., Sirtori, C., Capasso, F., Sivco, D.L., and Cho, A.Y. (1996) Quantum cascade disk lasers. *Appl. Phys. Lett.*, **69**, 2456–2458.
104. Gmachl, C., Capasso, F., Narimanov, E.E., Nockel, J.U., Stone, A.D., Faist, J., Sivco, D.L., and Cho, A.Y. (1998) High-power directional emission from microlasers with chaotic resonators. *Science*, **280**, 1556–1564.
105. Williams, B.S. (2007) Terahertz quantum-cascade lasers. *Nat. Photonics*, **1**, 517–525.
106. Scalari, G., Turcinková, D., Lloyd-Hughes, J., Amanti, M.I., Fischer, M., Beck, M., and Faist, J. (2010) Magnetically assisted quantum cascade laser emitting from 740 GHz to 1.4 THz. *Appl. Phys. Lett.*, **97**, 081110.
107. Chassagneux, Y., Palomo, J., Colombelli, R., Dhillon, S., Sirtori, C., Beere, H., Alton, J., and Ritchie, D. (2007) THz microcavity lasers with sub-wavelength mode volumes and thresholds in the milli-Ampere range. *Appl. Phys. Lett.*, **90**, 091113.
108. Dunbar, L.A., Houdré, R., Scalari, G., Sirigu, L., Giovannini, M., and Faist, J. (2007) Small optical volume terahertz emitting microdisk quantum cascade laser. *Appl. Phys. Lett.*, **90**, 141114.
109. Fasching, G., Benz, A., Unterrainer, C., Zobl, R., Andrews, A.M., Roch, T., Schrenk, W., and Strasser, G. (2005) Terahertz microcavity quantum cascade lasers. *Appl. Phys. Lett.*, **87**, 211112.

110. Colombelli, R., Srinivasan, K., Troccoli, M., Painter, O., Gmachl, C., Tennant, D.M., Sergent, A.M., Sivco, D.L., Cho, A.Y., and Capasso, F. (2003) Quantum cascade surface-emitting photonic-crystal laser. *Science*, **302**, 1374.
111. Xu, G., Colombelli, R., Braive, R., Beaudoin, G., Le Gratiet, L., Talneau, A., Ferlazzo, L., and Sagnes, I. (2010) Surface-emitting mid-infrared quantum cascade lasers with high-contrast photonic crystal resonator. *Opt. Express*, **18**, 11979–11989.
112. Krauss, T.F., Painter, O., Scherer, A., Roberts, J.S., and De La Rue, R.M. (1998) Photonic microstructures as laser mirrors. *Opt. Eng.*, **37**, 1143–1148, Invited paper in special issue on '30 years of integrated optics.
113. Raffaele, L., De La Rue, R.M., Roberts, J.S., and Krauss, T.F. (2001) Edge-emitting semiconductor microlasers with ultrashort-cavity and dry-etched high-reflectivity photonic microstructure mirrors. *IEEE Photonics Technol. Lett.*, **13**, 176–178.
114. Rennon, S., Klopf, F., Reithmaier, J.P., and Forchel, A. (2001) 12-micron-long edge-emitting quantum-dot microlaser. *Electron. Lett.*, **37**, 690.
115. Raffaele, L., De La Rue, R.M., and Krauss, T.F. (2002) Ultrashort in-plane semiconductor microlasers with high-reflectivity microstructured mirrors. *Opt. Quantum Electron.*, **34**, 101–111.
116. Erwin, G., Bryce, A.C., and De La Rue, R.M. (2005) Low-threshold oxide-confined compact edge-emitting semiconductor laser diodes with high-reflectivity 1D photonic crystal mirrors. *Proc. SPIE - Int. Soc. Opt. Eng.*, **5958**, 1–7.
117. Hvozdara, L., Lugstein, A., Finger, N., Gianordoli, S., Schrenk, W., Unterrainer, K., Bertagnolli, E., Strasser, G., Gornik, E. (2000) Quantum cascade lasers with monolithic air-semiconductor Bragg reflectors. *Appl. Phys. Lett.*, **77**, 1241.
118. Walker, C.L., Farmer, C.D., Stanley, C.R., and Ironside, C.N. (2004) Progress towards photonic crystal quantum cascade laser. *IEE Proc. Optoelectron.*, **151**, 502–507.
119. Semmel, J., Nähle, L., Höfling, S., and Forchel, A. (2007) Edge emitting quantum cascade microlasers on InP with deeply etched one-dimensional photonic crystals. *Appl. Phys. Lett.*, **91**, 071104.
120. Heinrich, J., Langhans, R., Seufert, J., Höfling, S., and Forchel, A. (2007) Quantum cascade microlasers with two-dimensional photonic crystal reflectors. *IEEE Photonics Technol. Lett.*, **19**, 1937.
121. Xu, G., Colombelli, R., Beaudoin, G., Largeau, L., Mauguin, O., and Sagnes, I. (2010) Loss-reduction in midinfrared photonic crystal quantum cascade lasers using metallic waveguides. *Opt. Eng.*, **49** (11), 111112.
122. Xu, G., Moreau, V., Chassagneux, Y., Bousseksou, A., Colombelli, R., Patriarche, G., Beaudoin, G., and Sagnes, I. (2009) Surface-emitting quantum cascade lasers with metallic photonic-crystal resonators. *Appl. Phys. Lett.*, **94**, 221101.
123. Xu, G., Chassagneux, Y., Colombelli, R., Beaudoin, G., and Sagnes, I. (2010) Polarized single-lobed surface emission in mid-infrared, photonic-crystal, quantum-cascade lasers. *Opt. Lett.*, **35**, 859–861.
124. Unterrainer, C., Colombelli, R., Gmachl, C., Capasso, F., Hwang, H., Sivco, D.L., and Cho, A.Y. (2002) Quantum cascade lasers with double metalsemiconductor waveguide resonators. *Appl. Phys. Lett.*, **80**, 3060.
125. Williams, B.S., Kumar, S., Callebaut, H., Hu, Q., and Reno, J.L. (2003) Terahertz quantum-cascade laser at lambda approximate to 100 μm using metal waveguide for mode confinement. *Appl. Phys. Lett.*, **83**, 2124–2126.
126. Dunbar, L.A., Moreau, V., Ferrini, R., Houdre, R., Sirigu, L., Scalari, G., Giovannini, M., Hoyler, N., and Faist, J. (2005) Design, fabrication and optical characterisation of quantum cascade lasers at terahertz frequencies using photonic crystal reflectors. *Opt. Express*, **13**, 8960–8968.
127. Benz, A., Fasching, G., Deutsch, C., Andrews, A.M., Unterrainer, K., Klang, P., Schrenk, W., and Strasser, G. (2007)

Terahertz photonic crystal resonators in double-metal waveguides. *Opt. Express*, **15**, 12418–12424.
128. Zhang, H., Dunbar, L.A., Scalari, G., Houdre, R., and Faist, J. (2007) Terahertz photonic crystal quantum cascade lasers. *Opt. Express*, **15**, 16818–16827.
129. Benz, A., Deutsch, C., Fasching, G., Unterrainer, K., Andrews, A.M., Klang, P., Schrenk, W., and Strasser, G. (2009) Active photonic crystal terahertz laser. *Opt. Express*, **17**, 941–946.
130. Chassagneux, Y., Colombelli, R., Maineult, W., Barbieri, S., Beere, H.E., Ritchie, D.A., Khanna, S.P., Linfield, E.H., and Davies, A.G. (2009) Electrically pumped photonic-crystal terahertz lasers controlled by boundary conditions. *Nature (London)*, **457**, 174.
131. Chassagneux, Y., Colombelli, R., Maineult, W., Barbieri, S., Khanna, S.P., Linfield, E.H., and Davies, A.G. (2009) Predictable surface emission patterns in terahertz photonic-crystal quantum cascade lasers. *Opt. Express*, **17**, 9491.
132. Srinivasan, K., Barclay, P.E., Painter, O.J., Chen, J., Cho, A.Y., and Gmachl, C. (2003) Experimental demonstration of a high quality factor photonic crystal microcavity. *Appl. Phys. Lett.*, **83**, 1915.
133. Chassagneux, Y., Colombelli, R., Maineult, W., Barbieri, S., Khanna, S.P., Linfield, E.H., and Davies, A.G. (2010) Graded photonic crystal THz quantum cascade lasers. *Appl. Phys. Lett.*, **96**, 031104.
134. Miyai, E. and Noda, S. (2005) Phase-shift effect on a two-dimensional surface-emitting photonic-crystal laser. *Appl. Phys. Lett.*, **86**, 111113.
135. Sevin, G., Fowler, D., Xu, G., Julien, F.H., Colombelli, R., Khanna, S.P., Linfield, E.H., and Davies, A.G. (2010) Optimized surface-emitting photonic-crystal THz quantum cascade lasers with reduced resonator size. *Appl. Phys. Lett.*, **97**, 131101.
136. Sevin, G., Fowler, D., Xu, G., Julien, F.H., Colombelli, R., Beere, H., and Ritchie, D. (2010) Continuous-wave operation of 2.7 THz photonic crystal quantum cascade lasers. *Electron. Lett.*, **46**, 1513.
137. Kumar, S., Williams, B.S., Qin, Q., Lee, A.W.M., Hu, Q., and Reno, J.L. (2007) Surface-emitting distributed feedback terahertz quantum-cascade lasers in metal-metal waveguides. *Opt. Express*, **15**, 113–128.
138. Mahler, L., Tredicucci, A., Beltram, F., Walther, C., Faist, J., Beere, H.E., and Ritchie, D.A. (2010) High-power surface emission from terahertz distributed feedback lasers with a dual-slit unit cell. *Appl. Phys. Lett.*, **96**, 191109.
139. G. Xu, R. Colombelli, S.P. Khanna, A. Belarouci, X. Letartre, L.H. Li, E.H. Linfield, A.G. Davies, H.E. Beere and D.A. Ritchie, Efficient power extraction in (THz) surface-emitting lasers using graded photonic heterostructures, *Nat. Commun.*, 3:952, doi: 10.1038/ncomms1958 (2012).
140. G. Xu, Y. Halioua, S. Moumdji, R. Colombelli, H.E. Beere, D.A. Ritchie, (2013) Stable single-mode operation of surface-emitting THz lasers with graded photonic heterostructure resonators, *Appl. Phys. Lett.*, **102**, 231105.
141. Amanti, M.I., Fischer, M., Scalari, G., Beck, M., and Faist, J. (2009) Low-divergence single-mode terahertz quantum cascade laser. *Nat. Photon.*, **3**, 586–590.
142. Amanti, M.I., Scalari, G., Castellano, F., Beck, M., and Faist, J. (2010) Low divergence THz photonic wire laser. *Opt. Express*, **18**, 6390–6395.
143. Kao, T.-Y., Hu, Q., and Reno, J.L. (2012) Perfectly phase-matched third-order distributed feedback terahertz quantum-cascade lasers. *Opt. Lett.*, **37**, 2070–2072.
144. Coccioli, R., Boroditsky, M., Kim, K.W., Rahmat-Samii, Y., and Yablonovitch, E. (1998) Smallest possible electromagnetic mode volume in a dielectric cavity. *IEE Proc. Optoelectron.*, **145**, 391.
145. Scheuer, J., Green, W.M.J., DeRose, G.A., and Yariv, A. (2005) Lasing from a circular Bragg nanocavity with an ultrasmall modal volume. *Appl. Phys. Lett.*, **86**, 251101.
146. Nozaki, K. and Baba, T. (2006) Laser characteristics with ultimate-small modal volume in photonic crystal slab

point-shift nanolasers. *Appl. Phys. Lett.*, **88**, 211101.

147. Hill, M.T., Oei, Y.-S., Smalbrugge, B., Zhu, Y., de Vries, T., van Veldhoven, P.J., van Otten, F.W.M., Eijkemans, T.J., Turkiewicz, J.P., de Waardt, H., Geluk, E.J., Kwon, S.-H., Lee, Y.-H., Nötzel, R., and Smit, M.K. (2007) Lasing in metallic-coated nanocavities. *Nat. Photonics*, **1**, 589–594.

148. Oulton, R.F., Sorger, V.J., Zentgraf, T., Ma, R.-M., Gladden, C., Dai, L., Bartal, G., and Zhangh, X. (2009) Plasmon lasers at deep subwavelength scale. *Nature*, **461**, 629–632.

149. Kühn, S., Håkanson, U., Rogobete, L., and Sandoghdar, V. (2006) Enhancement of single-molecule fluorescence using a gold nanoparticle as an optical nanoantenna. *Phys. Rev. Lett.*, **97**, 017402.

150. Noginov, M.A., Zhu, G., Belgrave, A.M., Bakker, R., Shalaev, V.M., Narimanov, E.E., Stout, S., Herz, E., Suteewong, T., and Wiesner, U. (2009) Demonstration of a spaser-based nanolaser. *Nature*, **460**, 1110–1112.

151. Cattoni, A., Ghenuche, P., Haghiri-Gosnet, A.M., Decanini, D., Chen, J., Pelouard, J.L., and Collin, S. (2011) $\Lambda^3/1000$ Plasmonic Nanocavities for Biosensing Fabricated by Soft UV Nanoimprint Lithography. *Nano Lett.*, **11**, 3557.

152. Bouchon, P., Koechlin, C., Pardo, F., Haidar, R., and Pelouard, J.-L. (2012) Wideband omnidirectional infrared absorber with a patchwork of plasmonic nanoantennas. *Opt. Lett.*, **37**, 1038–1040.

153. Todorov, Y., Tosetto, L., Teissier, J., Andrews, A.M., Klang, P., Colombelli, R., Sagnes, I., Strasser, G., and Sirtori, C. (2010) Optical properties of metal-dielectric-metal microcavities in the THz frequency range. *Opt. Express*, **18**, 13886.

154. Pendry, J.B., Holden, A.J., Robbins, D.J., and Stewart, W.J. (1999) Magnetisms from conductors and enhanced non linear phenomena. *IEEE Trans. Microwave Theory Tech.*, **47**, 2075–2084.

155. Smith, D.R. and Kroll, N. (1999) Negative refractive index in left-handed materials. *IEEE Phys. Rev. Lett.*, **85**, 2933–2936.

156. Balanis, C.A. (1996) *Antenna Theory: Analysis and Design*, 2nd edn, John Wiley & Sons, Inc., New York.

157. Tandaechanurat, A., Ishida, S., Guimard, D., Nomura, M., Iwamoto, S., and Arakawa, Y. (2011) Lasing oscillation in a three-dimensional photonic crystal nanocavity with a complete bandgap. *Nat. Photonics*, **5**, 91–94.

158. Wierer, J.J., David, A., and Megens, M.M. (2009) III-nitride photonic crystal light-emitting diodes with high extraction efficiency. *Nat. Photonics*, **3**, 163–169.

159. Machhadani, H., Kotsar, Y., Sakr, S., Tchernycheva, M., Colombelli, R., Mangeney, J., Bellet-Amalric, E., Sarigiannidou, E., Monroy, E., and Julien, F.H. (2010) Terahertz intersubband absorption in GaN/AlGaN step quantum wells. *Appl. Phys. Lett.*, **97**, 191101.

160. Belotti, E., Driscoll, K., Moustatkas, T.D., and Paiella, R. (2008) Monte Carlo study of GaN versus GaAs terahertz quantum cascade structures. *Appl. Phys. Lett.*, **92**, 10112.

161. Belotti, E., Driscoll, K., Moustatkas, T.D., and Paiella, R. (2009) Monte Carlo simulation of terahertz quantum cascade laser structures based on widebandgap semiconductors. *J. Appl. Phys.*, **105**, 113103.

162. Lončar, M., Lee, B.G., Diehl, L., Belkin, M., Capasso, F., Giovannini, M., Faist, J., and Gini, E. (2007) Design and fabrication of photonic crystal quantum cascade lasers for optofluidics. *Opt. Express*, **15**, 4499–4514.

163. Bahriz, M., Moreau, V., Colombelli, R., Crisafulli, O., and Painter, O. (2007) Design of mid-IR and THz quantum cascade laser cavities with complete TM photonic bandgap. *Opt. Express*, **15**, 5948–5965.

164. Ma, P., Kaspar, P., Fedoryshyn, Y., Strasser, P., and Jäckel, H. (2009) InP-based planar photonic crystal waveguide in honeycomb lattice geometry for TM-polarized light. *Opt. Lett.*, **34**, 1558.

165. Bai, Y., Gokden, B., Darvish, S.R., Slivken, S., and Razeghi, M. (2009) Photonic crystal distributed feedback quantum cascade lasers with 12 W output power. *Appl. Phys. Lett.*, **95**, 031105.
166. Capasso, F. (2010) High-performance midinfrared quantum cascade lasers. *Opt. Eng.*, **49**, 111102.
167. Gökden, B., Bai, Y., Bandyopadhyay, N., Slivken, S., and Razeghi, M. (2011) High peak power (34 W) photonic crystal distributed feedback quantum cascade lasers. *Proc. SPIE*, **7946**, 794606.

4
Photonic-Crystal VCSELs

Krassimir Panajotov, Maciej Dems, and Tomasz Czyszanowski

4.1
Introduction

Vertical-cavity surface-emitting lasers (VCSELs) are a generation of semiconductor lasers that differs considerably from the conventional edge emitting lasers (EELs). In contrast to the EELs where light propagates along the active region (AR) and is emitted from the edge, in VCSELs light propagates perpendicular to the quantum wells (QWs) inside very-short ($\sim\lambda$), high-finesse cavity built by very high reflectivity distributed Bragg reflectors (DBRs) [1, 2]. VCSELs are usually fabricated with a circular aperture defined by proton-implantation, air-post etching, or partial oxidation of an Al-rich layer (see Figure 4.1). This configuration of VCSELs provides several advantages, such as much smaller dimensions, inherent single longitudinal-mode emission, integration in two-dimensional arrays, circular beam shape to facilitate coupling to optical fibers, and on-wafer testing, which significantly reduces the production cost. However, the surface emission creates some problems, namely multitransverse-mode emission and polarization unstable behavior.

For many applications it is desirable to have single mode emission in the fundamental HE_{11} mode with an almost Gaussian shape. However, this is only possible for small aperture VCSELs that provide low power. Furthermore, by increasing the injection current and optical power in the fundamental mode, a spatial hole is burned in the carrier distribution, providing enhanced modal gain for the higher-order modes [3–5]. Such multitransverse-mode emission deteriorates the spectral purity of the single longitudinal regime of operation, which is important in applications such as spectroscopy, printing, high-speed data transmission over a single-mode optical fiber, and so on. Many efforts have been made to ensure single transverse-mode operation of VCSELs with large aperture and for a certain region of injection currents [6]. The general ideas are either to reduce (or even overcompensate) the transverse optical waveguiding or to introduce additional mode-selective losses for higher-order modes [6].

In order to reduce the transverse optical waveguiding one has to reduce the index step introduced by the air-post structure or by the partially oxidized AlAs layer. This

Compact Semiconductor Lasers, First Edition.
Edited by Richard M. De La Rue, Siyuan Yu, and Jean-Michel Lourtioz.
© 2014 Wiley-VCH Verlag GmbH & Co. KGaA. Published 2014 by Wiley-VCH Verlag GmbH & Co. KGaA.

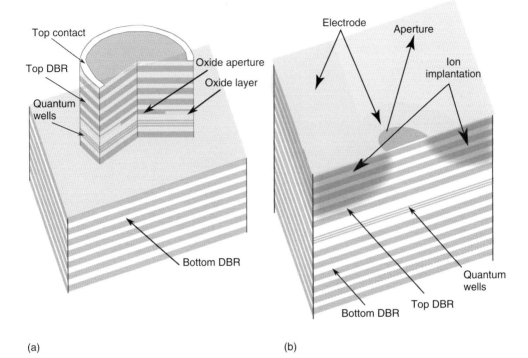

Figure 4.1 Typical vertical-cavity surface-emitting lasers (VCSELs). Transverse confinement of light is provided by either a partial lateral oxidation of an aluminum rich layer starting from the sides of the etched post (a) or ion implantation (b). (Please find a color version of this figure on the color plates.)

can be done by reducing the thickness of the AlAs layer to as little as 15–20 nm (such thin layers can still be oxidized reproducibly and can still prevent carrier tunneling through the oxidized area). The effective waveguiding, determined by the overlap of the standing wave longitudinal optical field distribution with the transverse distribution of the refractive index [7], can be even more reduced by placing the oxide layer in a node of the standing wave pattern of the VCSEL cavity (see Figure 4.2). Another approach is to make the index step gradual (i.e., tapered oxide aperture) by using several AlGaAs layers with different aluminum contents such that the center layer possesses the largest oxidization rate [8]. The idea of reduced contrast of the transverse effective refractive index has been pursued further, even to the situation when the sign of the index difference is reversed. Such antiguiding VCSEL structures can be realised by post-growth of air-post devices [9] or by including an additional AlGaAs layer which is oxidized in the center instead of the periphery [10]. A similar effect can be obtained by modifying radially the cavity resonance condition, e.g. by using etch and regrowth process to increase the cavity length around the laser aperture [11]. All these methods require growth interruption and eventually, using dielectric materials for the top

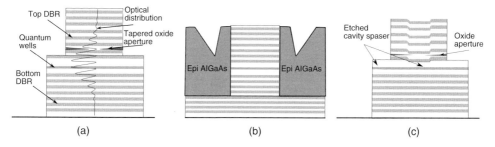

Figure 4.2 VCSELs with modified transverse optical waveguiding: (a) tapered oxide aperture; (b) antiguiding provided by regrowth of wide band-gap material surrounding the etched post; and (c) antiguiding by cavity resonance modification using etch and regrowth process. The longitudinal standing wave pattern is shown schematically in (a).

DBR. Introducing mode-selective losses for higher-order modes is another general approach for enhancing the single transverse-mode operation in VCSELs that makes use of the difference in the intensity patterns of the fundamental and higher-order modes. Higher-order modes have their intensity maximum at the periphery of the aperture and are emitted with a larger diffraction angle than the fundamental mode. The modal gain for the fundamental mode can be enhanced by adjusting the lateral gain profile to match its intensity profile, for example, by controlling the current injection profile [6] or by restricting lateral diffusion [12]. The optical losses can be made radially dependent by implementing saturable absorbers to act as a variable optical aperture [6], by applying externally a graded-index lens [13], a curved micromirror [14] or a Fabry–Perot etalon [15]. Furthermore, the p-contact can be used as an additional mode filter [16] or the reflectivity of the top Bragg mirror can be varied spatially, for example, by etching shallow local structures [17–19]. Introducing photonic crystal (PhC) in the VCSEL (see Figure 4.3) is an alternative approach to achieve enhanced single transverse-mode operation due to precise control of the strength of waveguiding and the modal losses and will be discussed in details in this chapter. Moreover, it could be easily adapted to simultaneously help avoiding polarization instabilities, typical for VCSELs.

The polarization properties of the light emitted by a semiconductor laser are determined by the polarization properties of: gain, that is, by the interband transition matrix element; and optical cavity. For QW active material commonly used in contemporary VCSELs because of their surface emission the transition matrix element is the same for linear polarization (LP) in any direction in the plane of the QW (perpendicular to the direction of light emission) [20]. Therefore, there is no intrinsic polarization gain anisotropy mechanism in conventional VCSELs. The optical confinement in VCSELs is usually achieved by a symmetric cylindrical or rectangular aperture defined by oxidization, air-post or proton-implantation, which does not provide waveguiding and reflectivity anisotropy contrary to EELs (where the lateral and transverse dimensions of the waveguide differ by orders of magnitude). As a result, even solitary VCSELs often suffer from polarization instabilities manifested as polarization switching between two orthogonal linearly

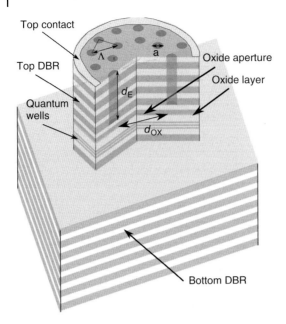

Figure 4.3 Photonic-crystal vertical-cavity surface-emitting laser (PhC-VCSEL). The PhC parameters: lattice pitch Λ, hole diameter a and etching depth d_E and the laser aperture diameter d_{Ox} are denoted. (Please find a color version of this figure on the color plates.)

polarized states or increased low frequency intensity noise because of polarization mode hopping [21]. Consequently, many approaches have been suggested to eliminate the current and temperature induced polarization switching. Similarly to the ideas of selective modal gain/losses for enhanced fundamental mode operation, one searches for polarization-dependent modal gain/loss. Anisotropic gain can be achieved through growth on a substrate with crystallographic orientation other than [001] [22, 23] or by in-plane uniaxial strain [21, 24]. Polarization selection may be imposed by the laser cavity, that is, by introducing either birefringence or dichroism. These include etching of rectangular mesas [25–28], etching of anisotropic apertures [29, 30] or by making the VCSEL top DBR polarization dependent by either using metal-semiconductor gratings [31–33] or directly etching a subwavelength grating [34]. All these approaches are usually rather difficult to implement and do not provide enhanced polarization stabilisation in the case of multitransverse-mode operation. It is therefore desirable to apply the technique for enhancing the fundamental mode regime for the sake of enhancing the single-polarization regime as well. Introducing PhC is a promising approach to achieving these two goals simultaneously.

Considerable amount of experimental work on incorporating PhC in the VCSEL structure has been carried out. In [35], 850-nm triangular-lattice PhC-VCSELs with a single-hole defect achieved a single-mode CW operation in the entire current range with side-mode suppression ratio (SMSR) of 35–40 dB (hereafter single-mode implies single (fundamental) transverse mode). The PhC lattice pitch is $\Lambda = 5$ µm, the hole diameter – $a = 3.5$ μm and the etching depth – $d_E = 14–17$ out 24 pairs of the top DBR. For an oxide aperture of diameter $d_{Ox} \sim 4\Lambda$ a maximum output power of 0.57 mW is achieved. Single-mode CW operation with 30 dB SMSR has also been achieved in 1-hole defect PhC-VCSELs in [36] while 19-hole PhC defect devices operate multimode. Both one-hole and seven-hole defect PhC-VCSELs have demonstrated single-mode operation in [37] however, for different PhC parameters: $\Lambda = 4$ μm, $a = 2.8$ μm and $\Lambda = 2$ μm, $a = 1$ μm, respectively. In [38] a 850 nm proton-implanted PhC-VCSEL with SMSR larger than 40 dB and CW (pulse current) output power of over 1 (2) mW has been demonstrated. Threefold improvement of the single-mode output power (SMSR of 30 dB) has been achieved in [39] for a single-defect PhC-VCSEL with $\Lambda = 4.4$ μm, $a = 2.2$ μm, $d_E = 16$ pairs and $d_{Ox} = 9$ μm.

A detailed study of more than 1600 single-hole defect 850 nm lasers in [40] has revealed that the expected "endlessly single-mode operation" with increasing Λ/λ for a suitably chosen a/Λ as observed for photonic-crystal fibers (PCFs) in [41] breaks down in PhC-VCSELs for $\Lambda/\lambda > 7$. The reason is that, besides the PhC, other contributions to the waveguiding and modal gain such as thermal lensing, carrier induced antiguiding, and spatial hole burning prevail. Endlessly single-mode operation occurs in PhC fibers because as Λ or λ are scaled, the change in the effective index in the PhC cladding exactly compensates for the normalised diameter of the core (D/λ) [41]. This property can be exploited to scale either Λ and therefore, the defect aperture allowing for greater single-mode output power or λ, which allows for a single design to maintain single-mode operation regardless of lasing wavelength. When etching the PhC holes in the oxide-confined PhC-VCSEL investigated in [40] the slope efficiency drops fourfold due to increased optical losses and, for the same reason, slightly increases with PhC defect width as less air-hole area remains in the oxide region. This also leads to a decrease of the threshold current. Similar results have been recently reported in [42] for commercial 850 nm VCSELs with oxide aperture of $d_{Ox} = 8.5$ μm and a focused ion beam drilled PhC with $\Lambda = 4$ μm and $a/\Lambda = 0.5$ and various hole depths from $d_E = 0$ to $d_E = 3.1$ μm (all 24 pairs of the top DBR). A significant decrease of the output power and an increase of the threshold current have been observed with increasing etch depth with single-mode operation for $d_E \geq 1.54$ μm. Oxide-aperture single-mode PhC-VCSELs with reduced divergence angle between 5.1° and 5.5° over the entire drive current range have been demonstrated in [43].

A similar study for proton-implanted PhC-VCSELs has been carried out in [44]. Contrary to oxide-confined PhC-VCSELs, in these devices the slope efficiency increases and the threshold current decreases when etching the PhC pattern because of the lack of index guiding, except for thermal lensing. An increase in slope efficiency and a decrease in threshold current are also observed for increasing defect diameters; this behavior, which is a direct consequence of decreasing diffraction loss, is similar to that obtained in oxide-confined PhC-VCSELs. For a given a/Λ, the defect diameter is increased by increasing the hole diameter a. However, the increased hole diameter results in a deeper etch depth, and is therefore, a better mode confinement, which is responsible for the decreased diffraction loss. Single-mode operation is obtained for defect diameters between 4 and 9 µm. Devices with defect diameters smaller than 4 µm do not lase and the ones with defect diameters larger than 9 µm lase multimode. A unique tradeoff between aperture diameter and etch depth exists in proton-implanted PhC-VCSELs: shallow etch depths provide single-mode emission if the defect diameter is enlarged, which allows scaling of the single-hole defect diameters greater than 10 µm. This is in contrast with oxide-confined PhC-VCSELs where a minimum d_E is needed to provide sufficient loss and to inhibit lasing of the strongly guided oxide modes [40].

Universal PhC designs that produce single-mode operation with >35dB SMSR and 1 mW output power over a wide range of wavelengths (780, 850, and 980 nm) have been demonstrated in [45]. The 780-, 850-, and 980-nm VCSELs have 30-, 22-, and 22-period p-type top DBR mirrors, respectively. Single-mode emission is achieved for $\Lambda = 4.5$ µm, $a/\Lambda = 0.7$, $d_E = 80\%$ of the top DBR and $d_{Ox} = 12$ µm.

The impact of optical losses introduced by the PhC for enhancing the single mode emission has been studied in [46, 47] for 850 nm VCSELs with 20 or 22 top p-type DBR periods, 35 bottom n-type DBR periods and AlGaAs/GaAs QW ARs. The optical loss induced by the PhC pattern was determined from the spectral splitting between the fundamental and first-order transverse modes measured for injection current below threshold (at $0.9 J_{th}$) in order to avoid thermal effects (see also Section 4.2). A careful examination of measurements revealed a clear cutoff between single-mode and multimode operation at approximately 5 cm^{-1} of loss difference. In order to study the PhC hole depth dependence, multiple etch depths for a specific PhC design were investigated. This study revealed that an increased etch depth corresponds to a decrease in optical loss, even for holes penetrating into the entire bottom DBR. Similar loss analysis of high-order modes in PhC-VCSELs has been carried out in [48].

PhC-VCSELs are capable of high-speed modulation. A small-signal modulation bandwidth of $f_{max} = 9$ GHz has been demonstrated in [49] for oxide-confined PhC-VCSELs with $a/\Lambda = 0.5$ and $d_E = 15$ out of 27 pairs of the top DBR. For proton-implanted single-mode PhC-VCSELs a maximum modulation bandwidth of $f_{max} = 15$ GHz was achieved for 11 µm implant diameter and PhC parameters $\Lambda = 4$ µm and $a/\Lambda = 0.7$. For multimode devices with implant diameter of 12 µm and $\Lambda = 4.5$ µm, $a/\Lambda = 0.4$, the bandwidth reached $f_{max} = 18$ GHz.

Single-mode 1.3 µm single-hole defect PhC-VCSELs have been realised in [50–52]. In [50] the AR contained 17 layers of 2.5 monolayer thick InAs quantum dots (QDs) overgrown by 5 nm $In_{0.15}Ga_{0.85}As$ QWs and separated by GaAs barriers. The PhC parameters were: $\Lambda = 5$ µm, $a = 2.5$ µm, $d_E = 18$ (out of 23) pairs and a combination of oxide- and proton-implanted carrier confinement was implemented. Only 0.2 mW CW output power has been achieved. In [51] the AR contained highly strained InGaAsN QWs while the PhC was etched in the dielectric top DBR. Single-mode operation with >30 dB SMSR was achieved for $\Lambda = 7.4$ µm, $a = 3.7$ µm, $d_E = 2$ µm and $d_{Ox} = 8$ µm. Submonolayer quantum-dot 1.3 µm PhC-VCSEL has been realised in [52]. A 3.8 mW single-mode output power with >35 dB SMSR was achieved at injection currents of $J \approx 30 \cdot J_{th}$ for a threshold current of $J_{th} = 0.9$ mA. The PhC parameters were: $\Lambda = 5$ µm, $a = 0.5 \Lambda$ and

$d_E = 16$ out 20 pairs of the top GaAs/AlGaAs DBR. The beam divergence angle remained very small and almost unchanged between 6.7° and 6.9° when increasing the current from 3 to 20 mA.

Polarization control induced by the PhC has been attempted in [53] and [54] for 850 nm VCSELs. A hexagonal lattice PhC with elliptical holes elongated along K and M directions has been used in [53] and 20 dB polarization-mode suppression-ration (PMSR) has been achieved, mostly due to an anisotropic current injection. In [54] square lattice PhC-VCSELs of various lattice constants have been demonstrated and a single-mode output power of 1 mW (SMSR of 20 dB) was obtained for $\Lambda = 5$ μm, $a = 0.7\,\Lambda$, $d_E = 1.7$ μm and a current aperture $d_{Ox} = 16$ μm. The threshold current increased from 2.5 to 4 mA when introducing the PhC pattern due to the increased modal losses. Polarization selectivity was achieved by decreasing the diameter of 2 air holes down to $a = 0.3\,\Lambda$ in a certain direction next to the single PhC defect, thus leading to lasing in a linearly polarized fundamental mode along this direction (see also Section 4.7). However, as the effective area of the cavity was increased, the lattice constant needed to be decreased to $\Lambda = 3$ μm for a single-mode operation in the whole region of injection currents.

An interesting application of PhC-VCSELs as micro-fluidic sensing devices has been demonstrated in [55, 56]. The oxide aperture extended under the PhC holes and when dissolved it formed a horizontal channel that was connected to the vertical PhC holes. When the oxide layer was removed, the PhC-VCSEL effective index contrast increased and multimode operation was supported. Then a reversible blueshift in the lasing wavelength of about 0.3 nm was observed with the introduction of deionized water accompanied by an increase in the divergence angle of the laser beam. Another sensor application has been recently reported in [57], namely hydrogen detection by a thin layer of palladium (Pd) deposited on the surface of etched air holes. As the H_2 reacts with Pd the resulting PdH_x induces a change in the complex refractive index leading to 60% output power increase and a 52 pm redshift of the lasing wavelength at 4% H_2 in [57].

In this chapter we present the efficient plane-wave admittance method (PWAM) for full-vectorial calculation of 3D photonic structures and a variety of numerical results for PhC-VCSELs. The chapter is organized as follows: Section 4.2 discusses the different numerical methods capable of modeling PhC-VCSELs; their shortcomings and advantages. Section 4.3 presents the mathematical basis of the PWAM. In Section 4.4 we investigate the impact of the PhC depth on VCSEL threshold characteristics and demonstrate the existence of an optimal depth that establishes the PhC waveguiding. In Section 4.5 we discuss the advantages of the bottom-emitting PhC-VCSELs compared to top-emitting ones with respect to the modal properties of the emitted light. Section 4.6 investigates the PhC parameter regions leading to an enhanced fundamental mode operation of PhC-VCSELs. Section 4.7 investigates PhC designs that lead to highly birefringent and dichroic PhC-VCSELs. In Section 4.8 we show that PhC-VCSELs with true photonic band gap are feasible and demonstrate truly band-gap waveguiding and low-index defect confinement in such photonic band gap (PBG)- VCSELs. Finally, in Section 4.9 we summarise the presented results with some prospects of further developments.

4.2
Numerical Methods for Modeling Photonic-Crystal VCSELs

The lack of axial symmetry and the complex 3D structure with very high index contrast between the PhC holes and the surrounding semiconductor material of PhC-VCSELs makes their optical modeling quite a challenging task. As a strong scattering takes place at the edges of the PhC holes the typical LP approximation is no longer valid preventing the application of popular simplified models [58]. Nevertheless, the first attempts to get insight into the light confinement mechanism of PhC-VCSELs relay on such simplified assumptions.

In [37] the band structure $k_0(k_\perp)$ of a perfect PhC structure without defect is first calculated for out-of-plane propagation using a 2D plane-wave expansion (PWE) technique. As the out-of-plane component $k_\perp$ of the normalised wavevector is larger than 3 for typical PhC-VCSEL parameters the light confinement effect is considered to be index guiding provided by the refractive index reduction in the periodic hole array with respect to the PhC defect aperture. Therefore, an equivalent index is deduced from the lowest PhC energy level available as $n_{eq} = k_\perp/k_0$ and then the transfer-matrix method is used to find the resonances and the effective indices of the different transverse sections of the entire PhC-VCSEL, that is, the effective-index method of Hadley [7] is applied. Finally, the PhC-VCSEL is mapped to 2D cylindrical step-index fiber, which allows calculating the resonant wavelength for a finite PhC hole etching depth. Later on, the same group of Choquette recognized that the PhC provides confinement not only by the effective index contrast but also by selective losses for different transverse modes and suggested a semiempirical approach to account for this effect [46]. To this aim they include an imaginary part of the effective refractive index of the PhC cladding region, which is estimated from the wavelength splitting of the fundamental and first-order transverse mode (this splitting is larger in a lossy structure). A very good agreement with experiment for the spectral positions of the higher-than-one transverse modes has been achieved in [46].

Similarly, the 3D PhC-VCSEL problem has been reduced by the effective-index method to a 2D Helmholtz equation, which has been solved either by finite difference in [59] or by finite element method in [60].

Three rigorous optical models and the effective PWE method have been compared in [58] for a benchmark PhC-VCSEL structure.

The coupled mode model was developed by Bava *et al.* in 2001 [61]. In this model, the electromagnetic field is approximated by a modal expansion of free space modes of a reference medium, that are common throughout all layers of the entire VCSEL structure. The optical field propagation is described by coupling of basis modes originating from perturbations of the refractive index compared to the reference medium; each different layer results in a different coupling coefficient matrix. Since the basis modes are common throughout the entire structure, no mode matching conversion that involves heavy matrix inversion calculations is required at layer interfaces. Furthermore, no transverse boundary conditions such as metallic boundary conditions with absorbing material coatings are needed.

The longitudinal boundary conditions at the topmost and bottommost interfaces of the VCSEL structure provide an eigenvalue equation. By solving the equation and requiring self consistency after a round trip propagation of the field over the VCSEL structure, the lasing wavelength and threshold material gain of a mode are obtained. Employing modal expansion rather than spatial discretization and the absence of mode matching conversion at layer interfaces lead to very high computational efficiency [58].

The Finite Element Method is used to solve either the vectorial or scalar Helmholtz equation in which the unknowns are the electric field profile $\mathbf{E}(r)$ and complex angular frequency (ω). One first selects an appropriate computational domain, encompassed by nonreflecting boundary conditions, that is, by perfectly matched layers (PML) (see Section 4.3). The unknown electric field is approximated as a linear combination of appropriately selected vectorial basis functions and an equivalent variational functional leads to an algebraic eigen problem [62]. The computational window is filled by prism elements. A triangular mesh is first created, which respects all lateral contours present in the PhC-VCSEL by projecting the electrical aperture, etched hole patterns, metallic contacts, and PML region into the same transverse plane. The triangularization of this domain results in a two-dimensional mesh, which is then extruded to the three-dimensional structure by filling it with prism elements with associated scalar unknowns to their vertices (or the mid-points of their vertical edges).

The plane-wave effective-index method is a simplified version of the rigorous PWAM [63] which will be presented in Section 4.3. It considers a planar structure and expands the lateral field distribution on a plane-wave basis and similar to the Hadley effective-index method [7] the electric field is separated into two components, which are determined separately for directions parallel and perpendicular to the device layers. However, in contrast to the Hadley method the optical field is considered a vectorial one, hence the polarization information is not lost [64]. The eigenmodes are determined as complex eigenvalues of a two-dimensional problem, which gives a significant performance boost with respect to computation time and memory requirements. However, because of optical field factorization, complex structures are not considered properly, especially in cases where the mode is not strongly confined [64].

3D finite-difference time-domain (FDTD) calculations of PhC-VCSEL have been attempted recently in [65]. However, the size of the structure prevents using a small enough mesh size that would give the same results in 1D computation as the transmission matrix method. Three-dimensional FDTD has been used to calculate the near field pattern and investigate the influences of the weakly guiding PhC waveguide on the divergence angle of a PhC-VCSEL in [43]. When the dimension of the exciting Gaussian source was chosen to be equal to 15 μm, the numerically obtained value of 5.2° of divergence angle was in good agreement with experiment.

We will now present in detail the efficient full-vectorial three-dimensional PWAM.

4.3
Plane-Wave Admittance Method

The PWAM was first introduced by Dems, Kotynski, and Panajotov in 2005, [63] for the specific usage of efficient modeling of 3D multilayered structures with high-contrast subwavelength features that need full-vectorial treatment [66]. It shares important ingredients with the plane-wave method (PWM), which has been widely used for PhC simulations [67, 68] and with the method of lines (MoL) [69, 70]. PWAM is a frequency-domain method, appropriate for the analysis of anisotropic, lossy, and even magnetic or negative refractive index photonic structures surrounded by arbitrarily shaped uniaxial perfectly matched layers (UPMLs) [71, 72]. The modeled structure is assumed to consist of parallel layers, and PWAM is especially efficient when certain layers appear multiple times within the stack, a notable example being the PhC-VCSELs for which the PWAM provides significant computational savings as compared to general purpose methods. Modeling of other photonic structures, such as PhC waveguides is also possible [64]. PWAM consists of two major steps. Firstly, each layer is analyzed with the PWE formulated with a general material model with a tensor-form, complex-valued permittivity and permeability. This part of the method taken alone [73] allows for efficient modal analysis of waveguide structures such as index guiding or photonic band-gap guiding anisotropic PCFs [41, 74] with UPML boundary conditions. Secondly, the PWE-based description of all layers is followed by a boundary matching scheme based on the admittance-matrix transfer method, which allows modeling fully three-dimensional structures. This part of PWAM is common with the MoL.

Consider an anisotropic material with permittivity and permeability represented as diagonal tensors and without any free charges ($\rho = 0$) or currents ($\mathbf{j} = 0$). The time-independent Maxwell equations for the electric $\mathbf{E}$ and magnetic $\mathbf{H}$ fields read

$$-\partial_z E_y + \partial_y E_z = -i\omega \mu_x \mu_0 H_x \tag{4.1}$$

$$\partial_x E_z - \partial_z E_x = -i\omega \mu_y \mu_0 H_y \tag{4.2}$$

$$-\partial_y E_x + \partial_x E_y = -i\omega \mu_z \mu_0 H_z \tag{4.3}$$

$$-\partial_z H_y + \partial_y H_z = i\omega \varepsilon_x \varepsilon_0 E_x \tag{4.4}$$

$$\partial_x H_z - \partial_z H_x = i\omega \varepsilon_y \varepsilon_0 E_y \tag{4.5}$$

$$-\partial_y H_x + \partial_x H_y = i\omega \varepsilon_z \varepsilon_0 E_z \tag{4.6}$$

where $i^2 = -1$, ω is the angular frequency of light and ∂_ξ means partial derivative in the ξ direction. From Equations 4.3 and 4.6 it is possible to express z components of the electric and magnetic vectors as

$$H_z = \begin{bmatrix} -\dfrac{i}{k_0 \eta_0 \mu_z} \partial_x & -\dfrac{i}{k_0 \eta_0 \mu_z} \partial_y \end{bmatrix} \begin{bmatrix} -E_y \\ E_x \end{bmatrix} \tag{4.7}$$

$$E_z = \begin{bmatrix} -\dfrac{i\eta_0}{k_0 \varepsilon_z} \partial_y & -\dfrac{i\eta_0}{k_0 \varepsilon_z} \partial_x \end{bmatrix} \begin{bmatrix} H_x \\ H_y \end{bmatrix} \tag{4.8}$$

where $k_0 = \omega/c = \omega(\mu_0 \varepsilon_0)^{1/2}$ is the normalized frequency and $\eta_0 = (\mu_0/\varepsilon_0)^{1/2}$ is the free space impedance. Substituting Equations 4.7 and 4.8 into Equations 4.1,

4.2, 4.4, and 4.5, we can obtain the generalized transmission line (GTL) equations

$$\partial_z \begin{bmatrix} -E_y \\ E_x \end{bmatrix} = -i\frac{\eta_0}{k_0} \begin{bmatrix} \partial_y \varepsilon_z^{-1} \partial_y + \mu_x k_0^2 & -\partial_y \varepsilon_z^{-1} \partial_x \\ -\partial_x \varepsilon_z^{-1} \partial_y & \partial_x \varepsilon_z^{-1} \partial_x + \mu_y k_0^2 \end{bmatrix} \begin{bmatrix} H_x \\ H_y \end{bmatrix} \quad (4.9)$$

$$\partial_z \begin{bmatrix} H_x \\ H_y \end{bmatrix} = -i\frac{1}{\eta_0 k_0} \begin{bmatrix} \partial_x \mu_z^{-1} \partial_x + \varepsilon_y k_0^2 & \partial_x \mu_z^{-1} \partial_y \\ \partial_y \mu_z^{-1} \partial_x & \partial_y \mu_z^{-1} \partial_y + \varepsilon_x k_0^2 \end{bmatrix} \begin{bmatrix} -E_y \\ E_x \end{bmatrix} \quad (4.10)$$

Now, consider a layer in the structure, invariant in the z direction and exhibiting a two-dimensional periodicity in the xy plane, that is,

$$\varepsilon(\mathbf{r}) = \varepsilon(\mathbf{r} + \mathbf{a}_i) \quad (i = 1, 2) \quad (4.11)$$

$$\mu(\mathbf{r}) = \mu(\mathbf{r} + \mathbf{a}_i) \quad (i = 1, 2) \quad (4.12)$$

with $\mathbf{a}_1$ and $\mathbf{a}_2$ being elementary lattice vectors in the xy plane. Because of this periodicity the solutions of the Maxwell equations obey the Bloch theorem, that is, the electric and magnetic fields can be represented as

$$\mathbf{E}(\mathbf{r}) = \overline{\mathbf{E}}(\mathbf{r}) \exp(-i\,\mathbf{k}_{xy} \cdot \mathbf{r}) \quad (4.13)$$

$$\mathbf{H}(\mathbf{r}) = \overline{\mathbf{H}}(\mathbf{r}) \exp(-i\,\mathbf{k}_{xy} \cdot \mathbf{r}) \quad (4.14)$$

where $\mathbf{k}_{xy}$ is a projection of the mode wavevector into the xy-plane and $\overline{\mathbf{E}}(\mathbf{r}), \overline{\mathbf{H}}(\mathbf{r})$ are periodic functions, that is, they can be expanded into two-dimensional Fourier (plane-wave) series as

$$\overline{\mathbf{E}}(\mathbf{r}) = \mathbf{E}^{\mathbf{g}} \exp(i\,\mathbf{g} \cdot \mathbf{r}) = \mathbf{E}^{\mathbf{g}} \,|\varphi_{\mathbf{g}}\rangle \quad (4.15)$$

$$\overline{\mathbf{H}}(\mathbf{r}) = \mathbf{H}^{\mathbf{g}} \exp(i\,\mathbf{g} \cdot \mathbf{r}) = \mathbf{H}^{\mathbf{g}} \,|\varphi_{\mathbf{g}}\rangle \quad (4.16)$$

where we use the Einstein summation convention[1] and $\mathbf{g}$ is the reciprocal lattice vector

$$\mathbf{g} = l_1 \mathbf{b}_1 + l_2 \mathbf{b}_2 \quad (4.17)$$

$$\mathbf{a}_i \cdot \mathbf{b}_j = 2\pi \delta_{ij} \quad (4.18)$$

with l_1 and l_2 being arbitrary integers and δ_{ij} the Kronecker symbol. Using the Bloch theorem together with Equations 4.15 and 4.16, we can represent the electric and magnetic fields as

$$\mathbf{E}(\mathbf{r}) = \mathbf{E}^{\mathbf{g}} \,|\varphi_{\mathbf{g} - \mathbf{k}_{xy}}\rangle \quad (4.19)$$

$$\mathbf{H}(\mathbf{r}) = \mathbf{H}^{\mathbf{g}} \,|\varphi_{\mathbf{g} - \mathbf{k}_{xy}}\rangle \quad (4.20)$$

The Fourier basis is orthonormal, that is,

$$\langle \varphi_{\mathbf{g}} | \varphi_{\mathbf{g}'} \rangle = \delta_{\mathbf{g}\mathbf{g}'} \quad (4.21)$$

where $\langle \cdot | \cdot \rangle$ denotes scalar product and $|\varphi_{\mathbf{g}}\rangle = \exp(i\,\mathbf{g} \cdot \mathbf{r})$. We introduce Equations 4.19 and 4.20 into the GTL Equations 4.9 and 4.10, left-multiply them

---

1) We use this convention in this whole chapter, unless explicitly stated otherwise.

by $\langle\varphi\mathbf{g}-\mathbf{k}_{xy}|$ and use the orthonormality condition Equation 4.21 to obtain

$$\partial_z \begin{bmatrix} E_y^{\mathbf{g}} \\ E_x^{\mathbf{g}} \end{bmatrix} = -i\frac{\eta_0}{k_0} \left\langle \varphi_{\mathbf{g}-\mathbf{k}_{xy}} \middle| \begin{matrix} \partial_y \varepsilon_z^{-1}\partial_y + \mu_x k_0^2 & -\partial_y \varepsilon_z^{-1}\partial_x \\ -\partial_x \varepsilon_z^{-1}\partial_y & \partial_x \varepsilon_z^{-1}\partial_x + \mu_y k_0^2 \end{matrix} \middle| \varphi_{\mathbf{g}'-\mathbf{k}_{xy}} \right\rangle \begin{bmatrix} H_x^{\mathbf{g}'} \\ H_y^{\mathbf{g}'} \end{bmatrix}$$
(4.22)

$$\partial_z \begin{bmatrix} H_x^{\mathbf{g}} \\ H_y^{\mathbf{g}} \end{bmatrix} = -\frac{i}{\eta_0 k_0} \left\langle \varphi_{\mathbf{g}-\mathbf{k}_{xy}} \middle| \begin{matrix} \partial_x \mu_z^{-1}\partial_x + \varepsilon_y k_0^2 & \partial_x \mu_z^{-1}\partial_y \\ \partial_y \mu_z^{-1}\partial_x & \partial_y \mu_z^{-1}\partial_y + \varepsilon_x k_0^2 \end{matrix} \middle| \varphi_{\mathbf{g}'-\mathbf{k}_{xy}} \right\rangle \begin{bmatrix} E_y^{\mathbf{g}'} \\ E_x^{\mathbf{g}'} \end{bmatrix}$$
(4.23)

where, in order to simplify notation, we use $E_y(\mathbf{r}) = -E_y^{\mathbf{g}}|\varphi_{\mathbf{g}-\mathbf{k}_{xy}}\rangle$ instead of directly using Equation 4.19. The matrices in the above equations can be computed easily when both permittivity and permeability are expanded in the Fourier basis, that is,

$$\varepsilon_i(\mathbf{r}) = \varepsilon_i^{\mathbf{g}} |\varphi_{\mathbf{g}}\rangle \quad (i = x, y) \tag{4.24}$$

$$\mu_i(\mathbf{r}) = \mu_i^{\mathbf{g}} |\varphi_{\mathbf{g}}\rangle \quad (i = x, y) \tag{4.25}$$

$$\varepsilon_z^{-1}(\mathbf{r}) = \kappa^{\mathbf{g}} |\varphi_{\mathbf{g}}\rangle \tag{4.26}$$

$$\mu_z^{-1}(\mathbf{r}) = \gamma^{\mathbf{g}} |\varphi_{\mathbf{g}}\rangle \tag{4.27}$$

Substituting this into (4.22) and (4.23), we obtain the GTL equations in the plane-wave basis

$$\partial_z \hat{\mathbf{E}} = -i\mathbf{R}_\mathbf{H}\hat{\mathbf{H}} \tag{4.28}$$
$$\partial_z \hat{\mathbf{H}} = -i\mathbf{R}_\mathbf{E}\hat{\mathbf{E}} \tag{4.29}$$

where

$$\hat{\mathbf{E}} = \begin{bmatrix} E_y^{\mathbf{g}} \\ E_x^{\mathbf{g}} \end{bmatrix}$$

$$\hat{\mathbf{H}} = \begin{bmatrix} H_x^{\mathbf{g}} \\ H_y^{\mathbf{g}} \end{bmatrix}$$

$$\mathbf{R}_\mathbf{E} = \frac{1}{\eta_0 k_0} \begin{bmatrix} -(g_x - k_x)(g_x' - k_x)\gamma^{\mathbf{g}-\mathbf{g}'} + k_0^2 \varepsilon_y^{\mathbf{g}-\mathbf{g}'} & -(g_x' - k_x)(g_y' - k_y)\gamma^{\mathbf{g}-\mathbf{g}'} \\ -(g_y - k_y)(g_x' - k_x)\gamma^{\mathbf{g}-\mathbf{g}'} & -(g_y - k_y)(g_y' - k_y)\gamma^{\mathbf{g}-\mathbf{g}'} + k_0^2 \varepsilon_x^{\mathbf{g}-\mathbf{g}'} \end{bmatrix}$$

$$\mathbf{R}_\mathbf{H} = \frac{\eta_0}{k_0} \begin{bmatrix} -(g_y - k_y)(g_y' - k_y)\kappa^{\mathbf{g}-\mathbf{g}'} + k_0^2 \mu_x^{\mathbf{g}-\mathbf{g}'} & (g_y - k_y)(g_x' - k_x)\kappa^{\mathbf{g}-\mathbf{g}'} \\ (g_x - k_x)(g_y' - k_y)\kappa^{\mathbf{g}-\mathbf{g}'} & -(g_x - k_x)(g_x' - k_x)\kappa^{\mathbf{g}-\mathbf{g}'} + k_0^2 \mu_y^{\mathbf{g}-\mathbf{g}'} \end{bmatrix}$$

Here k_x, k_y, g_x, and g_y are the corresponding components of the mode wavevector and the reciprocal lattice vector $\mathbf{g}$, respectively. In practical implementation the Fourier basis $\{\mathbf{g}\}$ is truncated at some point. The introduced error depends on two factors – the distribution of the electromagnetic field and the distribution of the material parameters. The latter can be convoluted with Gaussian window in order to improve the convergence.

Here, like in Ref. [63] we have assumed a diagonal form of the permittivity and permeability tensors. However, the PWAM may be formulated in the same way for a more general in-plane anisotropy: for the details of the corresponding PWE and the exact formulas for $\mathbf{R_E}$ and $\mathbf{R_H}$ operators considered below, please refer to Ref. [73].

By taking the z-derivative of each of the GTL equations in PWE basis Equations 4.28 and 4.29 and substituting the other one into it, we can write two second-order equations with the fields decoupled

$$\partial_z^2 \hat{\mathbf{E}} = -\mathbf{Q_E} \hat{\mathbf{E}} \tag{4.30}$$

$$\partial_z^2 \hat{\mathbf{H}} = -\mathbf{Q_H} \hat{\mathbf{H}} \tag{4.31}$$

where $\mathbf{Q_E} = \mathbf{R_H R_E}$ and $\mathbf{Q_H} = \mathbf{R_E R_H}$.

In the z-invariant layer, any of the above equations can be solved analytically, provided it is transposed into "diagonalized" coordinates. This is done by determination of all the eigenvalues of either $\mathbf{Q_E}$ or $\mathbf{Q_H}$ matrices. Because they are similar matrices in the linear algebra sense, that is, $\mathbf{Q_E} = \mathbf{R_E^{-1} Q_H R_E}$, they share the same set of eigenvalues and it is possible to write

$$\mathbf{Q_E} = \mathbf{T_E} \Gamma^2 \mathbf{T_E^{-1}} \tag{4.32}$$

$$\mathbf{Q_H} = \mathbf{T_H} \Gamma^2 \mathbf{T_H^{-1}} \tag{4.33}$$

where Γ^2 is a diagonal matrix of eigenvalues of $\mathbf{Q_E}$ and $\mathbf{Q_H}$ while $\mathbf{T_E}$ and $\mathbf{T_H}$ are the matrices constructed by their eigenvectors, respectively. In the following analysis we will use the fact that it is always possible to choose $\mathbf{T_E}$ and $\mathbf{T_H}$ such that

$$\mathbf{T_E} = \mathbf{R_H T_H} \Gamma^{-1} \tag{4.34}$$

Using the above relations we can write Equations 4.30 and 4.31 in their diagonalized forms

$$\partial_z^2 \tilde{\mathbf{E}} = -\Gamma^2 \tilde{\mathbf{E}} \tag{4.35}$$

$$\partial_z^2 \tilde{\mathbf{H}} = -\Gamma^2 \tilde{\mathbf{H}} \tag{4.36}$$

where $\tilde{\mathbf{E}} = \mathbf{T_E^{-1}} \hat{\mathbf{E}}$ and $\tilde{\mathbf{H}} = \mathbf{T_H^{-1}} \hat{\mathbf{H}}$. The mathematical operations we have just performed can also be explained as a representation of the electric (and similarly the magnetic) field via a linear combination of single-layer eigenmodes. Then the columns of the $\mathbf{T_E}$ ($\mathbf{T_H}$) represent the shape of each of such eigenmodes, the diagonal elements of Γ are their propagation constants and the elements of $\tilde{\mathbf{E}}$ ($\tilde{\mathbf{H}}$) – their amplitudes. Now Equations 4.35 and 4.36 have analytical solutions in the form of a propagating or standing wave. For the electric field, this solution can be written as

$$\tilde{\mathbf{E}}(z) = \cos(\Gamma z)\mathbf{A} + \sin(\Gamma z)\mathbf{B} \tag{4.37}$$

where $\mathbf{A}$ and $\mathbf{B}$ are some constant vectors that have to be determined from the boundary conditions. The application of the admittance transfer technique for this case is shown below.

Having found a solution of the GTL equations in a single z-invariant layer we need a numerically stable technique for a connection between separate layers and the final computation of eigenmodes. Originally, such a technique had been developed within the MoL using a finite-difference approximation [75] and is applied here to the case of PWE. For a uniform layer of thickness d, **A** and **B** can be expressed in terms of the diagonalised electric field, at $z = 0$ ($\tilde{\mathbf{E}}_0$) and at $z = d$ ($\tilde{\mathbf{E}}_d$), that is,

$$\mathbf{A} = \tilde{\mathbf{E}}_0 \tag{4.38}$$

$$\mathbf{B} = \frac{1}{\sin(\Gamma d)}\tilde{\mathbf{E}}_d - \cot(\Gamma d)\tilde{\mathbf{E}}_0 \tag{4.39}$$

Given $\tilde{\mathbf{E}}_0$ and $\tilde{\mathbf{E}}_d$ we would like to determine the magnetic field at both boundaries of the layer. For this purpose we will take Equation 4.28 and rewrite it in terms of $\tilde{\mathbf{E}}$ and $\tilde{\mathbf{H}}$:

$$\mathbf{T}_\mathbf{E}\partial_z\tilde{\mathbf{E}} = -i\mathbf{R}_\mathbf{H}\mathbf{T}_\mathbf{H}\tilde{\mathbf{H}}$$

which, after substituting Equation 4.34, gives

$$i\Gamma^{-1}\partial_z\tilde{\mathbf{E}} = \tilde{\mathbf{H}} \tag{4.40}$$

The left-hand-side of the above equation can be computed by taking the derivative of Equation 4.37,

$$\partial_z\tilde{\mathbf{E}}(z) = -\Gamma\sin(\Gamma z)\mathbf{A} + \Gamma\cos(\Gamma z)\mathbf{B} \tag{4.41}$$

Computing $\tilde{\mathbf{H}}_0$ at $z = 0$ and $\tilde{\mathbf{H}}_d$ at $z = d$ from Equation 4.40 with substitutions of Equations 4.38, 4.39, and 4.41, we have

$$\begin{bmatrix}\tilde{\mathbf{H}}_0\\ \tilde{\mathbf{H}}_d\end{bmatrix} = \begin{bmatrix}\mathbf{y}_1 & \mathbf{y}_2\\ -\mathbf{y}_2 & -\mathbf{y}_1\end{bmatrix}\begin{bmatrix}\tilde{\mathbf{E}}_0\\ \tilde{\mathbf{E}}_d\end{bmatrix} \tag{4.42}$$

where

$$\mathbf{y}_1 = -\cot(\Gamma d) \tag{4.43}$$

$$\mathbf{y}_2 = \frac{1}{\sin(\Gamma d)} \tag{4.44}$$

Equation 4.42 gives complete admittance relation for both sides of a single layer. Now let us consider the whole structure (Figure 4.4). We divide it into two parts and we derive the admittance relation in the matching plane between these two parts. To this aim we will use the iterative procedure for each part, starting at their outer edges, to determine the relation

$$\tilde{\mathbf{H}}_d^{(n)} = \mathbf{Y}^{(n)}\tilde{\mathbf{E}}_d^{(n)} \tag{4.45}$$

for every layer n. We assume $\tilde{\mathbf{E}}_0^{(1)}$, that is, the electric field at the beginning of the first layer, to be equal to zero. This is equivalent to putting our analyzed structure into a cavity limited by two perfect conductors. In order to correctly estimate the radiating field we need to use as the outer-most layers in the structure PMLs, that is, layers that do not reflect light for any incidence angle and polarization [71, 72, 76]. In PWAM we use the approach of Sacks *et al.* [71] in which the PMLs are physical mediums with uniaxially anisotropic permittivity and permeability (named uniaxial PMLs – UPMLs). When the vector normal to the UPML boundary is parallel to the $\hat{\mathbf{x}}$ unit vector and the permittivity and permeability of the attached simulation

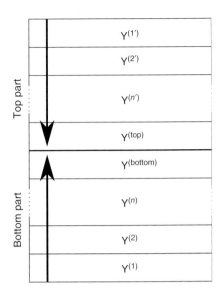

Figure 4.4 Admittance transfer in a multilayer Structure.

area are equal to ε_r and μ_r, UPML consists of the material with permittivity and permeability given by

$$\varepsilon_{\text{UPML}} = \varepsilon_r \begin{vmatrix} s^{-1} & 0 & 0 \\ 0 & s & 0 \\ 0 & 0 & s \end{vmatrix}, \quad \mu_{\text{UPML}} = \mu_r \begin{vmatrix} s^{-1} & 0 & 0 \\ 0 & s & 0 \\ 0 & 0 & s \end{vmatrix} \quad (4.46)$$

where s can be any arbitrary complex number different than 0. In particular if $s = \kappa - i\,\alpha/k_0$, then α is the ratio of damping in the PML and κ is responsible for coordinate stretching [77]. The UPML rotated counterclockwise by the angle θ with respect to a certain axis is defined as

$$\varepsilon_{\text{UPML}}(\theta) = R(\theta) \cdot \varepsilon_{\text{UPML}}(0) \cdot R(-\theta) \quad (4.47)$$

$$\mu_{\text{UPML}}(\theta) = R(\theta) \cdot \mu_{\text{UPML}}(0) \cdot R(-\theta) \quad (4.48)$$

where $R(\theta)$ is the respective rotation matrix in 3D. For instance, rotation around axis $\hat{c}z$ has the matrix

$$R(\theta) = \begin{bmatrix} \cos(\theta) & \sin(\theta) & 0 \\ -\sin(\theta) & \cos(\theta) & 0 \\ 0 & 0 & 1 \end{bmatrix} \quad (4.49)$$

It is easy to verify that the diagonal form of anisotropy allows surrounding the simulation area with a rectangular box of UPML, while the in-plane anisotropy is sufficient to build complex UPML geometries surrounding subsequent layers.

In PWAM, the PMLs at the boundary perpendicular to the z-axis (vertical PMLs) are treated differently from those at the other boundaries. Because of the analytical expansion in the z-direction, these vertical PMLs are matched to the host medium without any numerical dispersion or discretization-related errors.

Therefore it is possible to use a single absorbing layer with any arbitrary value of s and no reflections will occur. On the other hand, on the horizontal boundaries the material parameters are not represented precisely because of the truncated Fourier expansion. This may introduce undesired artificial reflections at the PML interface. To minimise this effect it is beneficial to apply gradual PMLs, that is, PMLs with s parameter changing gradually between unity and s_{max}.

Now, with the assumption that $\tilde{\mathbf{E}}_0^{(1)} = 0$, we can easily see from Equation 4.42 that

$$\mathbf{Y}^{(1)} = -\mathbf{y}_1^{(1)} \tag{4.50}$$

To find admittance in other layers we must use the constraint, that both the electric field $\hat{\mathbf{E}}$ and the magnetic one $\hat{\mathbf{H}}$ are continuous at all the layer boundaries. Thus we have

$$\mathbf{T}_{\mathbf{E}}^{(n)}\tilde{\mathbf{E}}_0^{(n)} = \mathbf{T}_{\mathbf{E}}^{(n-1)}\tilde{\mathbf{E}}_d^{(n-1)} \tag{4.51}$$

$$\mathbf{T}_{\mathbf{H}}^{(n)}\tilde{\mathbf{H}}_0^{(n)} = \mathbf{T}_{\mathbf{H}}^{(n-1)}\tilde{\mathbf{H}}_d^{(n-1)} \tag{4.52}$$

or

$$\tilde{\mathbf{E}}_0^{(n)} = \mathbf{t}_{\mathbf{E}}^{(n)}\tilde{\mathbf{E}}_d^{(n-1)} \tag{4.53}$$

$$\tilde{\mathbf{H}}_0^{(n)} = \mathbf{t}_{\mathbf{H}}^{(n)}\tilde{\mathbf{H}}_d^{(n-1)} \tag{4.54}$$

where $\mathbf{t}_{\mathbf{E}}^{(n)} = (\mathbf{T}_{\mathbf{E}}^{(n)})^{-1}\mathbf{T}_{\mathbf{E}}^{(n-1)}$ and $\mathbf{t}_{\mathbf{H}}^{(n)} = (\mathbf{T}_{\mathbf{H}}^{(n)})^{-1}\mathbf{T}_{\mathbf{H}}^{(n-1)}$. Now substituting this into Equation 4.42 we can state

$$\mathbf{t}_{\mathbf{H}}^{(n)}\mathbf{Y}^{(n-1)}\tilde{\mathbf{E}}_d^{(n-1)} = \mathbf{y}_1^{(n)}\mathbf{t}_{\mathbf{E}}^{(n)}\tilde{\mathbf{E}}_d^{(n-1)} + \mathbf{y}_2^{(n)}\tilde{\mathbf{E}}_d^{(n)} \tag{4.55}$$

$$\tilde{\mathbf{H}}_d^{(n)} = -\mathbf{y}_2^{(n)}\mathbf{t}_{\mathbf{E}}^{(n)}\tilde{\mathbf{E}}_d^{(n-1)} - \mathbf{y}_1^{(n)}\tilde{\mathbf{E}}_d^{(n)} \tag{4.56}$$

Removing $\tilde{\mathbf{E}}_d^{(n-1)}$ from these equations and doing some basic transformations, we can determine the admittance relation for nth layer

$$\tilde{\mathbf{H}}_d^{(n)} = \{\mathbf{y}_2^{(n)}\mathbf{t}_{\mathbf{E}}^{(n)}[\mathbf{y}_1^{(n)}\mathbf{t}_{\mathbf{E}}^{(n)} - \mathbf{t}_{\mathbf{H}}^{(n)}\mathbf{Y}^{(n-1)}]^{-1}\mathbf{y}_2^{(n)} - \mathbf{y}_1^{(n)}\}\tilde{\mathbf{E}}_d^{(n)} \tag{4.57}$$

which finally gives us

$$\mathbf{Y}^{(n)} = \mathbf{y}_2^{(n)}\mathbf{t}_{\mathbf{E}}^{(n)}[\mathbf{y}_1^{(n)}\mathbf{t}_{\mathbf{E}}^{(n)} - \mathbf{t}_{\mathbf{H}}^{(n)}\mathbf{Y}^{(n-1)}]^{-1}\mathbf{y}_2^{(n)} - \mathbf{y}_1^{(n)} \tag{4.58}$$

Using this relation we can determine the admittance matrix on the matching plane, with the aid of the iterative procedure for both top (we name the obtained matrix $\mathbf{Y}^{top}$) and bottom ($\mathbf{Y}^{bottom}$) parts of the structure. As both fields $\hat{\mathbf{E}}$ and $\hat{\mathbf{H}}$ must be continuous at all planes, then

$$\mathbf{T}_{\mathbf{H}}^{top}\tilde{\mathbf{H}}^{top} = \mathbf{T}_{\mathbf{H}}^{bottom}\tilde{\mathbf{H}}^{bottom}$$

and

$$\mathbf{T}_{\mathbf{E}}^{top}\tilde{\mathbf{E}}^{top} = \mathbf{T}_{\mathbf{E}}^{bottom}\tilde{\mathbf{E}}^{bottom} = \hat{\mathbf{E}}^{interface},$$

where $\tilde{\mathbf{E}}^{top/bottom}$ and $\tilde{\mathbf{H}}^{top/bottom}$ are the fields on the interface computed for the top/bottom part of the structure and $\mathbf{T}_{\mathbf{E}}^{top/bottom}$ and $\mathbf{T}_{\mathbf{H}}^{top/bottom}$ are the **T**-matrices of the layers directly adjacent to the interface. The above relations lead to the following eigenequation that must be satisfied for the whole structure

$$\{\mathbf{T_H}^{top}\mathbf{Y}^{top}[\mathbf{T_E}^{top}]^{-1} - \mathbf{T_H}^{bottom}\mathbf{Y}^{bottom}[\mathbf{T_E}^{bottom}]^{-1}\}\hat{\mathbf{E}}^{interface} = 0 \quad (4.59)$$

This equation can be only nontrivially satisfied when the matrix

$$\mathbf{M} = \mathbf{T_H}^{top}\mathbf{Y}^{top}[\mathbf{T_E}^{top}]^{-1} - \mathbf{T_H}^{bottom}\mathbf{Y}^{bottom}[\mathbf{T_E}^{bottom}]^{-1} \quad (4.60)$$

is singular. In the three-dimensional problem, this matrix is a function of the normalized frequency k_0 and the wavevector components k_x and k_y. In order to determine the eigenmodes we search for a combination of these variables in such a way that

$$\Phi(k_0, k_x, k_y) = \min|\det[\mathbf{M}(k_0, k_x, k_y)]| = 0 \quad (4.61)$$

In our numerical implementation, we fix two of these variables and search for the third one using the modified Broyden algorithm in a complex plane [78].

An open question is the choice of the matching plane. In general this choice can be arbitrary, however, the numerical error will be smallest if the electric field $\hat{\mathbf{E}}$ is large on the interface. Thus the best position of the matching plane depends on the expected field distribution in the structure.

Once an eigenfrequency of the analyzed device is known, the electromagnetic field distribution can be found in two steps: first the electric field at the matching interface is computed as the eigenvector of Equation 4.60 and then field distribution is determined with an iterative procedure in a reverse direction. As given by Equation 4.61, the physical eigenmodes are related to the situation where the matrix $\mathbf{M}$ (Equation 4.60) is singular, that is, the electric field at the interface $\hat{\mathbf{E}}_{interface}$ is an eigenvector of $\mathbf{M}$ corresponding to its eigenvalue $\lambda = 0$. Thus the determination of $\hat{\mathbf{E}}_{interface}$ is straightforward with any numerical algorithm for determination of eigenvectors.

To find the electric field distribution inside any uniform layer we must know the field in diagonalised coordinates $\tilde{\mathbf{E}}$ at both sides of the layer, namely $\tilde{\mathbf{E}}_0^{(n)}$ and $\tilde{\mathbf{E}}_d^{(n)}$. For the layer adjacent to the matching interface, the latter one is known and for any other layer it can be easily determined from Equation 4.51. To compute $\tilde{\mathbf{E}}_0^{(n)}$ let us consider Equation 4.42 which can be used to represent $\tilde{\mathbf{E}}_0^{(n)}$ as

$$\tilde{\mathbf{E}}_0^{(n)} = (\mathbf{y}_2^{(n)})^{-1}[\mathbf{y}_1^{(n)}\tilde{\mathbf{E}}_d^{(n)} + \tilde{\mathbf{H}}_d^{(n)}]$$

which using Equation 4.45 gives

$$\tilde{\mathbf{E}}_0^{(n)} = (\mathbf{y}_2^{(n)})^{-1}[\mathbf{y}_1^{(n)} + \mathbf{Y}^{(n)}]\tilde{\mathbf{E}}_d^{(n)} \quad (4.62)$$

An application of the above equation requires additional storage space to keep the matrix $\mathbf{Y}^{(n)}$ for each layer but provides numerical stability, lacking, for example, in the transfer-matrix method. Having both $\tilde{\mathbf{E}}_0^{(n)}$ and $\tilde{\mathbf{E}}_d^{(n)}$ it is possible to find the electric field in diagonalized coordinates in any layer using Equations 4.37–4.39 that together give

$$\tilde{\mathbf{E}}^{(n)}(z) = \cos(\Gamma^{(n)}z)\tilde{\mathbf{E}}_0^{(n)} + \sin(\Gamma^{(n)}z)\left[\frac{1}{\sin(\Gamma^{(n)}d^{(n)})}\tilde{\mathbf{E}}_d - \cot(\Gamma^{(n)}d^{(n)})\tilde{\mathbf{E}}_0\right] \quad (4.63)$$

where z is the relative position in the nth layer and the vector value in the brackets is constant and can be precomputed and stored in advance to save calculation time. The magnetic field can be determined directly from the electric one using the admittance-relation Equation 4.42 to get $\tilde{\mathbf{H}}_0^{(n)}$ and $\tilde{\mathbf{H}}_d^{(n)}$ and applying the following relation for finding $\tilde{\mathbf{H}}^{(n)}(z)$

$$\tilde{\mathbf{H}}^{(n)}(z) = \cos(\Gamma^{(n)}z)\tilde{\mathbf{H}}_0^{(n)} + \sin(\Gamma^{(n)}z)\left[\frac{1}{\sin(\Gamma^{(n)}d^{(n)})}\tilde{\mathbf{H}}_d - \cot(\Gamma^{(n)}d^{(n)})\tilde{\mathbf{H}}_0\right] \quad (4.64)$$

The x and y components of the physical fields can be now easily computed as $\hat{\mathbf{E}} = \mathbf{T}_E^{(n)}\tilde{\mathbf{E}}$ and $\hat{\mathbf{H}} = \mathbf{T}_H^{(n)}\tilde{\mathbf{H}}$ and to get the z-components, it is sufficient to use Equations 4.7 and 4.8, which in plane-wave basis can be expanded as

$$E_z^{\mathbf{g}} = \frac{\eta_0}{k_0}\kappa^{\mathbf{g}-\mathbf{g}'}[-(g_y' - k_y)H_x^{\mathbf{g}'} + (g_x' - k_x)H_y^{\mathbf{g}'}] \quad (4.65)$$

$$H_z^{\mathbf{g}} = \frac{1}{k_0\eta_0}\gamma^{\mathbf{g}-\mathbf{g}'}[(g_x' - k_x)E_y^{\mathbf{g}'} + (g_y' - k_y)E_x^{\mathbf{g}'}] \quad (4.66)$$

With this all the mathematics of PWAM is summarized and we will proceed with presenting numerous examples of how PWAM calculations provide an insight into and design guidelines for PhC-VCSELs.

4.4
Impact of Photonic-Crystal depth on VCSEL Threshold Characteristics

The present-day technology allows for a limited depth d_E of the PhC holes, depending on the hole diameter a an aspect ratio $d_E/a \approx 15$ is possible [79]. PWAM numerical studies on the impact of the hole depth on the modal behavior of PhC-VCSELs have been reported in [80, 81]. An 1.3 μm InAlGaAs/InP PhC-VCSEL is considered with $3 - \lambda$ InP cavity and four 10-nm-thick $Al_{0.0152}Ga_{0.495}In_{0.49}As$ QWs and 15-nm-wide $Al_{0.218}Ga_{0.25}In_{0.532}As$ barriers. The cavity is bounded by two $Al_{0.9}Ga_{0.1}As/GaAs$ DBRs with 27 and 35 pairs, respectively. Three rings of a hexagonal lattice PhC with a single defect provide the optical confinement with aperture radius $R_A = \Lambda - a/2$. Carrier confinement is assured by a proton-implanted tunnel junction (TJ) placed at a node position of the standing wave and, therefore, the imaginary part of the AR refractive index is defined by a radial step-like function. Two cases are considered: a passive one, which is defined by a zero imaginary part of the refractive index within the AR, and an active one which is defined by a material gain equal to 2000 $cm^{-1}(-1000\ cm^{-1})$ inside (outside) the optical aperture.

Two counter-acting optical mechanisms are introduced by etching the PhC in a VCSEL top DBR: (i) a light confinement mechanism by the PhC cladding effective index which decreases with increasing PhC hole depth and (ii) an effective decrease of the VCSEL top-DBR reflectivity. Hence the PhC acts effectively only if the light is well confined to the PhC defect, otherwise light leaks through the holes. Figure 4.5a shows the real and imaginary part of the emission wavelength for the passive structure as a function of the hole etching depth. Shallow holes do not provide waveguide mechanism: the mode is not confined to the optical aperture

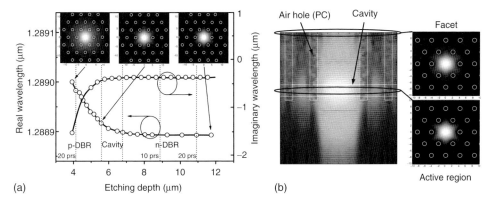

Figure 4.5 (a) Real and imaginary part of emission wavelength as a function of the etching depth in the case of the PhC-VCSEL passive structure. The insets show light intensity distribution within the AR for three different depths of etching. (b) Intensity distribution for optimal PhC hole depth: (left) logarithmic scaled vertical cross-section (the contrast is additionally scaled to emphasize the mode leakage) and (right) the horizontal cross-section of the AR and the facet. The PhC parameters are $a/\Lambda = 0.3$ and $R_A = 4$ μm. (After [80].)

of the PhC defect until $d_E \approx 4$ μm, which corresponds to about 20 pairs of the top DBR. Deeper etching causes stronger interaction with the PhC lattice and the light is squeezed into the optical aperture (see the insets in Figure 4.5a). Because of this shrinkage of the mode, an increase of the etching depth is accompanied by a blueshift of the emission wavelength. Consequently, the losses are reduced as the waveguiding mechanism overcomes the spreading of the mode. Further deepening of the holes reduces the losses to a constant level, which is achieved for $d_E^{opt} \approx 7.2$ μm (down to the fourth bottom-DBR pair). Such an etching depth is considered optimal for the given PhC parameters since it assures low modal losses with minimal etching depth. The mode distribution corresponding to d_E^{opt} is given in Figure 4.5b. Although the radial mode distributions show very good confinement in the AR and at the facet (right panel of Figure 4.5b), the logarithmically scaled vertical cross-section intensity distribution (left panel of Figure 4.5b) clearly shows that some mode leakage throughout the PhC holes still occurs.

The general tendencies with the hole depth presented in Figure 4.5a remain valid for other values of the PhC lattice parameters. The impact of the a/Λ ratio on the real and imaginary part of the wavelength is presented in Figure 4.6. Larger a/Λ ratio leads to larger modal gain because of better light confinement to the AR. The optimal PhC hole depth d_E^{opt} is therefore reduced for wider holes as such holes provide better confinement.

So far we have considered PhC-VCSEL with holes etched exactly at a certain number of DBR pairs, a study of etching depths with precision of a single layer has been carried out in [82]. An 1.3 μm QD PhC-VCSEL is considered. The current flow is defined by a selective oxidization of an AlAs layer placed at a node position in order not to alter the PhC interaction with the mode. Broad aperture of 16 μm and uniform gain distribution within the AR are assumed in order to reduce the

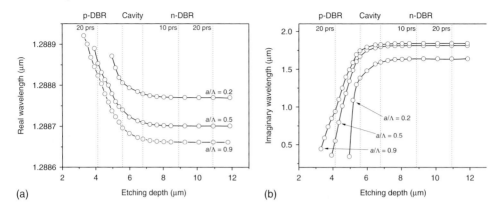

Figure 4.6 (a) Real and (b) imaginary part of emission wavelength as a function of the etching depth in the case of the PhC-VCSEL active structure and for different a/Λ ratios. After [80].

mode discrimination caused by the PhC-induced mode confinement. The AR is composed of five groups of three 8 nm thick $In_{0.15}Ga_{0.85}As$ QWs, each of them containing one InAs QD layer. In each group, located close to the successive antinode positions of the optical standing wave within the cavity, the QWs are separated by 32 nm thick GaAs barriers. Additional single InGaAs QWs containing single QD layers are placed at the beginning and end of the AR. The cavity is bounded by two $Al_{0.9}Ga_{0.1}As$/GaAs DBRs. Because the oxidation is placed at a node position of the standing wave, it impacts the waveguiding mechanism weakly. Therefore, the optical confinement is achieved mainly by means of a hexagonal air-hole PhC with a central single defect and by thermal focusing. The temperature distribution is determined at the threshold for a VCSEL without a PhC and is kept unchanged in the further analysis; neglecting it makes the mode confinement in a VCSEL without (or with a shallow) PhC impossible.

In Figure 4.7a,b we present, respectively, the fundamental mode wavelength and threshold gain as a function of the PhC etching depth revealing profound oscillations. Such oscillations are not observed for the above discussed case of etching of a certain number of DBR pairs (represented by the doted lines in Figure 4.7). There are two different mechanisms - for the case of shallow and deep etching - responsible for the appearance of the oscillations. For shallow etching, the mode is not confined by the PhC but by thermal focusing. This makes the mode broader than the aperture defined by the PhC and the overlap of the mode with the PhC holes is significant. If the bottom of the PhC hole is at an antinode position then light is partially reflected back to the resonator. If the hole bottom is at a node position then more light leaks through the holes. For deep etching, the optical field is confined by the PhC and consequently, it overlaps much less with the PhC holes. In this case, if the bottom of the PhC holes is at an antinode position the mode will be scattered more. If it is at a node position the scattering is decreased.

We now discuss the dependence of the threshold gain on the etching depth. As the gain aperture is broader than the mode distribution, the positive impact of the

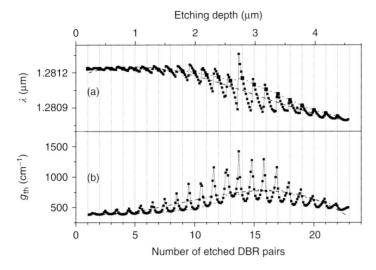

Figure 4.7 Wavelength of emission (a) and threshold gain (b) as a function of the PhC etching depth/number of the etched DBR pairs (lines with squares). The mean over DBR period values and the values for etching through a certain number of DBR pairs are represented by dashed and dotted lines, respectively. (After [82].)

PhC to improve their overlap is not observed here; rather, the PhC impact is now related to light scattering, diffraction, and light leakage, all of which contribute to an increase of the threshold gain. Hence, the lowest threshold gain is expected for a VCSEL without PhC, which is confirmed by Figure 4.7b. An increase of d_E deteriorates the DBR reflectivity gradually, which can be observed as an increase of the mean value of the threshold gain. For etching depth close to $d_E \approx 3$ μm, the holes are deep enough to significantly deteriorate the DBR reflectivity and too shallow to considerably confine the mode; therefore, the light leakage is the most intense and the mean value of the threshold gain is the largest. For deeper etching, the PhC confines the mode to the small region of the PhC defect. As a consequence, the overlap of the PhC holes with the mode is reduced and the light leakage, which is the main cause of light losses is reduced, too. Hence, the mean value of the threshold gain becomes smaller but never reaches the level of the one of a VCSEL without a PhC. The variations of the threshold gain within a single period of the DBR are governed by the light leakage in the case of shallow etching. It is the most pronounced process when the node of the mode coincides with the bottom of the PhC holes. An increase in the etching depth increases the light leakage and leads to larger oscillations of the threshold gain within one period of DBR. As the PhC starts to confine the mode, the light leakage is reduced, the oscillations are dumped. Now, the dominant loss mechanisms become diffraction and scattering, which are most pronounced when the antinode of the mode coincides with the bottom of PhC holes.

4.5
Top and Bottom-Emitting Photonic-Crystal VCSELs

Without loss of generality, we base our studies on the 1.3 μm InP PhC-VCSEL structure from the previous section. However, the light is now emitted either throughout the top DBR with PhC (as in Section 4.4) or throughout the bottom-DBR free of PhC (in that case the top and the bottom DBR are swapped) [83, 84]. The active case is considered only, that is, a material gain of 2000 (−1000) cm^{-1} within (outside) the PhC defect aperture, corresponding to carrier densities of $8 \times 10^{18}(8 \times 10^{17})$ cm^{-3}. The PhC parameters are $a/\Lambda = 0.3$ and $R_A = \Lambda - a/2 = 3.4$ μm (i.e. $\Lambda = 4$ μm, $a = 1.2$ μm) ensuring a single transverse-mode operation [85] (see also Section 4.6).

In the case of top-emitting PhC-VCSEL the minimum hole depth, which allows for confinement of the mode by the PhC single defect equals 5 μm (24 pairs). However, despite this confinement in the AR (Figure 4.8a AR) the intensity patterns at the laser facet and in the far field (Figure 4.8a NF and FF, respectively) evidence strong leakage. Apparently, the PhC does not efficiently confine the mode: it significantly overlaps with the PhC holes and light escapes through them leading to an increase of the optical losses. The mode pattern proves that the light escapes mostly through the first two rings. These results remain in agreement with the experimental

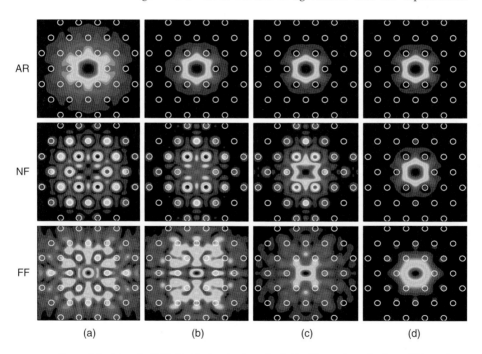

Figure 4.8 Intensity field distribution in the case of top-emitting PhC-VCSEL in three different cross-sections: in the middle of the active region (AR), of the facet (NF) and in far field - 10 μm from the facet (FF), and for four different etching depths: (a), (b), (c) and (d). (After [83].)

phenomena reported in [53]. Further deepening of the holes (Figure 4.8b) allows for better confinement of the mode within the AR. Light escapes through the first ring only but still its patterns at the facet and in the far field remain far from a Gaussian-like shape. When the holes are drilled throughout the upper DBR and the cavity (Figure 4.8c) the mode pattern improves considerably and the field is almost fully confined within the AR. Finally a Gaussian-like emission beam shape is achieved for the last analyzed etching depth (Figure 4.8d). This depth equals almost 11 µm and corresponds to PhC holes etched throughout the top DBR, the cavity, and down to the eighteenth pair of the bottom DBR.

Alternative view of the mode patterns is provided by the set of vertical cross-sections of the mode distributions (Figure 4.9). The consecutive figures (Figure 4.9a–d) demonstrate the increase of the mode confinement and reveal how light escapes throughout the PhC holes. The most pronounced mode leakage (Figure 4.9a) is observed for the case of weak confinement as already noticed from Figure 4.8a. This parasitic effect is diminished when stronger field confinement is provided by deeper holes (Figure 4.9b,c). The case of Figure 4.9d assures strong mode confinement and protects the light from penetration into the holes.

We now consider the case of bottom-emitting PhC-VCSEL. Emission through the DBR free from the PhC reveals significant improvement of the mode pattern at the facet and in the far field (Figure 4.10). At the same time, the distribution of the mode within the AR remains almost unchanged as compared with the top emission design. The field patterns for the whole range of analyzed etching depths reveal Gaussian-like shapes. The confinement of the mode at the facet follows the changes of the mode within the AR.

More quantitative comparison of top and bottom-emitting VCSELs is given in Figures 4.11, which present the confinement factors in the AR (Γ_A) and in the far field (Γ_{FF}) defined as the ratio of the light intensity integral over the optical aperture area to the intensity integral over the whole plane of the AR. The behavior of Γ_A for both designs confirms the earlier observations of the mode distributions. There is a minimal "set-on" etching depth, which allows for the mode to appear. The confinement factor increases with deepening of the PhC holes and also with

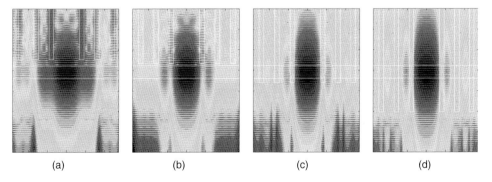

Figure 4.9 Vertical cross-section intensity distribution in a logarithmic scale: (a)–(d) correspond to the four different etching depths: (a), (b), (c) and (d). (After [83].)

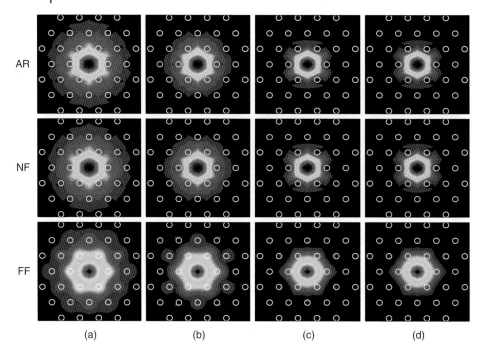

Figure 4.10 The same as Figure 4.8 but for a bottom-emitting PhC-VCSEL. (After [83].)

increasing the optical aperture. After a rapid growth of the confinement factor just after the set-on etching depth, it reaches a relative plateau – being almost constant for holes deeper than the bottom edge of the cavity. The set-on etching depth reduces when broadening the optical aperture R_A. As can be noticed from Figure 4.11a the set-on etching depths corresponding to the same R_A are different for the two VCSEL designs, however, the distances between the AR position and the set-on depth are the same in both designs. The major difference between the two designs is in the confinement factor Γ_{FF} in the far field (Figure 4.11b), a key parameter in telecommunication applications since it would determine the efficiency of coupling to optical fiber. As can be seen from Figure 4.11b, significantly better performance is achieved for the case of bottom emission design in a broad range of analyzed parameters with two- to fourfold improvement of confinement factor for small and medium size apertures. Despite the much thicker top DBR (35 pairs), for bottom emission design the etching depth even decreases for the same value of Γ_{FF}. The oscillations of Γ_{FF} for top-emitting design are related to the rapid changes of the mode distribution. The bottom-emitting (BE) structure is not affected in such a way as the mode patterns at the facet and in the far field remain close to the Gaussian-like shape for the whole range of etching depths. The oscillations of Γ_{FF} appear for narrow optical apertures, while for an aperture radius equal to 6 μm the confinement factor behavior is similar for both types of VCSELs. In that case the mode can be confined by shallow holes, and the leakage does not take place.

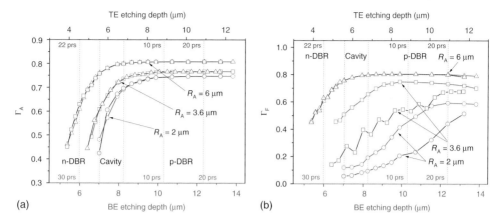

Figure 4.11 Confinement factors in the active region Γ_A (a) and in the far field Γ_{FF} (10 μm from the facet) and (b) as a function of the etching depth for different optical apertures. Top and bottom axis correspond to top-emitting (TE) and bottom-emitting (BE) designs. Upper (lower) curve for the same aperture size corresponds to BE (TE) design, respectively. (After [83].)

4.6
Enhanced Fundamental Mode Operation in Photonic-Crystal VCSELs

First, we discuss the 1.3 μm InP PhC-VCSEL structure as described in Section 4.4 [85]. Figure 4.12a presents the dependence of the emitted wavelength on the hole aperture for two guided modes (HE_{11} and HE_{21}) for the case of the passive structure. Broadening the holes causes narrowing of AR and strengthening of mode confinement. The impact of mode squeezing by the holes is observed in the typical blueshift of the emitted wavelength. The structure supports only fundamental mode for narrow holes in the range from 0.1 to 0.3 of the a/Λ ratio. From among the hybrid modes, only the fundamental and first-order modes are supported by the structure as the PhC aperture is too small to support higher-order modes. Figure 4.12b presents the imaginary part of the wavelength in the same domain as in Figure 4.12a. The main tendency is an increase of the modal losses with the increase of air-hole aperture, mainly because of the fact that expanding air holes deteriorates the DBR reflectivity and increases light scattering by the low-refractive index air columns.

Next, we investigate the influence of the gain region on the modal characteristics [85]. The gain profile has been introduced to the AR as a step-like function of the imaginary part of the refractive index: 2000 (−1000) cm^{-1} for $r \leq (>)3$ μm. The real wavelength analysis does not reveal any major change compared to the passive structure case, reflecting the fact that the introduction of the gain profile does not disturb the distribution of the real part of refractive index. More dramatic change is observed for the imaginary wavelength characteristics (Figure 4.12b). The modal losses of the modes have been affected positively by the existence of the gain region, revealing positive, or close to positive values of the imaginary part of the

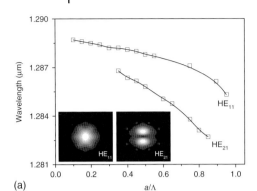
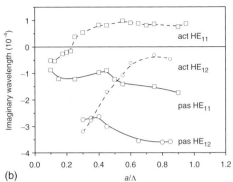

Figure 4.12 Real (a) and imaginary (b) wavelength of emission as a function of the a/Λ ratio for the fundamental and the first-order mode in a passive (pas) and active (act) structure. The insets show the distribution of the optical field within the active region cross-section. (After [85].)

wavelength. Narrow holes favor single mode action with the cost of higher modal losses because of the weak waveguiding effect and consequently, the penetration of the fundamental mode in the high loss region.

Figure 4.13a illustrates that indeed, the fundamental mode is weakly confined by holes of 0.1 Λ aperture. Further increase of the hole aperture (Figure 4.12b) leads to a larger modal gain of the fundamental mode, while HE_{12} remains cutoff for $a/\Lambda < 0.3$. That allows for a single mode operation, which is characterized by the highest modal gain for an aperture hole of 0.3 Λ. The fundamental mode becomes perfectly confined to the single PhC defect and simultaneously to the AR (Figure 4.13b). Hence optimal hole apertures supporting single mode action are in the range from 0.2 to 0.3 Λ. Further increase of the air holes diameter does not improve the fundamental mode characteristics (Figure 4.12a). Instead of that, the HE_{12} mode appears and tends to be gained, leading to a reduction of the fundamental mode discrimination. The difference between the modes in terms of the imaginary part of the wavelength reaches its minimum for 0.7Λ hole aperture. In that case the fundamental mode begins to suffer diffraction losses, which is visible in the mode distribution oscillations in the outer region (Figure 4.13c).

We now consider the same PhC-VCSEL design as in Section 4.5, for improved bottom emission [86]. The analyzed PhC parameters are varied within the limits and with steps as follows: R_A: 1–6 μm, step 1 μm; a/Λ: 0.1 - 0.9, step 0.1; etching depth d_E: 0–13.8 μm, step DBR pair thickness. Figure 4.14 illustrates how the area of a single mode operation (in gray) is limited: from below - by the minimum d_E for which well-confined mode in the PhC defect appears [10]; from above, by the cut off d_E for which the first higher-order mode appears. As it can be seen from Figure 4.14, the single-mode region is the largest for small a/Λ ratios as in this case it is not limited from above by the appearance of the first-order mode. This "extended" single-mode region is squeezed with the increase of R_A limited by: $0.2 < a/\Lambda < 0.5$ for $R_A = 1$ μm and by $0.1 < a/\Lambda < 0.2$ for $R_A = 6$ μm. The thickness of the remaining horizontal stripe in Figures 4.14 is almost constant,

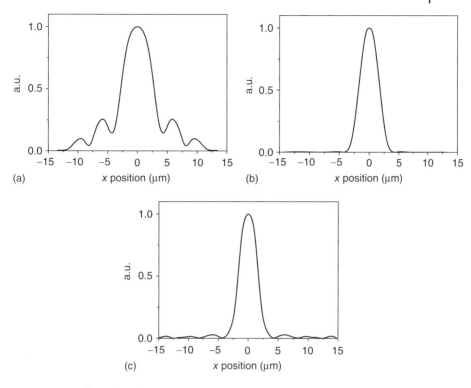

Figure 4.13 Profiles of the fundamental mode within active region for hole diameter/lattice pitch (a/Λ) ratios: 0.1 (a), 0.3 (b), and 0.7 (c). (After [85].)

approximately equal to 0.5 µm, and nearly independent on R_A (except for $R_A = 1$ µm when it is equal to 1 µm). The position of the single-mode region is shifted toward shallower etching depth with the increase of R_A and a/Λ. The described variation of the position and the area of single-mode region are governed mostly by the change of the PhC-induced waveguiding.

Figure 4.15 shows the modal gain difference between HE_{11} and HE_{12} modes. The largest difference is observed for PhC parameters close to the single mode region. This can be explained since the modal gain is related to the mode overlap with the AR. In the multimode region, but close to the single mode one, the first-order mode is weakly confined by the PhC, hence it suffers high losses, while the fundamental mode is relatively well localized into the AR. The reduction of the mode discrimination is because of the improvement of the confinement for both the HE_{11} and HE_{12} modes. Therefore the largest discrimination of the modes is observed for narrow optical apertures as well as for narrow and/or shallow holes [86].

PWAM has been combined with finite element electrical and thermal models, as well as with a gain model based on Fermi's golden rule in a self-consistent manner in [87–89] in order to determine the optimal PhC parameters assuring stable, single-mode operation in a broad range of driving currents. The optimal ratio of electrical aperture radius to one of the optical aperture is determined to be equal to

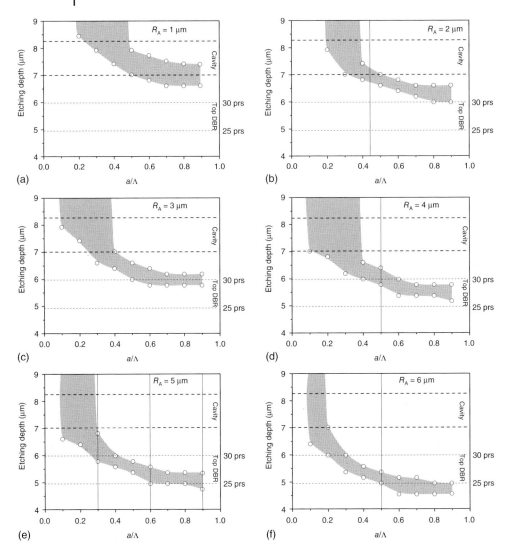

Figure 4.14 Region of single mode operation (gray fields) for six different optical apertures (from $R_A = 1$ μm to $R_A = 6$ μm) (a–f) mapped in the plane of etching depth and a/Λ ratio. The horizontal lines assign the PhC parameters, which have been chosen in the analysis presented in Figure 4.15. (After [86].)

$R_{PA}/R_A = 1/2(3/4)$ for PhC-VCSELs with TJ (TJ PhC-VCSELs) providing the lowest threshold current. The difference between PhC-VCSEL without and with TJ stems from the stronger current crowding effect for the first case. An increase of the mode discrimination can be achieved by a reduction of the electrical aperture at the cost of increasing the threshold current. A hole diameter of $a/\Lambda = 0.5$ causes stronger mode discrimination accompanied by a significant threshold current reduction in

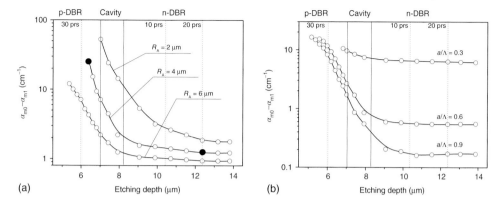

Figure 4.15 Mode gain difference between the HE_{11} and HE_{12} modes versus etching depth, under the change of (a) optical aperture and (b) a/Λ ratio. (After [86].)

comparison to VCSEL without PhC. This strong discrimination and low-threshold current can be observed for optical aperture radius smaller than 8 μm (6 μm) for PhC-VCSEL (TJ PhC-VCSEL). Furthermore, thanks to the reduction of the current crowding in TJ PhC-VCSEL, a lower temperature at threshold and a lower sensitivity to the ambient temperature have been observed than in PhC-VCSEL [87].

4.7
Highly Birefringent and Dichroic Photonic-Crystal VCSELs

As explained in the introduction, VCSELs have no a priori defined polarization selection mechanism and often suffer from polarization instabilities and switching between two orthogonal linearly polarized states [21]. In many applications, where the elements of the optical set-up are polarization dependent, polarization instabilities cannot be tolerated. For this reason, the polarization stabilization is an important issue in the design of modern VCSELs.

With the development of PhC-VCSELs, a new method of polarization stabilisation has emerged, namely an application of PhCs. For several years, PhCs have been successfully used for providing birefringence and dichroism in PhC fibers, while in the same time only few attempts were made to transfer such a PhC application to VCSELs. So far only two such attempts have been reported [53, 54] as described in the Introduction. Here, we perform a detailed analysis of the applicability of various PhC configurations for polarization stabilisation [90]. Contrary to Ref. [53], we investigate not only configuration with elliptical holes, but also some other designs that have been successfully applied to PhC fibers for providing large birefringence and single-polarization waveguiding.

An important issue to reflect on, when considering polarization control with PhCs, is breaking the C_6 point group symmetry. At first glance, a hexagonal lattice of cylindrical holes does not possess C_4 symmetry, that is, after rotating by 90°, the

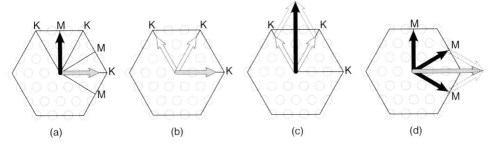

Figure 4.16 Orthogonal polarization-mode degeneracy for PhC-VCSEL lattice with C_6 symmetry. (a) x-polarized mode is oriented in Γ-K direction, while y-polarized one in Γ-M, (b) C_6 symmetry ensures that all the modes polarised in Γ-K are indistinguishable, (c) linear combination of two modes polarised in Γ-K gives a mode polarized in Γ-M, (d) linear combination of modes polarized in Γ-M gives a mode polarized in Γ-K.

PhC lattice will differ from the original one (Figure 4.16a). Thus one may assume that this is enough for differentiating the two orthogonal x- and y- polarizations. In fact, this is not true, [91]. The electric field vector of x-polarized LP mode (Figure 4.16a) is oriented in the K direction of PhC lattice and, because of the C_6 symmetry, this mode is indistinguishable from other modes marked with white arrows in Figure 4.16b. The solution of the Maxwell equations allows the existence of any linear combination of these modes, also the one in which the electric field is oriented along y-axis, that is, in the M direction of the PhC lattice (Figure 4.16c). Thus the fact, that there exists an x-polarized mode in C_6 lattice, induces the existence of the y-polarized mode degenerated with the original one. Obviously the opposite implication is also true (Figure 4.16d). This means, that in such a PhC lattice, there is neither frequency separation of the orthogonally polarised modes, nor any difference of their losses.

Hence, the efficient application of PhCs for polarization control requires breaking of the C_6 point group symmetry. This can be achieved by various methods, for example, by altering the shape of the holes or by changing their arrangement. In this section, we perform numerical analysis of a typical PhC-VCSEL. It is a gallium arsenide structure designed for operation at 980 nm wavelength, comprised of 24.5 pairs GaAs/AlGaAs top DBR and 29.5 pairs bottom DBR. The cavity has one wavelength optical length. The optical field confinement is provided by two rings of PhC with hexagonal lattice and single defect inside.

We consider two configurations for providing stable polarization of the emitted light. In a first step, we analyse a PhC structure with elliptical holes having the ratio of the ellipse radii d_y/d_x varying from 1 to 5 (Figure 4.17a). The second PhC structure has cylindrical holes of diameter d_c, however four of the holes have their diameter (d_h) increased (Figure 4.17b) and a small hole of diameter d_s is added inside the PhC defect in order to provide larger birefringence and dichroism [92].

Because the differences in the eigen frequencies for the two polarizations are very subtle it is necessary to increase the number of plane-waves used in calculations. We have found that satisfactory convergence is achieved with 53^2 plane-waves. The

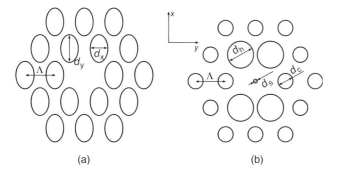

Figure 4.17 Photonic crystal configurations of analyzed structures: (a) elliptical holes, (b) holes of different diameters.

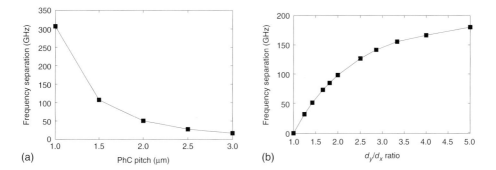

Figure 4.18 Resonant frequency separation of the x- and y-polarized modes (a) for the PhC shown in Figure 4.17a as a function of the PhC pitch and (b) for the PhC with elliptical holes as a function of the d_y/d_x ratio. (After [90].)

computation time with such a large basis is about two days for a single point on an AMD Opteron processor.

In order to ensure a relatively large polarization separation, we have chosen a small PhC pitch equal to $\Lambda = 1.0$ μm. This provides strong optical confinement of the mode inside the defect. For larger pitches the separation decreases quickly, as shown in Figure 4.18a. Similar effect can be observed for the modal dichroism.

In the first analyzed structure (Figure 4.17a) the average diameter of the holes has been kept constant and equal to $d_{avg} = (d_x \, d_y)^{1/2} = 0.4$ μm, while the ratio d_y/d_x has been varied. In order to avoid hole overlapping, their longer axes are always oriented along the ΓM direction. The computed resonant frequency splitting between the two orthogonally polarized fundamental modes is shown in Figure 4.18b. For ideally cylindrical holes, both x- and y-polarized modes have the same resonant frequencies (in fact they differ by about 2 GHz, which provides an estimate for the numerical precision at such choice of computational parameters). When increasing the hole ellipticity, the frequency splitting increases and the x-polarized mode has always a lower frequency than the y-polarized one. Contrary to the resonant frequencies, the modal losses for the two polarizations remain equal to each other within the numerical precision that we estimate to be around 3%. The field profiles are also almost identical, however the field for the x-polarized mode is slightly better confined inside the defect. Therefore, we can conclude that an elliptic hole PhC structure has a moderate impact on the VCSEL polarization properties, in agreement with experimental results reported in [53], where a largePMSR could only be achieved with the aid of asymmetric current injection.

Alternative approaches to provide good stabilisation of polarization have been extensively developed for birefringent PCF [93, 94]. The common practice to provide birefringence of the fiber is an asymmetric change in the radius of the PhC holes. An example of such a configuration is shown in Figure 4.17b. It comprises four holes of increased diameter out of six surrounding the defect. In addition, a small hole in the middle augments the birefringence and increases the polarization bandwidth [92].

Hence, we analyse a VCSEL with a PhC configuration based on the design of the highly birefringent microstructured optical fiber [92] as shown in Figure 4.17b. As we will show, such a PhC structure can provide a much larger frequency separation between the two orthogonal polarizations than the one with elliptical holes, reaching values as high as 400 GHz. Similar to the previous configuration, the PhC pitch is $\Lambda = 1$ µm. We have found that good results are obtained for a PhC where the majority of holes have a diameter $d_c = 0.4$ µm and the four larger holes have a diameter $d_h = 0.8$ µm. The size of the hole in the middle can be varied in a range $0.01 \leq d_s \leq 0.10$ µm.

The results of calculations are presented in Figure 4.19. Figure 4.19a shows the frequency splitting between the two orthogonally polarized fundamental VCSEL modes. In contrast to the case of the elliptic hole PhC-VCSELs, the x-polarized mode is now the high-frequency one. Also the frequency splitting is much enlarged: it ranges from 310 GHz at $d_s = 0.01$ µm up to almost 415 GHz for $d_s = 0.10$ µm. Figure 4.19b depicts the VCSEL relative loss dichroism, which is computed as $\delta_\alpha = (\alpha_y - \alpha_x)/\alpha_{\text{avg}}$, where α_x and α_y are the modal losses for the x- and y-polarized

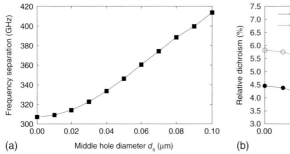
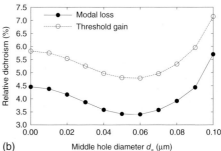

Figure 4.19 Resonant frequency separation of the x- and y-polarized modes (a) and relative loss and gain dichroism between x- and y-polarization (b) for the structure with four large holes (Figure 4.17b), as a function of d_s. (After [90].)

modes, respectively and α_{avg} is the average value of α_x and α_y. Surprisingly, δ_α is not a monotonic function of d_s, but possesses a minimum of around $d_s \approx 0.06$ μm.

A similar effect is observed when we consider the relative gain dichroism δ_g, defined analogously to the relative loss dichroism as $\delta_g = (g_y - g_x)/g_{avg}$, where g_x, g_y, and g_{avg} are the threshold gains of x, y-polarizations and their average value, respectively. The relative gain dichroism has the smallest value of 5% for $d_s = 0.06$ μm and reaches 7% for $d_s = 0.10$ μm. In addition, the large birefringence can increase the difference in threshold currents. Hence, with the aid of these two effects, a stable-polarization emission of the laser can be achieved. It is worth noticing that even for the structure without any hole in the middle ($d_s = 0$) the birefringence and dichroism are quite high.

4.8
Photonic-Crystal VCSELs with True Photonic Bandgap

PhCs have been envisioned as a way to funnel all spontaneous emission in the lasing mode by creating a PBG in lasers in a similar fashion as the BG for electrons in crystal lattices [95]. Since then, the rapid development of PhCs has led to the successful demonstration of such a spontaneous emission control in optically pumped [96, 97] and recently in electrically pumped [98] two-dimensional (2D) PhC slabs. In these PhC lasers, the PBG confines light to a defect in a 2D lattice while a total internal reflection confines light in the remaining third dimension. The development of PhCs with full 3D PBG remains a technological challenge despite several successful demonstrations [99, 100]. As discussed so far, incorporating PhC in VCSELs is a way to improve their optical properties, that is, to enhance their single-mode operation, to achieve stable polarization, and high-speed modulation. However, all PhC-VCSEL designs considered so far operate on index guiding inside the defect area of higher refractive index, which is different from the true bandgap confinement of short-period lattice surface-emitting lasers [101]. Recently, we have demonstrated using two different numerical methods that full PBG light confinement can actually be realised for the TE-like (with electric field lying in {ΓKM} plane) polarization, which is – as a matter of fact – the one of significance in VCSELs [102]. Such PhC-VCSELs would in fact be true PBG-VCSELs.

We first analyse a periodic stack of pairs of high (h) and low (l) refractive index layers with triangular two-dimensional lattice of air holes. The pair of layers is made of materials with refractive indices of high contrast as, for example, GaAs/AlO$_x$ or Si/SiO$_2$. Such a structure can be treated as a 3D PhC. Its unit cell and corresponding reciprocal lattice are shown in Figure 4.20. In our computations all units are normalised to the PhC lattice constant Λ. The diameter of the PhC holes is fixed at $a = 0.6\,\Lambda$. The optical thicknesses of h and l layers are quarter-wavelength ones at $\lambda = 4\,\Lambda$. Because of the linearity of Maxwell equations the value of Λ can be chosen arbitrarily in order to operate at any desired wavelength.

In the first step of our calculations – made with the popular MPB software [67] – we investigate the in-plane light propagation. In this case the modes can be

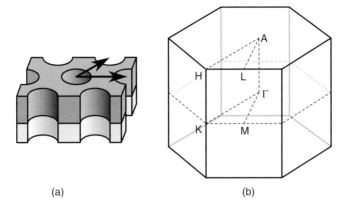

(a) (b)

Figure 4.20 (a) Unit cell of analzsed photonic crystal. (b) First Brillouin zone in its reciprocal lattice.

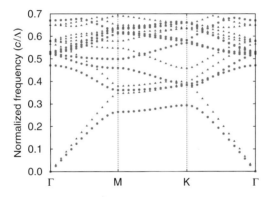

Figure 4.21 Band-diagram for in-plane modes. Red circles indicate TE-like modes and blue triangles TM-like ones. Here and in the rest of the paper we consider $n_H = 3.5$ and $n_L = 1.5$. (After [102].) (Please find a color version of this figure on the color plates.)

separated into purely TE- and TM-like [103]. They are both depicted in Figure 4.21 for $n_h = 3.5$ and $n_l = 1.5$. As can be seen from this figure a band gap for the TE-like modes exists ranging from 0.29 c/Λ to 0.36 c/Λ, where c is the speed of light in vacuum. At this point, it is important to consider the QW gain medium of a typical laser. As is well known for VCSELs, [21] the QW gain is determined by only the heavy-hole (HH) to conduction band (CB) transitions owing to the emission perpendicular to the surface of the QW. Considering QW in the {ΓKM} plane results in strongly anisotropic gain favoring the emission along the (ΓA) direction of the DBR with TE polarization as it couples to the HH-CB transitions. Emission of light with TM polarization will be strongly suppressed for properly designed QWs with large HH-LH band splitting, [20]. This means that, from a practical point of view, one does not need a band gap in TM-like modes because of the characteristics of the semiconductor junction laser.

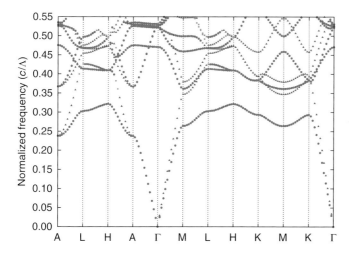

Figure 4.22 Three-dimensional band diagram. Red circles indicate the TE-like modes, blue triangles the TM-like modes and violet squares the modes with parity close to zero. (After [102].) (Please find a color version of this figure on the color plates.)

In surface-emitting lasers, the electromagnetic wave has a nonzero wavevector in the (ΓA) direction, which is perpendicular to the epitaxial layers. Hence, Figure 4.22 shows the three-dimensional band diagram with marked high-symmetry points indicated in Figure 4.20. In case of k-vector not lying in the {ΓKM}, plane the modes are characterized as partially TE- or TM- by considering the mode parity, that is, an expectation value of the mirror-flip operation through {ΓKM} plane. It is a number varying from -1 for purely TM-like modes to 1 for TE-like ones. The shape and the color of the markings in Figure 4.22 indicate the parity of the modes. As one can see, there is still a well-pronounced band gap in the TE-like (even) modes and hence, PBG confinement of the light is possible in any PhC-VCSEL utilising proposed DBRs.

As previously mentioned, of interest for VCSEL application is the existence of PBG for light with TE-like polarization emitted along the (ΓA) direction. In order to illustrate such a PBG confinement we create a low-refractive index line-defect in the periodic PhC-DBR structure by increasing the radius of a single air hole until it touches the neighboring ring of holes ($r_D = 0.7\ \Lambda$). The calculated electric field distributions of sample defect modes with frequency in the band gap is shown in Figure 4.23. As can be seen from this figure, a fundamental mode, well confined in the defect, is present in such PhC-VCSELs. Taking for example a wavelength of $\lambda = 1.55$ μm, the lattice constant for the mode shown in Figure 4.23a is $\Lambda \approx 0.5$ μm, which is feasible with the present-day technology.

The air-defect mode shown in Figure 4.23a clearly illustrates the feasibility of making PhC-VCSELs relying on true PBG light confinement. For efficient lasing however, the optical field has to be concentrated in the high-index QW material in order to provide high modal gain. Thus, the air-defect is unsuitable for application

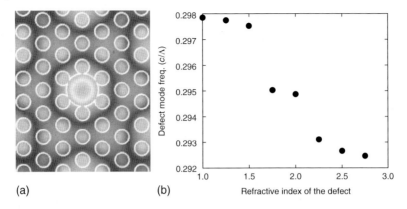

Figure 4.23 (a) Arrangement of the holes in a single-defect waveguide and electric field distributions of the defect modes at the symmetry point A and normalized frequency of 0.297853. (b) Normalized frequency of the fundamental defect mode as a function of the refractive index of the defect. (After [102].)

in a real VCSEL with a gain. However, the band-gap effects remain valid for both low- and high-index material inside the defect. In Figure 4.23b we prove the persistent existence of a confined mode when increasing the refractive index of the defect. As can be seen from this figure, the QW can be easily embedded in, for example, AlO_x/AlGaAs materials with an average refractive index of less than 3. In such a case, the defect mode would be confined by both effective-index guiding, as for present PhC-VCSELs, and by the lateral Bragg reflection, which should significantly improve the device optical properties like cavity Q-factor or threshold gain.

Finally, we consider a PhC cavity created by a low-index defect ($n_D = 1$) with radius $r_D = 0.5\Lambda$ and height $h_D = 2(h_l + h_h) + h_l$ surrounded by seven rings of holes and three (h/l) pairs with an additional high-index layer. We calculate the defect mode by the efficient PWAM using the periodic boundary conditions in {ΓKM} plane. The defect mode intensity distribution is shown in Figure 4.24. As can be seen from this figure, a PBG confined defect mode exists in such a PBG-VCSELs with modal distribution reflecting the sixfold symmetry of the PhC lattice and the fact that good vertical confinement is only achieved in the region $r_D > r > r_h$, where the vertical profile of the cavity differs from the one of the bulk PhC. The cavity quality factor of such a truly PBG-VCSEL is $Q = 43\,000$ for the case of Figure 4.24, that is, about three times higher than the one of the state-of-the-art oxide-confined VCSELs [104] and more than ten times higher than one of the typical PhC-VCSELs [63]. Similar to the waveguide above, the cavity may contain any high- or low-index material in which it may be possible to embed a QW. As shown here, one can make a cavity out of the same high- and low-index material as the rest of the structure, which would make its fabrication straightforward.

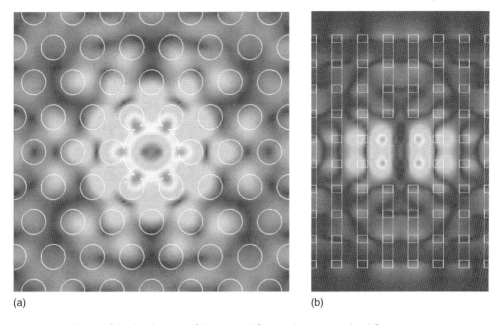

Figure 4.24 Electric field distributions of the cavity-defect modes at normalized frequency of 0.3112. (a) Cross-section in {ΓKM} plane of an undisturbed photonic crystal, and (b) cross-section in {ΓMA} plane. (After [102].)

4.9
Summary and Prospects

The VCSEL technology is fully mature, well established and commercialized. As a matter of fact, VCSELs are nowadays widely employed in short distance fiber interconnects, laser printing, computing, and sensing, and so on. Most importantly, the problem of efficient electrical current injection has been solved – the DBR grading and doping levels have been optimized and very small threshold currents and high slope and wall-plug efficiencies are common practice in contemporary VCSELs. This is in contrast to semiconductor PhC waveguide lasers emitting in the near infrared region, where the surface recombination in the air holes and the very low output power remain major problems [105]. Incorporating a PhC lattice of air holes in VCSELs is straightforward and indeed PhC-VCSELs have been experimentally demonstrated by many groups. The PhC serves to provide alternative and independent transverse optical confinement mechanism by index guiding. In commercial VCSELs transverse confinement is provided by either oxide-confined aperture, ion implantation or, structured TJ. High output power is only possible for large aperture sizes; however, this leads to multitransverse-mode operation with increased bandwidth. Multimode emission limits the speed of digital modulation through optical fiber owing to chromatic dispersion. Using PhC in VCSELs decouples the optical confinement from the electrical one and, as the PhC lattice parameters and hole depth can be precisely tuned in the wide region, it provides a tool to engineer the strength of the index guiding. The main goals pursued so far in PhC-VCSEL development have been ensuring high output power single fundamental mode emission and polarization stabilization.

In the Introduction we have reviewed numerous experimental realizations of PhC-VCSELs and have shown that they constitute a promising approach for achieving these two goals. Indeed, SMSR of 30–40 dB and 3.8 mW output power have been demonstrated. However, the expected "endlessly single-mode operation" with increasing the PhC pitch Λ for a suitably chosen hole diameter as observed for PhC fibers does not work for PhC-VCSELs. The reason is that, besides the PhC, other contributions to the waveguiding and modal gain such as thermal lensing, carrier induced antiguiding and spatial hole burning prevail. Oxide-confined PhC-VCSELs require a minimum etch depth to provide sufficient loss in order to inhibit lasing of the strongly guided oxide modes. This is not the case for proton-implanted PhC-VCSELs, which lack index guiding and where shallow etch depths can still provide single-mode emission. A clear cutoff between single-mode and multimode operations has been observed for about 5 cm^{-1} of loss difference between the fundamental and first-order transverse mode. Polarization control induced by the PhC has been demonstrated for elliptical hole triangular PhC lattice as well as for square lattice when decreasing the diameter of two air holes next to the PhC defect. Many of these experimental observations briefly reviewed in the introduction have been confirmed numerically in the following sections of this chapter.

In Section 4.2 we have briefly discussed different numerical methods capable of modeling PhC-VCSELs; their shortcomings and advantages. Next, in Section 4.3 we have presented in detail the mathematical basis of the PWAM developed for the specific usage of efficient modeling of 3D multilayered structures with high-contrast subwavelength features. The modeled structure is assumed to consist of parallel layers, and PWAM is especially efficient when certain layers appear multiple times within the stack. This is the case in PhC-VCSELs, for which the PWAM provides significant computational savings as compared to general purpose methods. PWAM uses the PWE method for the description of all layers, followed by a boundary matching scheme based on the admittance-matrix transfer method. In the following sections, we presented numerous applications of PWAM for full-vectorial 3D calculations of PhC-VCSELs.

In Section 4.4 we investigated the impact of the PhC depth on VCSEL threshold characteristics and demonstrated the existence of an optimal depth that establishes the PhC waveguiding without the need to etch throughout the whole VCSEL. In a broad range of PhC parameters, modal gain close to the maximal one can be assured by holes etched through the top DBR and the cavity. Only in the case of very narrow optical aperture ($R_A \leq 1$ μm) the etching should be considerably deeper. Moreover, to achieve PhC confinement and improve the modal performance with minimal technological efforts etching of only the top DBR is enough in most cases except for the smallest considered optical aperture. We also investigated the modal losses for broad aperture 1300 nm QD VCSEL and revealed that high precision PhC etching is necessary for optimising low-threshold performances; threshold gain is a periodic function of the etching depth and is low for optimal etch depths.

In Section 4.5 we discussed the advantages of the bottom-emitting PhC-VCSELs compared to the top-emitting ones with respect to the modal properties of the emitted light. Light emission throughout the mirror with incorporated PhC structure leads to significant distortion of the mode pattern at the laser facet. Alternatively, light emission throughout the DBR free from PhC improves drastically the emitted mode pattern. Such an improvement has been found for relatively narrow PhC apertures favoring single transverse-mode operation.

Section 4.6 investigated the PhC parameters leading to enhanced fundamental mode operation of PhC-VCSELs. We demonstrated that narrow PhC air holes cause severe optical field leakage while too broad holes cause diffraction losses, reduction of high-order mode discrimination, as well as reduction of DBR mirror reflection. Thus, the optimal hole diameter a for single mode operation and the highest fundamental mode gain is in the range $0.2 \leq a/\Lambda \leq 0.3$. We have mapped the regions of single mode operation on the plane: etching depth – a/Λ ratio for different optical apertures R_A. The position of the single mode region depends mostly on the optical aperture ($2R_A$) and on the a/Λ ratio. For some combinations of R_A and small a/Λ ratios, the first-order mode does not appear in the structure regardless of the etching depth. For most of the PhC parameters, this single mode region corresponds to holes etched only in the top DBR, which significantly reduces technological efforts. While the largest difference between the wavelengths of the fundamental and first-order modes is observed for PhC parameters assuring strong confinement, that is, deep, broad holes and narrow optical apertures, the largest modal gain difference is observed for shallow, narrow holes and narrow optical apertures.

Section 4.7 investigated PhC designs that lead to highly birefringent and dichroic PhC-VCSELs. We have considered two different PhC configurations in a GaAs/AlGaAs VCSEL: with elliptical holes and with selectively enlarged holes and a small central hole. Elliptic hole PhC structure can provide a large birefringence (of the order of 100 GHz in terms of frequency separation between modes of different polarizations). However, it has a moderate impact on the VCSEL dichroism (the modal losses for the two polarizations remain equal to each other within the numerical precision of about 3%). Very large birefringence of 400 GHz and large dichroism of 7% are demonstrated for the second PhC-VCSEL design using selectively enlarged holes and a small central hole.

In Section 4.8 we investigated the feasibility of PhC-VCSEL with true photonic band gap. Band-gap confinement for TE-polarized light has been demonstrated by calculating the 3D band diagrams for the VCSEL DBR structure. This band-gap effect, which is owing to the utilization of two materials with high refractive index contrast Δ in the DBR stack, only appears above a certain value of Δ. We illustrated the true PBG confinement for a low-index line-defect as well as for a low-index cavity mode. Such PBG-VCSELs are fully compatible with the mature VCSEL technology and do not require the complicated assembly of 3D PhCs.

Many of the theoretical and numerical results and PhC-VCSEL designs presented in this chapter remain to be proven experimentally and further developed. These concern the modal improvement by bottom emission, the impact of the hole depth on the threshold and modal characteristics, and the PhC designs for polarization stabilisation. Another interesting prospect for experimental and further theoretical work is along the lines of the true PBG confinement; the PBG-VCSEL suggested in [102]. Self consistency of electrical, thermal, gain, and optical PhC-VCSEL models is essential for proper numerical analysis as all these different physical phenomena are taking place in real-world VCSELs simultaneously. Detailed calibration and improvement of numerical methods by comparing with extensive experimental results still remain to be carried out. Such work would lead to new PhC designs, which would eventually prove the advantages of PhC-VCSELs for modal and polarization stabilization over other, alternative methods.

References

1. Li, H.E. and Iga, K. (eds) (2002) *Vertical-Cavity surface-Emitting Laser Devices*, Springer, Berlin.
2. Michalzik, R. (ed.) (2012) *VCSELs*, Springer Series in Optical Sciences, Vol. 166, Springer, Berlin.
3. Vakhshoori, D., Wynn, J.D., Zydzik, G.J., Asom, M., Kojima, K., Leibenguth, R.E., and Morgan, R.A. (1993) Top-surface emitting lasers with 1.9 V threshold voltage and the effect of spatial hole burning on their transverse mode operation and efficiencies. *Appl. Phys. Lett.*, **62**, 1448.
4. Valle, A., Sarma, J., and Shore, K.S. (1995) Spatial hole burning effect on the dynamics of vertical -cavity surface emitting lasers diodes. *IEEE J. Quantum Electron.*, **31**, 1423.
5. Law, J.Y. and Agrawal, G.P. (1997) Effects of spatial hole burning on gain switching in vertical -cavity surface emitting lasers. *IEEE J. Quantum Electron.*, **33**, 462.
6. Unold, H.J. et al. (2001) Large-area single-mode VCSELs and the self-aligned surface relief. *IEEE J. Sel. Top. Quantum Electron.*, **7**, 386.
7. Hadley, G.R. (1995) Effective index model for vertical - cavity surface - emitting lasers. *Opt. Lett.*, **20**, 1483.
8. Hegblom, E.R., Margalit, N.M., Fiore, A., and Coldren, L.A. (1999) High-performance small vertical-cavity lasers: a comparison of measured improvements in optical and current confinement in devices using tapered apertures. *J. Sel. Top. Quantum Electron.*, **5**, 553.
9. Choquette, K.D., Hong, M., Freud, R.S., Mannaerts, J.P., Wetzel, R.C., and Leibenguth, R.E. (1993) Vertical-cavity surface-emitting laser diodes fabricated by in situ dry etching and molecular beam epitaxial regrowth. *IEEE Photon. Technol. Lett.*, **5**, 284.
10. Oh, T.-H., McDaniel, M.R., Huffaker, D.L., and Deppe, D.G. (1998) Cavity-induced antiguiding in a selectively oxidized vertical-cavity surface-emitting laser. *Photon. Technol. Lett.*, **10**, 12.
11. Choquette, K.D., Hadley, G.R., Hou, H.Q., Geib, K.M., and Hammons, B.E. (1998) Leaky mode vertical cavity lasers using cavity resonance modification. *Electron. Lett.*, **34**, 991.
12. Naone, R.L., Floyd, P.D., Young, D.B., Hegblom, E.R., Strand, T.A., and Coldren, L.A. (1998) Interdiffused quantum wells for lateral carrier confinement in VCSEL's. *J. Sel. Top. Quantum Electron.*, **4**, 706.
13. Koch, B.J., Leger, J.R., Gopinath, A., and Wang, Z. (1997) Single-mode vertical cavity surface emitting laser by graded-index lens spatial filtering. *Appl. Phys. Lett.*, **70**, 2359.
14. Nikolajeff, F., Ballen, T.A., Leger, J.R., Gopinath, A., Lee, T.-C., and Williams, R.C. (1999) Spatial-mode control of vertical-cavity lasers with micromirrors fabricated and replicated in semiconductor materials. *Appl. Opt.*, **38**, 3030.

15. Chen, G., Leger, J.R., and Gopinath, A. (1999) Angular filtering of spatial modes in a vertical-cavity surface-emitting laser by a Fabry-Perot etalon. *Appl. Phys. Lett.*, **74**, 1069.
16. Ueki, N., Sakamoto, A., Nakamura, T., Nakayama, H., Sakurai, J., Otoma, H., Miyamoto, Y., Yoshikawa, M., and Fuse, M. (1999) Single-transverse-Mode 3.4-mW emission of oxide-confined 780-nm VCSEL's. *Photon. Technol. Lett.*, **11**, 1539.
17. Milster, T., Jiang, W., Walker, E., Burak, D., Claisse, P., Kelly, P., and Binder, R. (1998) A single-mode high power vertical cavity surface emitting laser. *Appl. Phys. Lett.*, **72**, 3425.
18. Claisse, P.R., Jiang, W., Kiely, P.A., Gable, B., and Koonse, B. (1998) Single high order mode VCSEL. *Electron. Lett.*, **34**, 681.
19. Martinsson, H., Vukusic, J.A., Grabherr, M., Michalzik, R., Jäger, R., Ebeling, K.J., and Larsson, A. (1999) Transverse mode selection in large area oxide-confined vertical-cavity surface-emitting lasers using a shallow surface relief. *Photon. Technol. Lett.*, **11**, 1536.
20. Coldren, L.A. and Corzine, S.W. (1995) *Diode Lasers and Photonic Integrated Circuits*, John Willey & Sons, Inc., New York.
21. Panajotov, K. and Prati, F. (2012) Polarization dynamics of VCSELs, *VCSELs*, Springer Series in Optical Sciences, Vol. 166, Chapter 6 (ed. Michalzik, R.), Springer, Berlin.
22. Uenohara, H., Tateno, K., Kagawa, T., Ohiso, Y., Tsuda, H., Kurokawa, T., and Amano, C. (1999) Investigation of dynamic polarization stability of 850 nm GaAs-based vertical-vavity surface-emitting lasers grown on (311)B and (100) substrates. *IEEE Photon. Technol. Lett.*, **11**, 400.
23. Niskiyama, N., Arai, M., Shinada, S., Azuchi, M., Miyamoto, T., Koyama, F., and Iga, K. (2001) Highly strained GaInAs-GaAs quantum-well vertical-cavity surface-emitting laser on GaAs (311)B substrate for stable polarization operation. *IEEE J. Sel. Top. Quantum Electron.*, **7**, 242.
24. Panajotov, K., Nagler, B., Verschaffelt, G., Georgievski, A., Thienpont, H., Danckaert, J., and Veretennicoff, I. (2000) Impact of in-plane anisotropic strain on the polarization behavior of vertical-cavity surface-emitting lasers. *Appl. Phys. Lett.*, **77**, 1590.
25. Choquette, K.D. and Leibenguth, R.E. (1994) Control of vertical - cavity laser polarization with anisotropic transverse cavity geometries. *IEEE Photon. Technol. Lett.*, **6**, 40.
26. Yoshikawa, T., Kawakami, T., Saito, H., Kosaka, H., Kajita, M., Kurihara, K., Sugimoto, Y., and Kasahara, K. (1998) Polarization-controlled single-mode VCSEL. *IEEE J. Quantum Electron.*, **34**, 1009.
27. Gayral, B., Gerard, J.M., Legrand, B., Gostard, E., and Thierry-Mieg, V. (1998) Optical study of GaAs/AlAs pillar microcavities with elliptical cross section. *Appl. Phys. Lett.*, **72**, 1421.
28. Ortsiefer, M., Shau, R., Zigldrum, M., Böhm, G., Köhler, F., and Amann, M.C. (2000) Submilliamp long-wavelength InP-based vertical-cavity surface-emitting laser with stable linear polarisation. *Electron. Lett.*, **36**, 1124.
29. Chua, C.L., Thotnton, R.L., Treat, D.W., and Donaldson, R.M. (1998) Anisotropic apertures for polarization-stable laterally oxidized vertical-cavity lasers. *Appl. Phys. Lett.*, **73**, 1631.
30. Mukaihara, T., Koyama, F., and Iga, K. (1993) Engineering polarization control of GaAs/ALGaAs surface-emitting lasers by anisotropic stress from elliptical etched substrate hole. *IEEE Photon. Technol. Lett.*, **5**, 133.
31. Mukaihara, T., Ohnoki, N., Hayashi, Y., Hatori, N., Koyama, F., and Iga, K. (1995) Polarization control of vertical-cavity surface emitting lasers using a birefringent metal/dielectric polarizer loaded on top distributed Bragg reflector. *IEEE J. Sel. Top. Quantum Electron.*, **1**, 667.
32. Ser, J.H., Ju, Y.G., Shin, J.H., and Lee, Y.H. (1995) Polarization stabilization of vertical-cavity top-surface-emitting lasers by inscription of fine metal-interlaced gratings. *Appl. Phys. Lett.*, **21**, 2769.

33. Berseth, C.A., Dwir, B., Utke, I., Pier, H., Rudra, A., Iakovlev, V.P., and Kapon, E. (1999) Vertical cavity surface emitting lasers incorporating structured mirrors patterned by electron-beam lithography. *J. Vac. Sci. Technol., B*, **17**, 3222.
34. Debernardi, P., Ostermann, J.M., Feneberg, M., Jalics, C., and Michalzik, R. (2005) Reliable polarization control of VCSELs through monolithically integrated surface gratings: a comparative theoretical and experimental study. *IEEE J. Sel. Top. Quantum Electron.*, **11**, 107.
35. Song, D.S., Kim, S.H., Park, H.G., Kim, C.K., and Lee, Y.H. (2002) Single-fundamental-mode photonic-crystal vertical-cavity surface-emitting lasers. *Appl. Phys. Lett.*, **80**, 3901.
36. Danner, A.J., Raftery, J.J., Yokouchi, N., and Choquette, K.D. (2004) Transverse modes of photonic crystal vertical-cavity lasers. *Appl. Phys. Lett.*, **84**, 1031.
37. Yokouchi, N., Danner, A.J., and Choquette, K.D. (2003) Two-dimensional photonic crystal confined vertical-cavity surface-emitting lasers. *IEEE J. Sel. Top. Quantum Electron.*, **9**, 1439.
38. Yang, H.P.D., Lai, F.I., Chang, Y.H., Yu, H.C., Sung, C.P., Kuo, H.C., Wang, S.C., Lin, S.Y., and Chi, J.Y. (2005) Single mode (SMSR > 40 dB) proton implanted photonic crystal vertical-cavity surface-emitting lasers. *Electron. Lett.*, **41**, 326.
39. Danner, A.J., Kim, T.S., and Choquette, K.D. (2005) Single fundamental mode photonic crystal vertical cavity laser with improved output power. *Electron. Lett.*, **41**, 325.
40. Danner, A.J., Raftery, J.J., Leisher, P.O., and Choquette, K.D. (2006) Single mode photonic crystal vertical cavity lasers. *Appl. Phys. Lett.*, **88**, 091114.
41. Birks, T.A., Knight, J.C., and Russell, P.S.J. (1997) Endlessly single-mode photonic crystal fiber. *Opt. Lett.*, **22**, 961.
42. Ivanov, P.S., Heard, P.J., Cryan, M.J., and Rorison, J.M. (2011) Comparative study of mode control in vertical-cavity surface-emitting lasers with photonic crystal and micropillar etching. *IEEE J. Quantum Electron.*, **47**, 1257.
43. Liu, A., Xing, M., Qu, H., Chen, W., Zhou, W., and Zhenga, W. (2009) Reduced divergence angle of photonic crystal vertical-cavity surface-emitting laser. *Appl. Phys. Lett.*, **94**, 191105.
44. Leisher, P.O., Danner, A.J., and Choquette, K.D. (2007) Parametric study of proton-implanted photonic crystal vertical-cavity surface-emitting lasers. *IEEE J. Sel. Top. Quantum Electron.*, **13**, 1292.
45. Kasten, A.M., Meng, P.T., Sulkin, J.D., Leisher, P.O., and Choquette, K.D. (2008) Photonic crystal vertical cavity lasers with wavelength-independent single-mode behavior. *IEEE Photon. Technol. Lett.*, **20**, 2010.
46. Siriani, D.F., Leisher, P.O., and Choquette, K.D. (2009) Loss-induced confinement in photonic crystal vertical-cavity surface-emitting lasers. *IEEE J. Quantum Electron.*, **45**, 762.
47. Siriani, D.F., Leisher, P.O., and Choquette, K.D. (2009) Mode control in photonic crystal vertical-cavity surface-emitting lasers and coherent arrays. *IEEE J. Sel. Top. Quantum Electron.*, **15**, 909.
48. Alias, M.S. and Shaari, S. (2010) Loss analysis of high order modes in photonic crystal vertical-cavity surface-emitting lasers. *IEEE J. Lightwave Technol.*, **28**, 1556.
49. Kim, T.S., Danner, A.J., Grasso, D.M., Young, E.W., and Choquette, K.D. (2004) Single fundamental mode photonic crystal vertical cavity surface emitting laser with 9 GHz bandwidth. *Electron. Lett.*, **40**, 21.
50. Yang, H.P.D., Chang, Y.H., Lai, F.I., Yu, H.C., Hsu, Y.J., Lin, S.Y., Hsiao, R.S., Kuo, H.C., Wang, S.C., and Chi, J.Y. (2005) Single mode InAs quantum dot photonic crystal VCSELs. *Electron. Lett.*, **41**, 1130.
51. Leisher, P.O., Danner, A.J., and Choquette, K.D. (2006) Single-mode 1.3 μm photonic crystal vertical-cavity surface-emitting laser. *IEEE Photon. Technol. Lett.*, **18**, 2156.

52. Lai, F.-I., Yang, H.P.D., Lin, G., Hsu, I.-C., Liu, J.-N., Maleev, N.A., Blokhin, S.A., Kuo, H.C., and Chi, J.Y. (2007) Quantum-dot photonic crystal vertical-cavity surface-emitting lasers. *IEEE J. Sel. Top. Quantum Electron.*, **13**, 1318.
53. Song, D.S., Lee, Y.J., Choi, H.W., and Lee, Y.H. (2003) Polarization-controlled, single-transverse-mode, photonic-crystal, vertical-cavity, surface-emitting lasers. *Appl. Phys. Lett.*, **82**, 3182.
54. Lee, K.-H., Baek, J.-H., Hwang, I.-K., Lee, Y.-H., Lee, G.-H., Ser, J.-H., Kim, H.-D., and Shin, H.-E. (2004) Square-lattice photonic-crystal vertical-cavity surface-emitting lasers. *Opt. Express*, **12**, 4136.
55. Samakkulam, K., Sulkin, J., Giannopoulos, A., and Choquette, K.D. (2006) Micro-fluidic photonic crystal vertical cavity surface emitting laser. *Electron. Lett.*, **42**, 809.
56. Kasten, A.M., Sulkin, J.D., Leisher, P.O., McElfresh, D.K., Vacar, D., and Choquette, K.D. (2008) Manufacturable photonic crystal single-mode and fluidic vertical-cavity surface-emitting lasers. *IEEE J. Sel. Top. Quantum Electron.*, **14**, 1123.
57. Griffin, B.G., Arbabi, A., Kasten, A.M., Choquette, K.D., and Goddard, L.L. (2012) Hydrogen detection using a functionalized photonic crystal vertical cavity laser. *IEEE J. Quantum Electron.*, **48**, 160.
58. Dems, M., Chung, I.-S., Nyakas, P., Bischoff, S., and Panajotov, K. (2010) Numerical methods for modeling photonic-crystal VCSELs. *Opt. Express*, **15**, 16042.
59. Ivanov, P.S., Unold, H.J., Michalzik, R., Maehnss, J., Ebeling, K.J., and Sukhoivanov, I.A. (2003) Theoretical study of cold-cavity single-mode conditions in vertical-cavity surface-emitting lasers with incorporated two-dimensional photonic crystals. *J. Opt. Soc. Am. B*, **20**, 2442.
60. Ivanov, P.S. and Rorison, J. (2009) Theoretical optimization of transverse waveguiding in oxide-confined VCSELs with internal photonic crystals. *J. Opt. Soc. Am. B*, **26**, 2461.
61. Bava, G.P., Debernardi, P., and Fratta, L. (2001) Three-dimensional model for vectorial fields in vertical-cavity surfaceemitting lasers. *Phys. Rev. A*, **63**, 23816.
62. Nyakas, P. (2007) Full-vectorial three-dimensional finite element optical simulation of vertical-cavity surface-emitting lasers. *IEEE J. Lightwave Technol.*, **25**, 2427.
63. Dems, M., Kotynski, R., and Panajotov, K. (2005) Plane-wave admittance method – a novel approach for determining the electromagnetic modes in photonic structures. *Opt. Express*, **13**, 3196.
64. Dems, M. and Panajotov, K. (2007) Modeling of single- and multimode photonic-crystal planar waveguides with planewave admittance method. *Appl. Phys. B*, **89**, 19.
65. Ivanov, P.S., Ho, Y.-L.D., Cryan, M.J., and Rorison, J. (2012) Modelling investigations of DBRs and cavities with photonic crystal holes for application in VCSELs. *J. Opt.*, **14**, 125103.
66. Dems, M., Czyszanowski, T., and Panajotov, K. (2007) Numerical analysis of high Q-factor photonic-crystal VCSELs with plane-wave admittance method. *Opt. Quantum Electron.*, **39**, 419.
67. Johnson, S. and Joannopoulos, J. (2001) Block-iterative frequency-domain methods for Maxwell equations in a planewave basis. *Opt. Express*, **8**, 173.
68. Joannopoulos, J.D., Johnson, S.G., Winn, J.N., and Meade, R.D. (2008) *Photonic Crystals: Molding the Flow of Light*, Princeton University Press, Princeton, NJ.
69. Dreher, A. and Pregla, R. (1992) Analysis of planar waveguides with the method of lines and absorbing boundary conditions. *IEEE Microw. Guid. Wave Lett.*, **1**, 239.
70. Helfert, S.F., Barcz, A., and Pregla, R. (2003) Three-dimensional vectorial analysis of waveguide structures with the method of lines. *Opt. Quantum Electron.*, **35**, 381.
71. Sacks, Z.S., Kingsland, D.M., Lee, R., and Lee, J.F. (1995) A perfectly matched anisotropic absorber for use

as an absorbing boundary condition. *IEEE Trans. Antennas Prop.*, **43**, 1460.
72. Berenger, J.P. (1994) A perfectly matched layer for the absorption of electromagnetic waves. *J. Comput. Phys.*, **114**, 185.
73. Kotynski, R., Dems, M., and Panajotov, K. (2007) Waveguiding losses of microstructured fibres-plane wave method revisited. *Opt. Quantum Electron.*, **39**, 469.
74. Knight, J.C. and Russell, P.S.J. (2002) Applied optics: new ways to guide light. *Science*, **296**, 276.
75. Conradi, O., Helfert, S.F., and Pregla, R. (2001) Comprehensive modeling of vertical-cavity laser-diodes by the method of lines. *IEEE J. Quantum Electron.*, **37**, 928.
76. Taflove, A. and Hagness, S.C. (2000) *Computational Electrodynamics: The Finite-Difference Time-Domain Method*, 2nd edn, Artec House Inc., Boston, MA.
77. Chew, W.C., Jin, J.M., and Michielssen, E. (1997) Complex coordinate stretching as a generalized absorbing boundary condition. *Microw. Opt. Technol. Lett.*, **15**, 363.
78. Press, W., Teukolsky, S.A., Vetterling, W.T., and Flannery, B.P. (1992) *Numerical Recipes in C: The Art of Scientific Computing*, 2nd edn, Cambridge University Press, New York.
79. www.oxfordplasma.de (accessed on November 30, 2013).
80. Czyszanowski, T., Dems, M., and Panajotov, K. (2007) Impact of the hole depth on the modal behaviour of long wavelength photonic crystal VCSELs. *J. Phys. D: Appl. Phys.*, **40**, 2732.
81. Czyszanowski, T., Dems, M., and Panajotov, K. (2007) Optimal parameters of photonic crystals vertical-cavity surface-emitting diode lasers. *IEEE J. Lightw. Technol.*, **25**, 2331.
82. Czyszanowski, T., Dems, M., Sarzala, R.P., Nakwaski, W., and Panajotov, K. (2011) Precise lateral mode control in photonic crystal vertical-cavity surface-emitting lasers. *IEEE J. Quantum Electron.*, **47**, 1291.
83. Czyszanowski, T., Dems, M., Thienpont, H., and Panajotov, K. (2008) Modal gain and confinement factors in top- and bottom-emitting photonic-crystal VCSEL. *J. Phys. D: Appl. Phys.*, **41**, 085102.
84. Czyszanowski, T., Sarzala, R., Piskorski, L., Dems, M., Wasiak, M., Nakwaski, W., and Panajotov, K. (2008) Threshold characteristics of bottom-emitting long wavelength VCSELs with photonic-crystal within the top mirror. *Opt. Quantum Electron.*, **40**, 149.
85. Czyszanowski, T., Dems, M., Thienpont, H., and Panajotov, K. (2007) Optimal radii of photonic crystal holes within DBR mirrors in long wavelength VCSEL. *Opt. Express*, **15**, 1301.
86. Czyszanowski, T., Dems, M., and Panajotov, K. (2007) Single mode condition and modes discrimination in photonic-crystal 1.3 μm AlInGaAs/InP VCSEL. *Opt. Express*, **15**, 5604.
87. Czyszanowski, T., Sarzaa, R.P., Dems, M., Nakwaski, W., Thienpont, H., and Panajotov, K. (2009) Optimal photonic-crystal parameters assuring single-mode operation of 1300 nm AlInGaAs vertical-cavity surface-emitting laser. *J. Appl. Phys.*, **105**, 093102.
88. Czyszanowski, T., Sarzaa, R.P., Dems, M., Nakwaski, W., Thienpont, H., and Panajotov, K. (2009) Strong modes discrimination and low threshold in cw regime of 1300 nm AlInGaAs/InP VCSEL induced by photonic crystal. *Phys. Status Solidi A*, **206**, 1396.
89. Sarzala, R., Czyszanowski, T., Wasiak, M., Dems, M., Piskorski, L., Nakwaski, W., and Panajotov, K. (2012) Numerical self-consistent analysis of VCSELs advances in optical technologies. *Adv. Opt. Technol.*, **2012**, Article ID 689519, doi: 10.1155/2012/689519.
90. Dems, M., Czyszanowski, T., Thienpont, H., and Panajotov, K. (2008) Highly birefringent and dichroic photonic crystal VCSEL design. *Opt. Commun.*, **281**, 3149.
91. Steel, M.J., White, T.P., Martijn de Sterke, C., McPhedran, M.C., and Botten, L.C. (2001) Symmetry and degeneracy in microstructured optical fibers. *Opt. Lett.*, **26**, 488.

92. Szpulak, M., Olszewski, J., Martynkien, T., Urbanczyk, W., and Wojcik, J. (2004) Polarizing photonic crystal fibers with wide operation range. *Opt. Commun.*, **239**, 97.
93. Ortigosa-Blanch, A., Knight, J.C., Wadsworth, W.J., Arriaga, J., Mangan, B.J., Birks, T.A., and Russell, P.S.J. (2000) Highly birefringent photonic crystal fibers. *Opt. Lett.*, **25**, 1325.
94. Hansen, T.P., Broeng, J., Libori, S.E.B., Knudsen, E., Bjarklev, A., Jensen, J.P., and Simonsen, H. (2001) Highly birefringent index-guiding photonic crystal fibers. *IEEE Photon. Technol. Lett.*, **13**, 588.
95. Yablonovich, E. (1987) Inhibited spontaneous emission in solid-state physics and electronics. *Phys. Rev. Lett.*, **58**, 2059.
96. Painter, O., Lee, R.K., Sherer, A., Yariv, A., O'Brien, J.D., Dapkus, P.D., and Kim, I. (1999) Two-dimensional photonic band-gap defect mode laser. *Science*, **284**, 1819.
97. Hwang, J.K., Ryu, H.Y., Song, D.S., Han, I.Y., Park, H.K., Jang, D.H., and Lee, Y.H. (2000) Continuous room-temperature operation of optically pumped two-dimensional photonic crystal lasers at 1.6 µm. *IEEE Photon. Technol. Lett.*, **22**, 1295.
98. Park, H.-G., Kim, S.-H., Kwon, S.-H., Ju, Y.-G., Yang, J.-K., Baek, J.-H., Kim, S.-B., and Lee, Y.-H. (2004) Electrically driven single-cell photonic crystal laser. *Science*, **305**, 1444.
99. Ogawa, S., Imada, M., Yoshimoto, S., Okano, M., and Noda, S. (2004) Control of light emission by 3D photonic crystals. *Science*, **305**, 227.
100. Aoki, K., Guimard, D., Nishioka, M., Nomura, M., Iwamoto, S., and Arakawa, Y. (2008) Coupling of quantum-dot light emission with a three-dimensional photonic-crystal nanocavity. *Nat. Photon.*, **2**, 688.
101. Noda, S., Yokoyama, M., Imada, M., Chutinan, A., and Mochizuki, M. (2004) Polarization mode control of two-dimensional photonic crystal laser by unit cell structure design. *Science*, **293**, 1123.
102. Panajotov, K. and Dems, M. (2010) Photonic crystal vertical-cavity surface-emitting lasers with true photonic bandgap. *Opt. Lett.*, **45**, 829.
103. Johnson, S.G., Fan, S., Villeneuve, P.R., Joannopoulos, J.D., and Kolodziejski, L.A. (1999) Guided modes in photonic crystal slabs. *Phys. Rev. B*, **60**, 5751.
104. Leisher, P., Raftery, J. Jr., Kasten, A., and Choquettea, K. (2006) Etch damage and deposition repair of vertical-cavity surface-emitting lasers. *J. Vac. Sci. Technol., B*, **24**, 104.
105. Checoury, X., Colombelli, R., and Lourtioz, J.-M. Electrically pumped photonic crystal lasers, Chapter 3 in this book.

5
III–V Compact Lasers Integrated onto Silicon (SOI)

Geert Morthier, Gunther Roelkens, and Dries Van Thourhout

5.1
Introduction

Silicon is emerging as a very promising platform for realizing highly integrated photonic integrated circuits (PICs), which can be fabricated using standard tools from the electronics industry, but an efficient source is lacking. Several solutions can be considered to overcome this problem. In some cases a fully independent external light source, for example, connected to the chip through an optical fiber, has been proposed. This source could then be considered as an optical supply line, in the same way as current electronic chips now have electrical supply lines. One could either use a single-wavelength source or a mode-locked laser (MLL) for WDM-applications (wavelength division multiplexing). In many cases, such a solution does not provide sufficient flexibility, and it is difficult to route the optical supply to any desired location on the chip. Furthermore it also does not help to overcome on-chip losses. Therefore several groups have proposed to place preprocessed III–V optoelectronic devices (DFB-lasers (distributed feedback), SOAs (semiconductor optical amplifiers), etc.) closer to the silicon chip, using well-known flip-chip integration techniques. The III–V devices can then either be located next to the silicon chip and edge-coupled [1–3] or on top of the silicon chip and coupled to grating couplers (e.g., Luxtera "lamp" solution [4]). Methods that use epitaxial lift off have also been developed [5]. These approaches have several advantages. Compared to the off-chip source described above they allow for larger flexibility (multiple channels, integration within the laser cavity, amplification, etc.). Compared to the wafer bonding approach described in this chapter, they have the advantage that the III–V devices can be prescreened and selected for performance. They are also based on well-known technological steps, and so they are currently closer to the market. Their main drawback lies in the accurate alignment required between the III–V chip and the silicon chip (typically ±0.5 to ±1 μm), which is time consuming and costly. In addition, they still do not allow for sufficient flexibility and integration density for some applications and, in particular, they do not allow the realization of low-power compact sources such as the ones discussed in this book.

Compact Semiconductor Lasers, First Edition.
Edited by Richard M. De La Rue, Siyuan Yu, and Jean-Michel Lourtioz.
© 2014 Wiley-VCH Verlag GmbH & Co. KGaA. Published 2014 by Wiley-VCH Verlag GmbH & Co. KGaA.

For these reasons several groups began to develop an alternative technology that used a wafer bonding technique to integrate the III–V semiconductor material onto the silicon waveguides. This technique, in principle, overcomes all the problems mentioned above; the critical, part-by-part alignment step needed in flip-chip integration approaches is now replaced by a wafer-scale lithography step, allowing for much faster and more accurate alignment. Furthermore the III–V devices can now be integrated everywhere on the chip, without limits on integration density and flexibility. Finally the wafer bonding approach (or heterogeneous integration process, as we refer to it further on in this chapter) has also allowed the demonstration of very promising microscale devices, which simply cannot be realized using the standard flip-chip integration approach.

Indeed, while the heterogeneous integration of III–V lasers onto silicon (silicon-on-insulator or SOI) was motivated by the near impossibility of fabricating electrically injected laser diodes in silicon, it turns out that the III–V membranes lend themselves excellently to implementation of very compact laser sources. The III–V layer stack (typically InP-based for telecom wavelength and with a thickness from 0.5 to 1 µm) has a refractive index of over 3 and is, on either side, surrounded by material with refractive index below 1.5 (air at the top and polymer and/or silica on the bottom side). This refractive index contrast (very similar to that of the silicon waveguides) gives extremely strong optical confinement in the waveguides, which is exactly what makes the compact laser diodes possible.

Moreover, in this approach the compact laser diodes can be integrated in a natural way with the complex and compact passive optical-waveguide circuits that can be fabricated with great accuracy and reliability using the SOI platform. Silicon photonics, because of the large refractive index contrast and due to the fact that mature complementary metal-oxide semiconductor (CMOS) fabrication technology can be used, has emerged in the past decade as a photonics platform of high potential. PICs implemented using passive waveguide circuits in the SOI material system, combined with heterogeneously integrated laser diodes, amplifiers, or photodetectors are thus extremely well suited for the realization of low-footprint and low-power photonic circuits for applications in optical interconnects and optical logic. Such photonic circuits are increasingly becoming a requirement, for example, for high-performance computing.

The heterogeneous integration approach, apart from allowing compact, potentially low-power sources, also poses some challenges. For example, the membrane structure inevitably means that metal contacts are much closer to the active layer and care has to be taken to avoid excessive loss because of that fact. Thermal management is another challenge. Especially in the case of adhesive bonding, heat sinking through the bonding layer is inefficient and alternative heat-sinking provision has often to be made.

In the following sections, we discuss in detail the bonding and fabrication technology, as well as different implementations of heterogeneously integrated edge-emitting lasers and microring and microdisk lasers. PICs with such heterogeneously integrated compact lasers have many potential applications, for example, in optical interconnects (in particular on-chip interconnects), optical sensing (such as biosensing), and in all-optical logic. Several of these applications are addressed in this chapter.

5.2
Bonding of III–V Membranes on SOI

The bonding of the III–V membrane onto the SOI can be performed using either molecular bonding (bonding based on Van der Waals forces) or adhesive bonding (using an intermediate adhesive polymer). In principle, III–V wafers can be bonded with SOI wafers, but so far mainly III–V dies have been bonded to SOI dies. Alignment is not necessary during bonding, as the III–V processing is carried out after bonding.

Figure 5.1 shows the general outline of the process. The starting point (a) is the processed SOI wafer, containing the silicon waveguide circuits. In most cases these are relatively simple passive circuits but recently integration with silicon modulators has also been demonstrated [6]. Next, the surface is prepared for the bonding process (b). Depending on the approach chosen (see below) a suitable bonding agent can be applied, the wafer can be planarized using oxide or the surface can be activated, for example, using oxygen plasma (or any combination thereof). Then the III–V material is bonded on top of the silicon wafer, with the

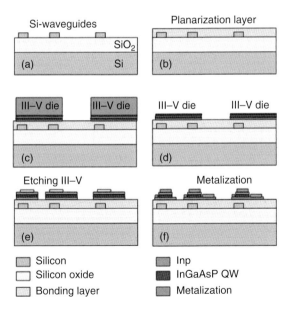

Figure 5.1 Schematics of heterogeneous integration process, with (a) processed SOI wafer or die, (b) planarized SOI (using oxide or BCB), (c) after bonding of unprocessed III–V material, (d) after III–V substrate removal, (e) after lithography and dry etching, and (f) after metalization. (Please find a color version of this figure on the color plates.)

epitaxial layers toward the silicon waveguides (c). The III–V material can be either in the form of a full wafer (wafer-to-wafer bonding) or in the form of (multiple) smaller dies (die-to-wafer bonding). Next, the substrate of the III–V material is removed, up to the epitaxial layers, using a chemical etching step (d) (sometimes preceded by a mechanical grinding step). This process requires incorporation of a suitable sacrificial etch-stop layer in the III–V layer stack, in most cases an InGaAs layer (assuming InP-based active layers). Once the sacrificial etch-stop layer has also been removed, a silicon substrate becomes available with thin III–V epitaxial layers on top, which can now be further processed using standard wafer-scale processing techniques, such as lithography and dry etching, to form the desired laser and/or amplifier structures (e). Finally metalization steps connect the lasers, and possibly the silicon modulators to the electrical supply lines (f). Devices fabricated using this approach is discussed further in this chapter. We first go into greater detail on the bonding process, which forms the basis of this integration method and the necessary post-processing steps (substrate removal, characterization methods).

Bonding and, in particular, wafer bonding is a process that is well known in the micro-electronics and MEMS industries. Different types of bonding have been developed, including adhesive bonding (using bonding agents such as BCB (benzocyclobutene), Polyimides, Photoresists, etc.), anodic bonding, and direct or fusion-bonding. In particular, most of the SOI wafers used in silicon photonics are realized through the so-called UNIBOND [7] process whereby two silicon wafers are bonded with an intermediate oxide layer in between. However, the bonding of III–V epitaxial material imposes some additional challenges to the bonding process. The III–V material is typically more fragile than silicon-based materials and has a very different thermal expansion coefficient. The latter imposes the requirement that the *temperatures* involved in the bonding process should be limited to 250–400 °C (compared to the temperatures in the range of 800–1000 °C often used in SOI-wafer manufacturing). Preserving the doping profiles in the preprocessed silicon wafers may also restrict the maximum temperature budget available.

Of course the *thermal properties* and the *transparency* of the bonding layer (which excludes anodic bonding and metallic bonding processes) are also important. Another issue is related to the *size-mismatch* between the silicon wafers and the III–V epitaxial material. While the silicon waveguides are processed on 150, 200 mm (mostly) or, in the future, even 300 mm wafers, the InP wafers are typically only between 2 and 4 in. (however 150 mm (6 in.) has also been demonstrated [8]). So one has either to sacrifice part of the silicon wafer or resort to die-to-wafer bonding techniques, whereby multiple InP-based dies are bonded on the silicon wafer, at the locations where the lasers are needed (e.g., see Figure 5.5b). However, die-to-wafer bonding techniques require significant effort in order to be developed toward a fully industrialized process: careful preparation of the dies is required and the bonding process is more prone to edge-effects.

The *bonding-layer thickness* is also a parameter that requires careful control. Depending on the laser design, the requirements on the bonding layer may differ.

For traditional lasers (DFB, DBR (distributed Bragg reflector), etc.) with large output power, the bonding layer typically has to be as thin as possible to allow for a good coupling between the light in the III–V material and the silicon waveguide. For microsources, on the other hand, one wants to limit the amount of light coupled out of the cavity and a somewhat thicker bonding layer may be required (100–300 nm). However, in many cases evanescent coupling is used and the coupling rate depends exponentially on the bonding-layer thickness. In other words, very good control and repeatability in the process are required.

5.2.1
Adhesive Bonding

In adhesive bonding, a suitable agent is applied to one or both substrates and serves as a "glue" to connect them. For an exhaustive review see Niklaus *et al.* [9]. The bonding agent can provide planarization of the substrate surface and is typically somewhat tolerant to surface roughness, defects, and contamination. Bonding using thermoplastic, elastomeric, and thermosetting materials has been reported in the literature. Given that the first two agents respectively remelt or become rubbery at elevated temperatures, they cannot be used for our envisaged application, which requires subsequent processing at temperatures >250 °C. Epoxies, spin-on-glasses (SOGs), polyimides, and DVS-BCB (divinylsiloxane-bis-benzocyclobutene) are the most common thermosetting materials used for bonding. Polyimides are known to be not very practical, because they result in large voids at the interface that are related to the creation of by-products during the bonding process. SOGs can be used, but require careful selection to avoid large film stresses leading to cracks [10]. DVS-BCB is a thermosetting polymer, which has been widely used because it has several properties that make it very suitable for use in bonding applications (e.g., see [11]). It exhibits almost no outgassing and very limited shrinkage. Following curing, it is resistant to most chemicals and shows a high bond strength (see [11] for comparison). It has very good planarizing properties, making it also possible to bond structured wafers. Table 5.1 shows its main properties. The main drawback for our present application is its very low thermal conductivity.

DVS-BCB has been commercialized by Dow Chemical. Depending on the temperature used, it cures completely in a few seconds at 320 °C, up to a few hours at 175 °C. Typical curing temperatures used in our research are around 250 °C. For a description of the full polymerization process, we refer the reader to the literature (e.g., see [13]). The glass transition temperature, T_g, after curing is around 350 °C.

While the use of DVS-BCB for bonding has been widely described in the literature, the bonding-layer thickness was mostly on the order of several micrometers – which, as discussed above, is not compatible with efficient coupling of light between a III–V laser and the underlying silicon. Therefore, research has focused on decreasing the bonding-layer thickness. Commercial DVS-BCB formulations give a layer thickness from 1 to 25 µm. By adding mesitylene a custom solution

Table 5.1 Main properties of DVS-BCB.

Optical properties	
Refractive index	1.54 at 1.55 μm
Optical loss	< 0.1 dB cm^{-1} at 1.55 μm
Mechanical properties	
Tensile modulus	2.9 GPa
Intrinsic stress	28 MPa
Tensile strength	89 MPa
Poisson ratio	0.34
Shrinkage after cure	0.05
Thermal properties	
Glass transition temperature	>350 °C
Thermal expansion coefficient	42 ppm K^{-1}
Thermal conductivity	0.29 W mK^{-1}
Other properties	
Moisture uptake	Very low

From [12].

allowing layer thicknesses down to a few tens of nanometers can be formulated. These form the basis for our bonding process.

The first step in the bonding process is cleaning of the respective wafer surfaces. While the adhesive bonding process can accommodate some nonplanarity, the requirements become more stringent with decreasing bonding-layer thickness. For good adhesion it is also important that organic contamination is removed. A typical cleaning process for the SOI wafers consists of a Piranha step, removing organics, and an SC1 solution step, removing both particles and organic materials. The InP die is typically "cleaned" by removing a pair of sacrificial InP/InGaAs layers, which lifts off contaminants and particles remaining on the surface.

Several variants of the bonding process have been developed over the past few years [14, 15]. One variant is outlined below and is illustrated in Figure 5.2. The process is carried out in a commercial wafer bonder.

1) Clean and dry wafer.
2) Spin-coat a mesitylene diluted DVS-BCB layer on the SOI substrate (50–70 nm target thickness).
3) Partially cure the DVS-BCB layers at 150 °C for 15 min.
4) Slowly cool down the substrate to room temperature.
5) Attach the III–V wafer or dies to the substrate at room temperature, either using vacuum tweezers or using a flip-chip machine for high-accuracy alignment, in the case of dies.
6) Load the bond fixture with the wafers into the bond chamber, after which the chamber is pumped-down to 10^{-3} mbar at room temperature.
7) Ramp up the temperature of the bond tool to 150 °C in the vacuum environment (ramp rate 5 °C min^{-1}).

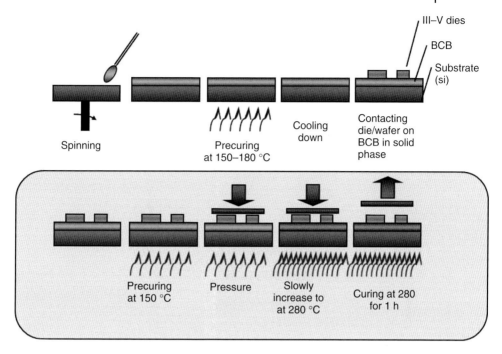

Figure 5.2 Adhesive bonding process. (Picture courtesy S. Keyvaninia.)

8) Apply a pressure of 0.3–1.5 MPa to the wafer stack, using a sheet of graphite between the top chuck and the wafer stack.
9) Ramp up the temperature of the bond tool to 280 °C in the vacuum environment (ramp rate 1.5 °C min^{-1}).
10) Release the bond force and retain temperature at 280 °C for 60 min in a pressurized chamber with nitrogen flow, followed by a natural cool down.

In [15] an alternative method was described whereby spacers are used between the dies and substrate, prior to loading into the bonding tool. In that case the substrates are brought into contact after evacuating the bonding chamber and heating up the substrates. The advantage of the so-called cold bonding method outlined in Figure 5.2 is that no dedicated spacers are needed, and dies can be brought into contact before moving them to the bonding chamber. Figure 5.3a shows an SOI substrate with four InP dies bonded on top (after substrate removal). The bonding-layer thickness was less than 50 nm. Figure 5.3b shows a cross-section illustrating how the adhesive bonding process is able to cope with nonplanar substrates.

5.2.2
Direct Bonding

In a direct bonding process, substrates are held together through strong interfacial bonds, without an intermediate adhesive layer. As already mentioned above, most

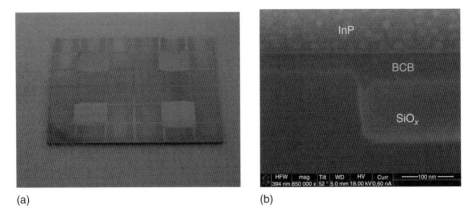

Figure 5.3 (a) InP dies bonded to SOI after chemical substrate removal and (b) cross-section of InP-die bonded on SOI waveguide. (Pictures courtesy of S. Keyvaninia.)

SOI wafers now available are fabricated through such a process, demonstrating that this is a manufacturable and potentially high-yield process.

However, when bonding III–V dies on a silicon-based substrate, several new problems arise. Often the III–V surface quality is inferior to that of commercial silicon wafers. This possible problem can be overcome through careful process control. Edge roughness from dicing or cleaving the dies may also hamper the bonding process. More fundamental is the difference in thermal expansion coefficients between silicon and InP. The conventional direct bonding process includes a high temperature anneal stage (>600 °C), which is not allowable when bonding different types of material, where preferably the temperature is kept at a much lower level (<400 °C or even <250 °C). Therefore plasma assisted bonding processes have been developed [13, 16] that make it possible to keep the temperature at an acceptable level. One such process, from [13] is illustrated in Figure 5.4. The starting point is again a thorough cleaning step. Next, an O_2 plasma step renders the surfaces hydrophilic by growing a thin layer of highly reactive native oxide. Then the surfaces are brought into contact at room temperature and initially held together through weak Van der Waals forces. Finally the substrates are heated and strong covalent bonds start to form. In this process, gaseous by-products of H_2O and H_2 are formed and can lead to interfacial voids. Removing these by-products is therefore key to obtaining successful bonding. Several strategies toward achieving this have been demonstrated in the literature. The authors of [17] use an amorphous SiO_2 layer deposited on one or both substrates to absorb the gaseous by-products. This method has the advantage that it is generic: the same type of surface preparation processes can be used independent of the exact substrates involved in the process. The drawback is that it requires a few extra preparation steps, possibly including a CMP step (chemical-mechanical polishing) and that it imposes a minimum intermediate layer thickness from a few tens up to hundreds of nanometers, which may also be undesirable from an optical or thermal point of view.

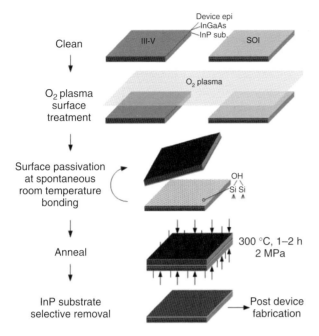

Figure 5.4 O$_2$-assisted low temperature bonding of III–V to Si. (After [13].) (Please find a color version of this figure on the color plates.)

The alternative approach is to produce evacuation channels that run over the wafer, which allow the gases to escape, and use horizontal channels running over the entire length of the substrate. However, it is difficult to envisage provision of such outgassing channels with sufficient density, without restricting design flexibility. For these reasons, Liang et al. [8, 18] developed what they call vertical outgassing channels (VOCs). These channels are etched through the top silicon device layer and connect to the SiO$_2$ buried oxide. After mating the two substrates, the gaseous by-products and possibly trapped air can then migrate toward the VOCs and eventually be absorbed in the buried oxide layer. The size of the VOCs is typically on the order of 0.3–3 μm. More important is the spacing between the VOCs, which should be below 50 μm to guarantee void-free bonding. The results of this bonding process are shown in Figure 5.5a.

5.2.3
Substrate Removal

After bonding the InP wafer or dies, the substrate needs to be removed, which can be achieved by using a complete wet etching process, stopping on a sacrificial InGaAs layer embedded below the active layers. High etch rates are obtained by using pure HCl, but to avoid the formation of gas bubbles, a mixture of HCl and H$_2$O is often used. Although straightforward in die bonding processes, the wet etching process has the drawback that ramps are often formed on both sides of

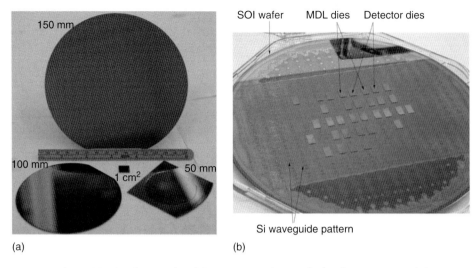

Figure 5.5 (a) Photographs of thin InP epi-layers of 1 cm², 50, 100, and 150 mm in diameter bonded onto an SOI substrate. (From [13].) (b) Picture of multiple die-to-wafer bonding process, including dies with different epitaxial layers (MDL = microdisk laser). (After [19], © IEEE 2009.)

the bonded substrate, because HCl does not etch an exposed (01-1) plane [13]. One approach for reducing the size of these ramps is to start the substrate removal with a mechanical substrate removal step, leaving only a few tens of micrometers of the bonded dies. Afterwards a wet chemical etch is used to remove the remainder of the substrate.

5.3
Heterogeneously Integrated Edge-Emitting Laser Diodes

While the micrometer-scale lasers discussed in the next section are small-footprint devices with output-power levels compatible with intra-chip optical interconnect applications, these devices are less suitable for off-chip communication. Higher-output power devices are typically required for that purpose. In this section we describe the realization of III–V-on-silicon light sources that are suitable for off-chip optical communication. Both DVS-BCB bonded laser diodes and directly bonded laser diodes are described. In these devices, thin layers of III–V semiconductors are bonded to silicon. The laser cavity gets its gain from the III–V layers, but couples its output light into a silicon waveguide. Often part of the cavity structure is implemented by means of patterning in silicon, thereby taking advantage of the resolution and accuracy of lithography tools in CMOS fabs. In that sense, these hybrid III–V/silicon lasers take the best of both worlds.

In spite of rapid progress in this field since about 2006 [20], the design and fabrication of hybrid silicon lasers present specific challenges and trade-offs.

There are many choices to be made in terms of the cavity structure, the optical coupling between the silicon waveguide and the III–V waveguide, and the technological approach. A relatively large variety of approaches has been reported over the years. In particular, in the design a choice needs to be made as to whether the optical mode is predominantly confined in the silicon waveguide or in the III–V overlay. Both options have their advantages and disadvantages and they are not even mutually exclusive, given the fact that, for example, adiabatic tapers inside the laser cavity can be used to change the amount of optical confinement in the respective layers. Confining the optical mode predominantly to the silicon waveguide layer has the advantage of making the coupling to a passive silicon waveguide straightforward. Moreover, wavelength-selective features can easily be defined in the silicon waveguide layer using CMOS fabrication techniques, which provides an accurate mechanism for controlling the emission wavelength of the laser. However, a drawback of this silicon-confined approach is that only a small fraction of the optical mode interacts with the gain material, resulting in the need for longer laser cavities and higher power consumption. On the other side of the design space, one can find III–V-on-silicon laser geometries where the optical mode is completely confined in the III–V waveguide layer and where the optical cavity is defined in the III–V semiconductor. A representative example of such a laser geometry is the III–V microdisk laser heterogeneously integrated onto a silicon waveguide layer, which is discussed in Section 5.4. This laser geometry allows the realization of ultracompact light sources, given the high confinement of the optical mode in the gain material, while the coupling with the silicon waveguide layer is achieved through an intracavity evanescent coupling scheme. The emission wavelength is determined by the diameter of the III–V microdisk, which makes it less straightforward to control the emission wavelength.

Many different types of heterogeneously integrated lasers have been demonstrated in recent years. In the subsequent sections we elaborate on the different device types that have been reported in the literature.

Besides the approach based on the bonding of III–V semiconductor epitaxial layer stacks with a silicon waveguide circuit, monolithic integration techniques have been developed in recent years, including the hetero-epitaxial growth of GaSb-based epitaxial layer stacks on silicon with sufficiently low defect density for laser operation [21] and the realization of Ge-on-silicon lasers [22]. While these devices are currently not yet on a par with the heterogeneously integrated laser devices, it will be definitely interesting to see how these technologies will evolve.

5.3.1
Fabry-Perot Lasers

The simplest device geometry is the Fabry–Perot laser, where two facets define the laser cavity. Even for this simple device geometry, many coupling schemes to the underlying silicon waveguide circuit have been developed. In [23], a spot-size converter external to the cavity was used to couple the edge-emitted light efficiently into the underlying silicon waveguide circuit. In [15, 24], a hybrid optical mode structure was used, with predominant confinement of light in the silicon waveguide layer. In [25, 26], an intracavity spot-size converter was used to transform the optical mode adiabatically from being strongly confined in the III–V gain material to a mode that is completely confined in the silicon waveguide layer. These different coupling concepts are schematically outlined in Figure 5.6.

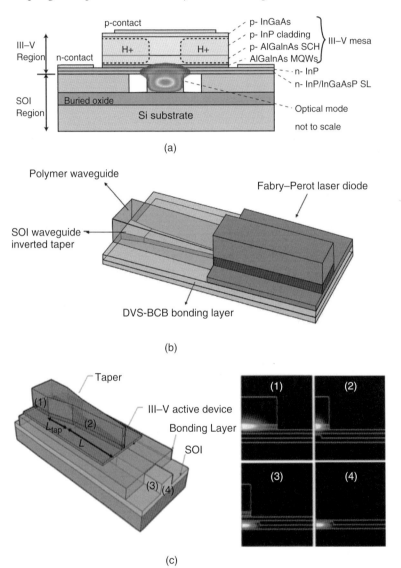

Figure 5.6 Overview of heterogeneously integrated Fabry–Perot type laser structures: (a) evanescent hybrid laser, (b) inverted taper coupler outside the laser cavity, and (c) intracavity spot-size converter. (Please find a color version of this figure on the color plates.)

The fabrication is carried out using the wafer-to-wafer or die-to-wafer bonding techniques described earlier. The definition of the facets is made either by dicing and polishing or by dry etching. Dry etching of the facets has the advantage of being CMOS-compatible and does not restrict the length of the silicon waveguide to the length of the laser cavity. A proton implantation on both sides of the mesa can be used to prevent lateral current leakage [15, 24].

5.3.2
Mode-Locked Lasers

In principle a Fabry–Perot laser could be used as a multiwavelength light source, were it not for the fact that instability in the power levels, phases, and wavelengths of the individual modes make this approach unfeasible. However, these instabilities can be reduced significantly in a MLL, which emits an array of continuous-wave (CW) wavelengths that are phase locked to each other. Stability is especially improved if optical injection locking is implemented [13].

Semiconductor MLLs can be made by introducing a small absorbing section into a standard laser cavity. Fabrication of these lasers requires only one additional process step to isolate a section of the active material from the rest. This isolated section can then be reverse-biased to act as a saturable absorber (SA), while the rest of the active material is forward-biased as a gain section to generate light. With the proper choices of length for the two regions and the proper bias conditions, it becomes favorable for all laser modes to interfere constructively when passing through the absorber section. This means that all laser modes are in-phase with each other and, as a result, pulses are generated in the cavity. The mode spacing and pulse repetition rate are determined by the cavity length of the MLL.

A typical III–V/SOI MLL is depicted schematically in Figure 5.7 [27]. It has a 4 mm-long cavity composed of an 80 μm-long SA section placed next to the backside mirror and a large gain section that extends along the rest of the cavity. Laser mirrors are formed by the end facets, each of which provides approximately 33% power reflectivity. The SA section is made up of the same active material as the gain section, but it is electrically isolated from the gain section via a proton implantation between the sections. Based on this device geometry, several demonstrations of heterogeneously integrated MLLs have been made. Repetition frequencies up to 40 GHz have been demonstrated.

5.3.3
Racetrack Resonator Lasers

Racetrack resonator lasers have been presented as a solution for defining laser cavities without dicing and facet polishing. They can be considered as a scaled-up version of the microdisk lasers that will be discussed in Section 5.4 of this chapter. A schematic view of the device is shown in Figure 5.8. The optical

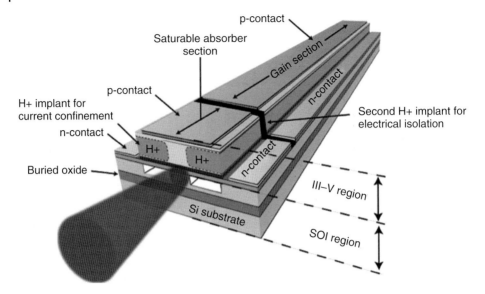

Figure 5.7 Typical mode-locked laser geometry. (After [27], © OSA 2007.)

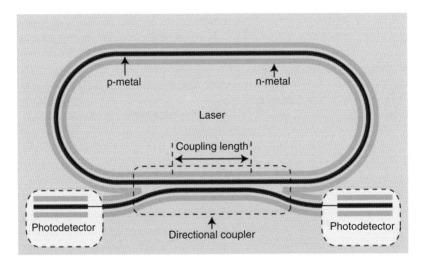

Figure 5.8 The layout of the racetrack laser resonator and the photodetectors. (After [28], © OSA 2007.)

design is based on the evanescent coupling scheme developed for Fabry–Perot lasers, where the optical mode is predominantly confined to the silicon waveguide layer. The light is coupled out of the racetrack region by a directional coupler placed on the bottom arm [28]. Still using this cavity geometry, a mode-locked racetrack resonator laser (7 ps pulses at 30 GHz repetition-rate) was also demonstrated [29].

5.3.4
DFB and Tunable Lasers

5.3.4.1 Distributed Feedback Lasers

DFB lasers are lasers where the whole resonator consists of a periodic structure that has a small corrugation level and acts as a distributed reflector in the wavelength range of possible laser action, and also contains a gain medium. Typically, the periodic structure is realized with a specific phase shift in its middle, to support single-mode operation. DFB lasers are attractive for single-wavelength emission as they have a single longitudinal mode output and their short cavity lengths allow for low threshold currents while still producing output powers in the milliwatt regime. The first-order distributed gratings can be defined in several ways. Either the grating can be incorporated directly into the III–V semiconductor film [30], as is the case in classical III–V semiconductor DFB lasers, or it can be incorporated into the silicon waveguide layer, where it can be defined using either state-of-the-art deep UV lithographic techniques or electron-beam lithography (EBL) [13]. Both alternative types of structures are represented schematically in Figure 5.9.

5.3.4.2 Distributed Bragg Reflector Lasers

DBR lasers consist of an active gain region surrounded by two passive DBRs, which provide the wavelength-selective feedback. In contrast to DFB lasers, the longer lasing cavity that is typical of DBR lasers reduces their thermal impedance and consequently enhances their output-power levels. An example of such a heterogeneous III–V/silicon DBR laser implementation is shown in Figure 5.10. The device layout includes two passive Bragg-reflector mirrors placed 600 μm apart to form an optical cavity. Two 80 μm-long tapers sandwich the 440 μm-long silicon evanescent gain region. The back and front mirror lengths are 300 and 100 μm, respectively [13].

5.3.4.3 Sampled Grating DBR Lasers

A tunable DBR laser has also been demonstrated. The device is fabricated using quantum-well intermixing, where atomic-scale disordering of the quantum well

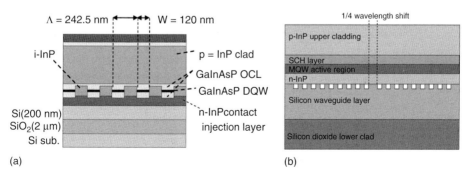

Figure 5.9 III–V DFB laser bonded on silicon, (a) with the DFB grating located in the active layer and (b) the DFB grating is located in the silicon waveguide layer.

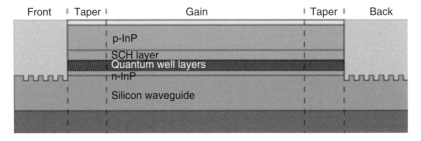

Figure 5.10 DBR silicon evanescent laser layout.

active region is performed before bonding, in order to shift the bandgap of the as-grown III–V material. The device utilizes two material bandgaps that have photoluminescence peaks at 1520 and 1440 nm. Current injection into the different III–V regions provides tuning over a range of 13 nm [31].

5.3.4.4 Ring-Resonator Based Tunable Laser

By using silicon wavelength-selective feedback elements that can be tuned, a (widely) tunable, heterogeneously integrated, III–V/silicon laser can be realized. The schematic of a particular implementation of such a device is shown in Figure 5.11. Here the laser cavity is formed by two broadband first-order Bragg reflectors, while the intracavity ring resonator serves as an intracavity etalon. The resonance wavelengths of the ring can be tuned by implementing a heater on top of the silicon waveguide circuit. The coupling between the silicon waveguide circuit and the III–V semiconductor amplifier is realized by using a spot-size converter, as shown in Figure 5.6c.

In this way, a single-wavelength tunable laser (>30 dB side-mode suppression ratio), with a 9 nm tuning range was realized recently [32].

5.3.4.5 Heterogeneously Integrated Multiwavelength Laser

Besides using MLLs as multiwavelength light sources, other options can be considered that provide a higher optical output power per laser line. A multiwavelength laser can be realized by interconnecting an array of SOAs to an arrayed waveguide grating wavelength multiplexer, inside a waveguide facet-defined laser resonator.

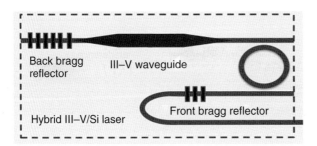

Figure 5.11 Heterogeneously integrated tunable laser. (After [32], © OSA 2012.)

While this is a classical approach in III–V semiconductor laser structures, it has only recently been demonstrated on a silicon platform [33].

5.3.5
Proposed Novel Laser Architectures

The evanescent laser designs rely on a tradeoff between modal gain and coupling efficiency into the SOI waveguide, as discussed before. This tradeoff requires the use of a sufficiently thin bonding layer (below 100 nm) between the III–V stack and the SOI waveguide, as the modal overlap with the neighboring waveguide decreases exponentially with increasing distance between the waveguides. Nevertheless, thicker bonding layers would be beneficial to the integration process as they would facilitate the encapsulation of the SOI waveguide. Furthermore, in the case of direct bonding, a thick oxide layer limits the formation of interface defects and, in the case of adhesive bonding, a thick polymer layer can better accommodate the wafer topography.

5.3.5.1 Exchange Bragg Coupling Laser Structure
Contradirectional coupling between the III–V and the SOI eigenmodes (so-called exchange Bragg coupling) has been proposed to allow for DFB, as well as for a more equal distribution of the laser field between the III–V and SOI waveguides [34].

5.3.5.2 Resonant Mirrors
To obtain a high-level, narrowband reflection using a short grating structure, one can opt for a laser design based on resonant mirrors [35]. Again, the laser mode inside the gain section is completely confined to a III–V mesa-shaped waveguide. At the ends of that gain section, silicon gratings underneath the III–V waveguide will provide optical feedback, as shown in Figure 5.12. The silicon grating acts as a periodic perturbation of the III–V waveguide, but because the overlap of the III–V waveguide mode with the grating is weak, this perturbation is typically very small, yielding very long grating reflectors.

To solve this problem, a quarter-wavelength phase-shifting section is introduced near the center of the silicon grating, turning it into a resonant cavity at the

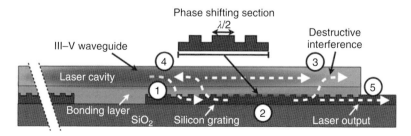

Figure 5.12 Side view of a hybrid laser with resonant mirrors.

Bragg wavelength. A limited fraction of the light in the III–V waveguide couples to the silicon grating resonator (Figure 5.12, 1) and power will build up inside that grating cavity (Figure 5.12, 2). Eventually, a significant amount of light will couple back into the III–V waveguide. The light that couples codirectionally to the light incident from the III–V laser cavity will interfere destructively with the latter (Figure 5.12, 3) and result in zero overall transmission through the III–V waveguide. The counter-directionally coupled light (Figure 5.12, 4) will propagate back into the laser cavity and hence provide optical feedback. Because this reflection mechanism is based on a resonance phenomenon inside the silicon grating, its spectral response will be narrow-banded.

The resonant mirror also offers an elegant mechanism for coupling of the laser output to a silicon waveguide (Figure 5.12, 5). By engineering the position of the quarter-wavelength phase-shifting section, one can tune the amount of optical power that leaks from the silicon grating cavity into an output waveguide.

The advantage of this approach is that it does not require very thin bonding layers, as the coupling from the III–V waveguide to the silicon grating cavity can be very small, but because the coupling is distributed along the length of the grating, the III–V waveguide mode has to be phase-matched to the resonant grating mode for the reflector to work. This requires careful design of both waveguide geometries.

5.3.6
Heat Sinking Strategies for Heterogeneously Integrated Lasers

Compared with III–V semiconductor lasers, heterogeneously integrated laser diodes have the advantage that they can leverage from the well-developed silicon processing infrastructure for the fabrication of the passive and electro-optic waveguide circuits – either defining the laser cavity or connecting the heterogeneously integrated lasers to the outside world. On the other hand, heterogeneously integrated lasers are defined on an SOI waveguide circuit, which inherently provides less efficient heat sinking compared to the same laser realized on its native III–V semiconductor substrate. Therefore, developing an efficient heat-sinking strategy is of paramount importance for high power and elevated temperature operation of these devices. Generally it is considered that adhesively bonded laser diodes have a worse thermal resistance in comparison with directly bonded devices, because of the low thermal conductivity of the adhesive bonding layer. However, in practice it turns out that, in particular, a buried SiO_2 layer contributes substantially to the overall thermal impedance. In Figure 5.13, the optical power versus injection current curves are plotted for an adhesively bonded Fabry–Perot laser. This curve clearly illustrates that adhesively bonded lasers can deliver milliwatt optical output-power levels up to 60 °C, in a continuous-wave operation [25].

Several strategies can be followed to alleviate the self-heating issues of bonded lasers. A first strategy is to mount the lasers up-side down onto a ceramic heat spreader. While this provides the best heat sinking, it makes the electrical and optical interfacing to the device more difficult. Instead of using a ceramic heat spreader, the SiO_2 buried oxide layer can be replaced by a more thermally conductive material,

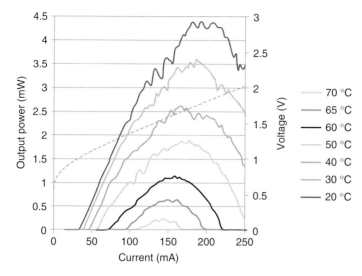

Figure 5.13 Optical output power versus drive current for an adhesively bonded Fabry–Perot laser (950 μm active region length). (After [25], © OSA 2012.)

such as nanocrystalline diamond, deposited using a chemical vapor deposition technique [36]. When relying on the classical SOI waveguide substrates, thermal vias can be implemented to provide a still more efficient heat-sinking route. Two strategies have been followed along this direction: either the p-type metal contact is used as a thermal via, in order to sink the heat to the silicon substrate [13], or poly-silicon shunts are utilized [37]. These heat sink strategies are depicted in Figure 5.14.

5.4
Microdisk and Microring Lasers

Most of the microlasers that have been reported on in the literature are either microdisk lasers or photonic-crystal lasers. While photonic-crystal lasers theoretically allow much more compact devices than microdisk lasers, their electrical pumping is much more difficult. Many authors have reported on optically pumped photonic-crystal lasers, but only a few electrically pumped photonic-crystal lasers have been presented. In [38], an ultralow threshold photonic-crystal laser (with a threshold of 287 nA at 150 K) is described, but it was only operated at low temperatures up to 150 K. More information on photonic-crystal lasers, both optically and electrically pumped, is given in other chapters of this book.

Microdisk lasers on the other hand were demonstrated as early as 1992 [39]. In [40], an electrically injected microdisk laser with 2 μm diameter was experimentally demonstrated, under pulsed operation, with a threshold current of 0.2 mA. The whispering gallery mode (WGM) in such lasers makes it possible to implement

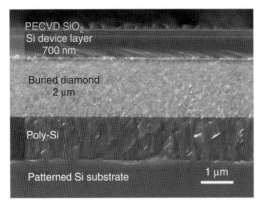
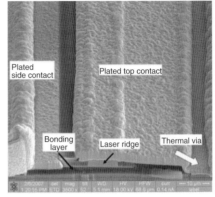

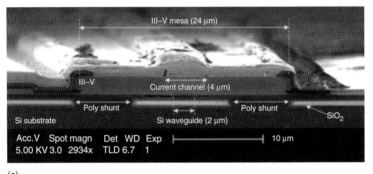

Figure 5.14 Strategies to alleviate the self-heating of heterogeneously integrated laser diodes: (a) silicon-on-diamond substrate, (b) metal thermal vias, and (c) poly-silicon thermal shunts.

electrical contacting away from the modal energy distribution. Especially for the case of membrane lasers, the contact metallization should then not cause excessive losses.

Ring lasers, which are similar to disk lasers, with rather large radius have been demonstrated to have significant advantages over other lasers. In the EU-project IOLOS (www.iolos.org), unidirectional operation was demonstrated, as well as the possibility of switching between clockwise (CW) and counter clockwise (CCW) by means of injected optical pulses. The ring lasers had typical diameters of several hundred micrometers and threshold currents of several tens of milliamperes [41]. In addition to applications as all-optical flip-flops, the ring geometry also allows other optical signal processing operations such as signal regeneration and all-optical label swapping [42, 43]. However, the large dimensions of the InP ring lasers limit the maximum switching speed in these applications. Ring lasers with smaller diameters have meanwhile been fabricated, and some recent, previously unpublished, results are described in Chapter 6.

Finally, it was thought for many years that ring lasers generally have much lower sensitivity to external reflections [44], although this was never really proven. We show later in this paragraph that ring and disk lasers, operating in a unidirectional regime, are indeed less sensitive to external reflections. Moreover, if biased just below threshold and for relatively low optical powers, the ring/disk lasers can even be used as isolators, giving over 10 dB isolation, at the resonance wavelengths.

5.4.1
Design of Microdisk Lasers

In order to obtain a small threshold current (and low-power consumption) for a small-footprint laser, it is necessary to minimize the different losses. Microdisk lasers with a small diameter may exhibit relatively high bending losses and relatively high scattering losses, due to the sidewall surface roughness that is experienced by the WGM. This sidewall surface roughness is also the main factor responsible for the coupling between CW and CCW modes and must be minimized in order to obtain a bistable operation. The calculated bending losses are shown as a function of the disk radius in Figure 5.15 for the fundamental transverse electric (TE)-polarized WGM and for different bottom contact layer thicknesses. Our microdisk lasers are surrounded by BCB; this allows isolation/separation of the n- and p-type contacts. Further reductions in the bending loss are possible if the surrounding BCB is replaced by a polymer with a lower refractive index.

The specific thin membrane structure that is used in the heterogeneous integration of the microdisks can be an additional cause of loss. The top contact is very

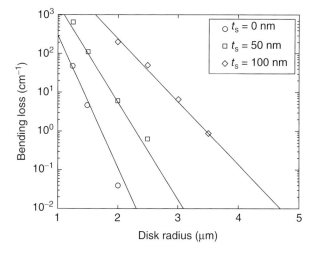

Figure 5.15 Calculated bending losses versus disk radius for the fundamental TE-polarized whispering gallery mode, for thin microdisks, and for different bottom contact layer thicknesses t_s. The curves represent eigenmode-expansion calculations, whereas the markers show the results of 2D-FDTD modeling. (After [43].)

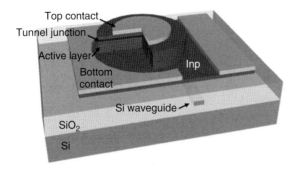

Figure 5.16 Schematic representation of the heterogeneous SOI-integrated microdisk laser. (Please find a color version of this figure on the color plates.)

close to the active layer. To avoid excessive optical losses due to the top contact, care must be taken to avoid, as much as possible, overlap between the top contact and the WGM. This can be achieved by placing the top contact centrally, as shown in Figure 5.16. The thin membrane structure also implies that the p++ contact layer, with its large intervalence band absorption (IVBA), is very close to the active layer. To avoid the large loss caused by IVBA, tunnel junctions are preferred for the injection of holes [43].

The heterogeneous integration of the microdisk on SOI circuits via BCB or silica layers limits the possibilities of heat sinking significantly and heating remains a major problem with these lasers. The first fabricated devices therefore only operated under pulsed current injection or showed a roll-off in the $L-I$ characteristic at relatively low CW currents. In more recent designs [45], extra heat sinking was provided by making the gold layer of the top contact extra thick (600 nm).

There have also been experiments using BCB doped with diamond nanoparticles to give the bonding layer a better thermal conductivity and thus provide better heat sinking downwards [46].

Other potential measures for improving the heat sinking are to deposit an oxide on the InP die and use a very thin BCB layer (of e.g., 50 nm). This gives a bonding yield of close to 100% and the oxide thickness can be controlled accurately. A somewhat thicker oxide (of e.g., 200 nm) in combination with a very thin BCB layer still gives much better heat sinking that a purely BCB bonding layer. So far, we have mainly used silica as the insulating oxide, but alumina has higher thermal conductivity and we have recently begun to deposit alumina on the InP die. Finally, the heat sinking also improves with the thickness of the n-type bottom contact layer, but this improvement comes at the expense of modal leakage.

Coupling between the WGM of the microdisk and the straight silicon wire waveguide occurs through evanescent coupling. The degree of coupling depends on the thickness and composition of the bonding layer. Some simulation results are shown in Figure 5.17 for the coupling between a 500 nm wide SOI wire waveguide and 7.5 μm diameter microdisks with thickness of 0.5 μm (thin) or 1 μm (thick).

So far, current injection into the active layer has been performed with the help of tunnel junctions. Heavily doped p-type contact layers would indeed cause excessive

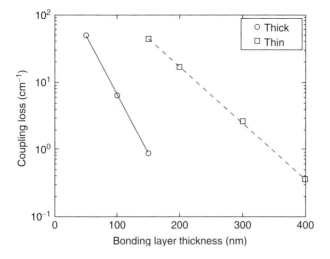

Figure 5.17 Coupling loss versus bonding-layer thickness for thin and thick microdisks with 7.5 μm diameter and a 500 nm wide SOI wire waveguide. (After [43].)

losses for the WGM. Nevertheless, tunnel junctions lead to a relatively high series resistance. It should in principle also be possible to fabricate efficient microdisk lasers without tunnel junctions, for example, by etching away the highly absorbent p-type contact layers at the microdisk edges, where the WGM overlaps significantly with the contact layer. Since part of the injected current produces unnecessary carriers in the center of the disk (where no gain is needed), a higher efficiency could be obtained by etching a part of the disk center (and, for instance, filling it with BCB).

5.4.2
Static Operation

Making use of tunnel junctions in the thin (580 nm thick) InP epi-stack, with a thick gold layer on the top contact for heat sinking and minimizing the scattering losses due to sidewall surface roughness, has led to microdisk lasers with threshold currents just below 0.5 mA and with maximum output-power levels of several tens of microwatts. Typical L–I and V–I characteristics for a 7.5 μm diameter microdisk laser are shown in Figure 5.18a. The power is coupled out from the silicon wires through surface grating couplers that have an efficiency of 21% at 1554 nm and 30% at 1584 nm.

The small dimensions of this microdisk laser imply a large mode spacing or free spectral range (FSR). The optical spectrum for the 7.5 μm diameter microdisk laser is shown in Figure 5.18b. The fundamental mode has an FSR of about 30 nm and the side-mode rejection ratio is 35 dB. However, the small dimensions also imply that the exact lasing wavelength is very sensitive to small variations in the disk diameter, which makes the exact lasing wavelength unpredictable. For several

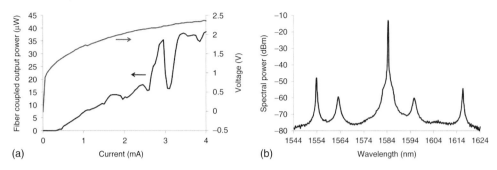

Figure 5.18 (a) Output power and voltage versus current for a 7.5 μm diameter microdisk laser and (b) optical spectrum of the same microdisk laser.

applications, it is important to have a predefined lasing wavelength, for example, when several microdisk lasers have to be locked to each other. To compensate for the unpredictability of the lasing wavelength, thermal tuning can be implemented using integrated ring heaters, as shown in Figure 5.19. The gap between the ring and the central disk is a critical parameter in this situation; it should be small enough to allow for efficient heating and large enough to avoid optical leakage of the cavity WGM to the heater electrode. An optimum value for the gap appears to be between 1 and 2 μm. Experimentally (see Figure 5.19b), a tuning rate of 0.31 nm mW^{-1} was measured for this type of thermal tuning. A 2.2 nm shift could be realized without severe impact on the laser output power.

The evanescent coupling of microdisk lasers to straight Si waveguide stripes is ideal for the implementation of laser arrays and the realization of, for example, multiwavelength transmitters. Microdisk lasers with different diameters have different emission wavelengths (which can be tuned thermally as explained above) and can thus be used to fabricate WDM sources, for example, for application to optical interconnects. Figure 5.20 shows a picture of a fabricated multiwavelength laser,

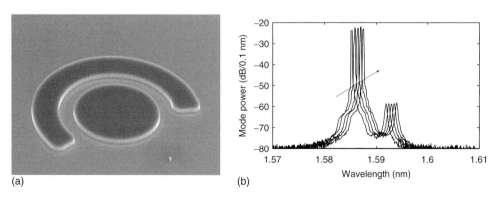

Figure 5.19 (a) SEM image of a microdisk cavity with an integrated ring heater and (b) tuning of the lasing wavelength for heating power levels of 0, 1.6, 3.5, and 7.1 mW, respectively. (After [47], © IEEE 2008.)

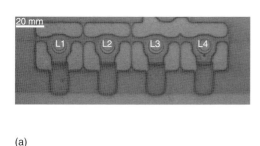
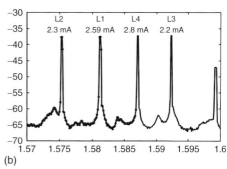

Figure 5.20 (a) Fabricated multiwavelength laser (before metalization) and (b) emission spectrum of a multiwavelength laser with 6 nm channel spacing. (After [47], © IEEE 2008.)

as well as the emission spectrum of a multiwavelength laser with 6 nm channel spacing. The bias currents of the microdisk lasers have been adjusted so as to provide the same output power in each channel [47].

One problem under static operation, and for use as transmitters in optical interconnect, may be that the lasing can be either in the clockwise or the CCW direction or even bidirectional. This problem could in principle be solved by terminating one side of the bus (silicon) waveguide with a Bragg reflector. Such a reflector would force the microdisk laser to operate and emit in the direction of the Bragg reflection. First experiments using such configurations tend to confirm this behavior, but owing to the use of a very strong Bragg reflection, the results obtained are not fully conclusive.

5.4.3
Dynamic Operation and Switching

5.4.3.1 Direct Modulation

Injection-locked ring lasers can provide wide direct-modulation bandwidths [48, 49]. Modulation of the master laser allowed a 3 dB bandwidth over 40 GHz in [49]. Theoretically, direct modulation of a microdisk laser under CW injection locking should also enable the achievement of such wide modulation bandwidths. Modulation at 20 Gbps has been demonstrated at only 1 mA bias and for 190 mV data signals, thereby corresponding to 50 fJ per bit [50].

However, even without injection locking, small-signal direct-modulation bandwidths of 11 GHz have been obtained with non-optimized microdisk lasers biased at current levels of a few milliamperes [51].

5.4.3.2 All-Optical Set-Reset Flip-Flop

As has been shown for InP ring lasers, microdisk lasers can be operated in a unidirectional mode (either CW or CCW) if the coupling between the CW and CCW modes, because of scattering from sidewall surface roughness or other reflections, is sufficiently small and if the gain suppression is sufficiently large to overcome this coupling [52]. The unidirectional regime typically starts above the threshold at a current that depends on the coupling between the CW and CCW modes. This coupling obviously favors bidirectional operation. However, the gain suppression is such that different gain levels occur for the clockwise and CCW modes, as soon as the clockwise and CCW modes have different power levels and this effect favors unidirectional operation. More specifically, the gain for the clockwise mode can be expressed as $g_{CW} = g_0(N)(1 - \varepsilon_S P_{CW} - \varepsilon_C P_{CCW})$ and that for the CCW mode as $g_{CCW} = g_0(N)(1 - \varepsilon_S P_{CCW} - \varepsilon_C P_{CW})$ with the self gain suppression ε_S being twice the cross gain suppression ε_C, that is, $\varepsilon_S = 2\varepsilon_C$. The power levels P_{CW} and P_{CCW} of clockwise and CCW modes and, therefore, the gain suppressions increase with increasing current, and unidirectional operation is possible for those current levels for which the gain suppression $\varepsilon_S P$ is large enough to counter the coupling between the CW and CCW modes.

Once a bistable, unidirectional regime is achieved, switching between CW and CCW modes can be performed with the injection of set and reset optical pulses. Using microdisks of 7.5 µm diameter, we have succeeded in switching the emission mode with optical pulse energies of only 1.8 fJ [45]. The disks were biased at 3.5 mA, giving a total operating power level of less than 6 mW and the switching time was found to be 60 ps. The dynamic, high-speed extinction ratio was 11 dB. Unidirectional behavior and switching potential for these lasers were demonstrated by optimizing the etching process (carried out at the Technical University of Eindhoven) in laser fabrication, giving very low sidewall surface roughness, as can be seen in the secondary electron microscope (SEM) image of Figure 5.21. Further reduction of the coupling between CW and CCW modes, and thus a lower threshold for unidirectional behavior, is probably still possible by attenuating and, ideally, eliminating parasitic reflections from the grating couplers, fiber facets, and so on. This threshold can also be reduced by optimizing the active layer material for large gain suppression.

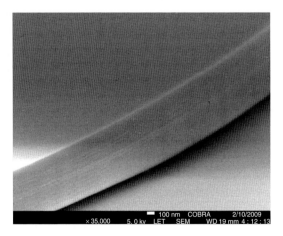

Figure 5.21 SEM image showing the sidewall surface roughness (on a micrometer scale, e.g., the thickness of the disk is about 0.6 µm). (After [44].)

5.4.3.3 Gating and Wavelength Conversion

Microdisk lasers biased either below or above threshold are also useful for performing wavelength conversion and gating/time-demultiplexing. Both operations are based on the same principle: a high power pump beam (e.g., as emitted from a short pulse laser source) operating at one microdisk resonance wavelength and carrying the signal information, causes a modulation of the effective index of the microdisk and therefore of the resonance wavelengths. A weaker CW probe beam at a second microdisk resonance wavelength is then intensity modulated by the resonance wavelength modulation. A typical transmission characteristic of an unbiased microdisk resonator is shown in Figure 5.22. The pump beam is typically located at the shorter wavelength resonance (with larger absorption) and the probe beam at the longer wavelength resonance (with higher Q-factor).

Gating/wavelength conversion/demultiplexing has recently been demonstrated at 10 Gbps using unbiased microdisk lasers. The time trace and eye diagram are shown in Figure 5.23. The pump and probe power levels were 1.5 and 0.375 mW, respectively. The rise and fall times indicate that data rates of at least 20 Gbps should be feasible [53]. The rise and fall times are determined by the effective carrier lifetime. They can be further reduced, for example, by ion-implantation but also by adding a DC beam (holding beam) to the short pulses. Such a holding beam would reduce the effective carrier lifetime after the pump pulse. The gating/wavelength conversion scheme also regenerates the incoming signal, that is, the extinction ratio at the output can be considerably larger than at the input.

Using the same principles as for the gating process, the unbiased microdisks can also be applied for the realization of time–domain demultiplexing and signal format conversion, in particular from nonreturn-to-zero, on-off-keying (NRZ-OOK) to return-to-zero, on-off-keying (RZ-OOK) [54]. Signal format conversion is shown

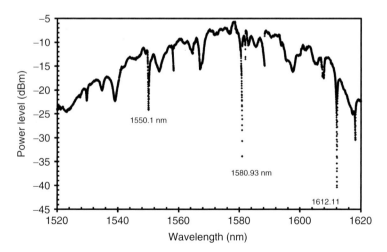

Figure 5.22 Transmission spectrum of an unbiased microdisk, showing three resonances. (After [53], © OSA 2011.)

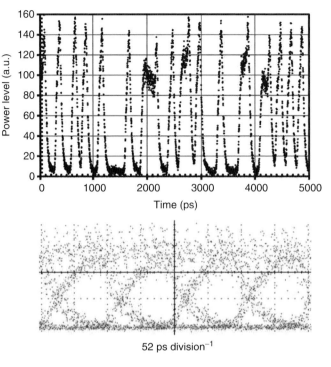

Figure 5.23 Time trace and eye diagram after all-optical wavelength conversion of a 10 Gbps pseudo-random nonreturn-to-zero bit sequence. (After [53], © OSA 2011.)

in Figure 5.24 at 5 Gbps, in which (a) shows the NRZ sequence 0011001100, (b) is a clock signal at 5 GHz, and (c) is the resulting RZ sequence 0011001100.

5.4.3.4 Narrowband Optical Isolation and Reflection Sensitivity of Microdisk Lasers

Stable unidirectional behavior in microdisk lasers is only possible because the gain suppression is twice as large for the non-lasing direction as for the lasing direction. Making use of this property, one can also use microdisk resonators (i.e., lasers just below threshold) as narrowband optical isolators. For example, in the configuration shown in Figure 5.25a, with the input signal injected from the lower waveguide, a reflection level of 10% or less at the output (upper waveguide) can be isolated by more than 10–20 dB, depending on the input power level. Figure 5.25b gives the isolation-ratio obtained numerically, as a function of the input power level, for different coupling between the straight waveguides and the microdisk. For the typically low output-power levels of microdisk lasers (below 100 μW), isolation of over 10 dB would be obtained. Higher-output power levels deplete the carrier density and gain and therefore dampen the resonance for the transmission between input and drop port, so that the influence of gain suppression becomes weak. However, this kind of isolation still needs to be demonstrated experimentally.

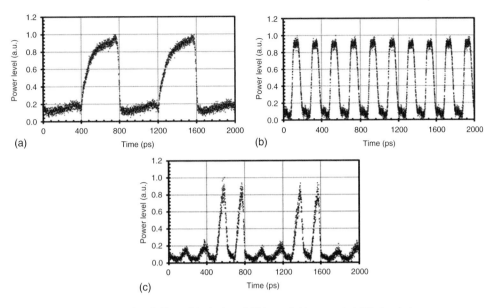

Figure 5.24 (a) NRZ signal at 5 Gbps, (b) clock at 5 GHz, and (c) converted RZ signal at 5 Gbps. (After [54], © OSA 2011.)

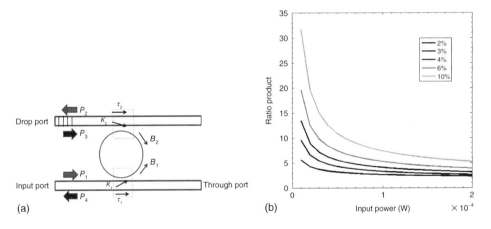

Figure 5.25 (a) Configuration used for the optical isolation and (b) simulated isolation ratio in linear scale. The microdisk laser is biased just below the threshold current.

For the same reason that microdisk resonators can act as narrowband optical isolators, microdisk lasers (and semiconductor ring lasers) in general are much less sensitive to external reflections than other laser diodes, especially if they are operated at high power levels in a unidirectional mode [55]. Unidirectional operation can be induced by connecting the microdisk lasers on one side with a Bragg (or other) reflector. Smaller external, parasitic reflections of the unidirectional mode (and coming from the opposite side of the Bragg reflector) do not in this case couple to the laser mode – or do so very weakly – and therefore do not affect it either (with the possible exception of a small reduction in lasing power).

5.4.3.5 Phase Modulation

Resonances in the transmission spectrum of microdisks biased below threshold correspond to phase changes that are as large as 360°, if the current is sufficiently high, but the current is still smaller than the lasing threshold. This situation is illustrated in Figure 5.26, which gives the experimentally measured amplitude (gain) and phase shift of the transmission versus wavelength [56].

As can be seen in Figure 5.26, it is not difficult to find two current values that, for a specific wavelength, result in an equal transmission level but also in a phase difference of 180°. Modulating the injection current between the two values then results in phase modulation. In [55], such a phase modulation was demonstrated at 1.25 Gbps with a power imbalance of 0.6 dB. The bit rate was limited by the low current injection and could be extended to one or a few tens of gigabytes per second with an optimized microwave design where parasitic effects in the metal contacts were minimized.

5.4.3.6 Microwave Photonic Filter

The aforementioned phase shifts are also found for single-sideband modulated optical carriers. A minimum modulation frequency of 18 GHz was required to obtain a phase shift of 2π: at lower frequencies a full 2π phase shift was not

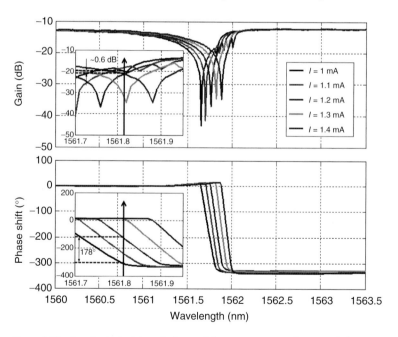

Figure 5.26 Amplitude (gain) and phase of the transmission of a microdisk resonator versus wavelength and (inset) detail of the optical carrier spectral placement. (After [56], © OSA 2012.)

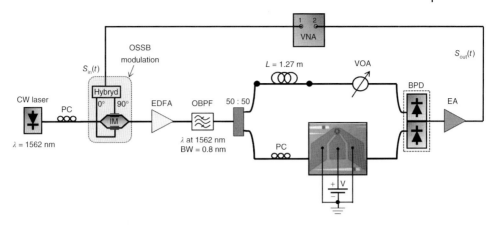

Figure 5.27 Experimental setup of the complex-valued two-tap microwave photonic tunable filter. (After [57], © OSA 2012.)

achievable. These microwave phase shifts can be used to implement microwave photonic filters, such as the one shown in Figure 5.27. In this scheme, the single-sideband modulated optical carrier is split into two equal parts, which undergo different propagation delays (with an upper branch that is 1.26 m longer) and phase shift (via the microdisk filter in the lower branch). The two parts of the modulated optical carrier are detected and subtracted in the balanced photodetector (BPD). The scheme acts as a Mach–Zehnder interferometer for the microwave sideband, with the phase shift in one of the arms being adjustable via control of the current injection in the microdisk. The microwave filtering characteristic is shown in Figure 5.28 [57].

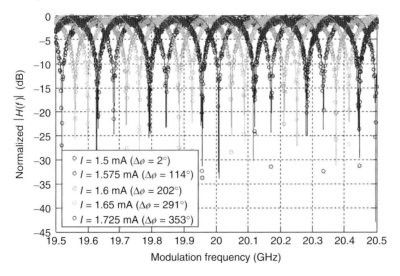

Figure 5.28 Normalized frequency response of the microwave photonic (MWP) tunable filter for different injection currents. (After [57], © OSA 2012.)

5.5
Summary and Conclusions

The heterogeneous integration of III–V semiconductor membrane lasers and amplifiers on SOI circuits has matured greatly in recent years. Molecular as well as adhesive bonding of (multiple) III–V dies on SOI dies or wafers is now possible with very high yields, reproducibility, and uniformity. In the case of adhesive bonding, very thin bonding layers can be obtained.

Using standard III–V semiconductor processing, a variety of laser types has been fabricated from bonded III–V membranes. The range includes both micrometer-scale lasers with low threshold and low output power that are compatible with intra-chip optical interconnect applications and longer, higher-output devices for off-chip communication. The latter are typically edge-emitting devices, such as DFB/DBR lasers or MLLs that have been demonstrated with milliwatt-level output powers, good single-mode behavior and, in some cases, broad wavelength tuning ranges. Heterogeneously integrated on SOI, they can also readily be integrated with modulators and passive devices implemented in SOI.

InP-based membrane microdisk lasers are very suitable for the implementation of a variety of all-optical logic devices and transmitters that can be operated with low-power levels and low energy per bit. Because they can be heterogeneously integrated onto SOI waveguide passive circuits and are very compact, they allow the fabrication of compact PICs. Crucial for the operation of microdisk lasers and their further developments are: (i) the reduction of the scattering because of the sidewall surface roughness, (ii) the reduction of metal losses and other losses in highly doped p-type materials, and (iii) the reduction of temperature increases with the simultaneous improvement of the heat sinking. Several methods that potentially result in better heat sinking have been described in this chapter, as well as possibilities for reducing bending losses and losses because of metal contacts and IVBA.

References

1. Fujioka, N., Chu, T., and Ishizaka, M. (2010) Compact and low power consumption hybrid integrated wavelength tunable laser module using silicon waveguide resonators. *IEEE J. Lightwave Technol.*, **28** (21), 3115–3120.
2. Shimizu, T. *et al.* (2011) High density hybrid integrated light source with a laser diode array on a silicon optical waveguide platform for inter-chip optical interconnection. 8th IEEE International Conference on Group IV Photonics (GFP), 2011, pp. 181–183.
3. Mitze, T., Schnarrenberger, M., Zimmerman, L., Bruns, J., Fidorra, F., Janiak, K., Kreissl, J., Fidorra, S., Heidrich, H., and Petermann, K. (2006) CWDM transmitter module based on hybrid integration. *IEEE J. Sel. Top. Quantum Electron.*, **12** (5), 983–987.
4. Mekis, A., Gloeckner, S., Masini, G., Narashimha, A., Pinguet, T., Sahni, S., and De Dobbelaere, P. (2011) A grating-coupler-enabled CMOS photonics platform. *IEEE J. Sel. Top. Quantum Electron.*, **17** (3), 597–608.

5. Jokerst, N.M. (2005) A thin-film laser, polymer waveguide, and thin-film photodetector cointegrated onto a silicon substrate. *IEEE Photonics Technol. Lett.*, **17** (10), 2197–2199.
6. Koch, B. et al. (2011) A 4x12.5 Gb/s CWDM Si photonics link using integrated hybrid silicon lasers. Proceedings of CLEO 2011.
7. SOITEC's Unibond(R) process. (1996) *Microelectron. J.*, **27** (4-5), R36.
8. Liang, D., Bowers, J.E., Oakley, D.C., Napoleone, A., Chapman, D.C., Chen, C.L., Juodawlkis, P.W., and Raday, O. (2009) High-quality 150 mm InP-to-silicon epitaxial transfer for silicon photonic integrated circuits. *Electrochem. Solid-State Lett.*, **12** (4), H101–H104.
9. Niklaus, F., Stemme, G., Lu, J.-Q., and Gutmann, R.J. (2006) Adhesive wafer bonding. *J. Appl. Phys.*, **99** (3), 031101.
10. Brouckaert, J. (2010) *Integration of Photodetectors on Silicon Photonic Integrated Circuits (PICs) for Spectroscopic Applications*, Ghent University.
11. Niklaus, F., Andersson, H., Enoksson, P., and Stemme, G. (2001) Low temperature full wafer adhesive bonding of structured wafers. *Sens. Actuators, A*, **92** (1–3), 235–241.
12. Garrou, P. et al. (1993) Rapid thermal curing of BCB dielectric. *IEEE Trans. Compon. Hybrid. Manuf. Technol.*, **16** (1), 46–52.
13. Roelkens, G., Liu, L., Liang, D., Jones, R., Fang, A., Koch, B., and Bowers, J. (2010) III-V/silicon photonics for on-chip and inter-chip optical interconnects. *Laser Photonics Rev.*, **4** (6), 751–779.
14. Liu, L., Roelkens, G., Van Campenhout, J., Brouckaert, J., Van Thourhout, D., and Baets, R. (2010) III-V/silicon-on-insulator nanophotonic cavities for optical network-on-chip. *J. Nanosci. Nanotechnol.*, **10** (3), 1461–1472.
15. Stankovic, S., Jones, R., Heck, J., Sysak, M., Van Thourhout, D., and Roelkens, G. (2011) Die-to-die adhesive bonding procedure for evanescently-coupled photonic devices. *IEEE Photonics Technol. Lett.*, **23** (23), 1781–1783.
16. Pasquariello, D. and Hjort, K. (2002) Plasma-assisted InP-to-Si low temperature wafer bonding. *IEEE J. Sel. Top. Quantum Electron.*, **8** (1), 118–131.
17. Kostrzewa, M., Di Cioccio, L., Zussy, M., Roussin, J., Fedeli, J.M., Kernevez, N., Regreny, P., Lagahe-Blanchard, C., and Aspar, B. (2006) InP dies transferred onto silicon substrate for optical interconnects application. *Sens. Actuators, A*, **125** (2), 411–414.
18. Liang, D. and Bowers, J.E. (2008) Highly efficient vertical outgassing channels for low-temperature InP-to-silicon direct wafer bonding on the silicon-on-insulator substrate. *J. Vac. Sci. Technol., B*, **26**, 1560–1568.
19. Van Campenhout, J., Binetti, P., Romeo, P., Regreny, P., Seassal, C., Leijtens, X., de Vries, T., Oei, Y., Van Velthoven, P., Notzel, R., Di Cioccio, L., Fedeli, J.M., Smit, M., Van Thourhout, D., and Bates, R. (2009) Low-footprint optical interconnect on an SOI chip through heterogeneous integration of InP-based microdisk lasers and microdetectors. *IEEE Photonics Technol. Lett.*, **21** (8), 522–524.
20. Liang, D. and Bowers, J. (2010) Recent progress in lasers on silicon. *Nat. Photonics*, **4**, 511.
21. Reboul, J., Cerutti, L., Rodriguez, J.B., Grech, P., and Tournie, E. (2011) Continuous-wave operation above room-temperature of GaSb-based laser diodes grown on silicon. *Appl. Phys. Lett.*, **99** (12), 121113.
22. Liu, J., Sun, X., Camacho-Aguilera, R., Kimerling, L., and Michel, J. (2011) Ge-on-Si laser operating at room temperature. *Opt. Lett.*, **35** (5), 679–682.
23. Roelkens, G., Van Thourhout, D., Baets, R., Notzel, R., and Smit, M. (2006) Laser emission and photodetection in an InP/InGaAsP layer integrated on and coupled to a silicon-on-insulator waveguide circuit. *Opt. Express*, **14** (18), 8154–8159.
24. Fang, A.W., Park, H., Cohen, O., Jones, R., Paniccia, M.J., and Bowers, J.E. (2006) Electrically pumped hybrid AlGaInAs-silicon evanescent laser. *Opt. Express*, **14**, 9203–9210.

25. Lamponi, M., Keyvaninia, S., Jany, C., Poingt, F., Lelarge, F., de Valicourt, G., Roelkens, G., Van Thourhout, D., Messaoudene, S., Fedeli, J.M., and Duan, G.H. (2012) Low-threshold heterogeneously integrated InP/SOI lasers with a double adiabatic taper coupler. *IEEE Photonics Technol. Lett.*, **24** (1), 76–78.
26. Ben-Bakir, B., Descos, A., Olivier, N., Bordel, D., Grosse, P., Augendre, E., Fulbert, L., and Fedeli, J.M. (2011) Electrically driven hybrid Si/III-V fabry-perot lasers based on adiabatic mode transformers. *Opt. Express*, **19** (11), 10317–10325.
27. Koch, B.R., Fang, A.W., Cohen, O., and Bowers, J.E. (2007) Mode-locked silicon evanescent lasers. *Opt. Express*, **15**, 11225–11233.
28. Fang, A.W., Jones, R., Park, H., Cohen, O., Raday, O., Paniccia, M.J., and Bowers, J.E. (2007) Integrated AlGaInAs-silicon evanescent racetrack laser and photodetector. *Opt. Express*, **15**, 2315–2322.
29. Fang, A.W., Koch, B.R., Gan, K.-G., Park, H., Jones, R., Cohen, O., Paniccia, M.J., Blumenthal, D.J., and Bowers, J.E. (2008) A racetrack mode-locked silicon evanescent laser. *Opt. Express*, **16**, 1393–1398.
30. Okumura, T., Maruyama, T., Yonezawa, H., Nishiyama, N., and Arai, S. (2008) Injection type GaInAsP/InP/Si DFB lasers directly bonded on SOI substrate. IPRM 2008 Proceedings.
31. Sysak, M.N., Anthes, J.O., Liang, D., Bowers, J.E., Raday, O., and Jones, R. (2008) A hybrid silicon sampled grating DBR laser integrated with an electroabsorption modulator using quantum well intermixing. Proceedings of ECOC 2008.
32. Duan, G.-H., Jany, C., Le Liepvre, A., Provost, J.-G., Make, D., Lelarge, F., Lamponi, M., Poingt, F., Fedeli, J.-M., Messaoudene, S., Bordel, D., Brision, S., Keyvania, S., Roelkens, G., Van Thourhout, D., Thomson, D.J., Gardes, F.Y., and Reed, G.T. (2012) 10Gb/s integrated tunable hybrid III-V/Si laser and silicon Mach-Zehnder modulator. Submitted to ECOC 2012.
33. Kurczveil, G., Heck, M., Peters, J., Garcia, J., Spencer, D., and Bowers, J. (2011) An integrated hybrid silicon multiwavelength AWG laser. *IEEE J. Sel. Top. Quantum Electron.*, **17** (6), 1521–1527.
34. Dupont, T., Grenouillet, L., Chelnokov, A., and Viktorovitch, P. (2010) Contradirectional coupling between III-V stacks and silicon-on-insulator corrugated waveguides for laser emission by distributed feedback effect. *IEEE Photonics Technol. Lett.*, **22**, 1413–1415.
35. De Koninck, Y., Roelkens, G., and Baets, R. (2010) Cavity enhanced reflector based hybrid silicon laser. IEEE Photonics Society Annual Meeting, 2010.
36. Liang, D., Fiorentino, M., Todd, S., Kurczveil, G., and Beausoleil, R. (2011) Fabrication of silicon-on-diamond substrate and low-loss optical waveguides. *IEEE Photonics Technol. Lett.*, **23** (10), 657–659.
37. Sysak, M., Liang, D., Jones, R., Kurczveil, G., Piels, M., Fiorentino, M., and Beausoleil, R. (2011) Hybrid silicon laser technology a thermal perspective. *IEEE J. Sel. Top. Quantum Electron.*, **17** (6), 1490–1498.
38. Ellis, B., Mayer, M.A., Shambat, G., Sarmiento, T., Harris, J., Haller, E.E., and Vuckovic, J. (2011) Ultralow-threshold electrically pumped quantum-dot photonic-crystal nanocavity laser. *Nat. Photonics*, **5** (5), 297–300.
39. McCall, S.L., Levi, A.F.J., Slusher, R.E., Pearton, S.J., and Logan, R.A. (1992) Whispering gallery mode microdisk lasers. *Appl. Phys. Lett.*, **60**, 289–291.
40. Baba, T., Fujita, M., Sakai, M., Kihara, M., and Watanabe, R. (1997) Lasing characteristics of GaInAsP-InP strained quantum-well microdisk injection lasers with diameter of 2-10 μm. *IEEE Photonics Technol. Lett.*, **9**, 878–880.
41. Mezosi, G., Strain, M.J., Furst, S., Wang, Z., Yu, S., and Sorel, M. (2009) Unidirectional bistability in AlGaInAs microring and microdisk semiconductor lasers. *IEEE Photonics Technol. Lett.*, **21**, 88–90.

42. Li, B., Memon, M.I., Mezosi, G., Wang, Z.R., Sorel, M., and Yu, S. (2009) Characterization of all-optical regeneration potentials of a bistable semiconductor ring laser. *J. Lightwave Technol.*, **27**, 4233–4240.
43. Van Campenhout, J. (2007) Thin-film microlasers for the integration of electronic and photonic integrated circuits. PhD thesis. Ghent University, http://www.photonics.intec.ugent.be/publications/phd.asp?ID=159 (accessed 11 Ocotber 2013).
44. Siegman, A.E. (1986) *Lasers*, Chapter 13, University Science Books, Mill Valley, CA.
45. Liu, L., Kumar, R., Huybrechts, K., Spuessens, T., Roelkens, G., Geluk, E., De Vries, T., Regreny, P., Van Thourhout, D., Morthier, G., and Baets, R. (2010) An ultra-small, low-power, all-optical flip-flop memory on a silicon chip. *Nat. Photonics*, **4** (3), 182–187.
46. Bazin, A., Halioua, Y., Monnier, P., Bordas, F., Karle, T., Perruchas, S., Gacoin, T., Girard, H., Sagnes, I., Roelkens, G., Raj, R., and Raineri, F. (2010) Thermal improvement of InP wire photonic crystal laser on silicon by addition of diamond nanoparticles in polymer bonding layer. Proceedings of ECOC 2010, Torino, Italy.
47. Van Campenhout, J., Liu, L., Romeo, P., Van Thourhout, D., Seassal, C., Regreny, P., Di Sioccio, L., Fedeli, J.M., and Baets, R. (2008) A compact SOI-integrated multiwavelength laser source based on cascaded InP microdisks. *IEEE Photonics Technol. Lett.*, **20**, 1345–1347.
48. Bochove, E.J. (1997) Theory of modulation of an injection-locked semiconductor laser diode with applications to laser characterization and communications. *J. Opt. Soc. Am.*, **14**, 2381–2391.
49. Memon, M.I., Li, B., Mezosi, G., Wang, Z., Sorel, M., and Yu, S. (2009) Modulation bandwidth enhancement in optical injection-locked semiconductor ring laser. *IEEE Photonics Technol. Lett.*, **21** (24), 1792–1794.
50. Raz, O., Dorren, H., Kumar, R., Morthier, G., Regreny, P., and Rojo-Romero, P. (2011) 50 fJ-per-bit, high speed, directly modulated light sources for on-chip optical data communications. Proceedings of OFC 2011.
51. Jung, H.D., Kumar, R., Regreny, P., Dorren, H., Koonen, T., and Raz, O. (2011) Analogue modulation characteristics of InP membrane microdisc lasers for in-building networks. *Electron. Lett*, **47**, 192–193.
52. Van der Sande, G., Gelens, L., Tassin, P., Scire, A., and Danckaert, J. (2008) Two-dimensional phase-space analysis and bifurcation study of the dynamical behaviour of a semiconductor ring laser. *J. Phys. B*, **41**, 1–8.
53. Kumar, R., Spuesens, T., Mechet, P., Kumar, P., Raz, O., Olivier, N., Fedeli, J.-M., Roelkens, G., Baets, R., Van Thourhout, D., and Morthier, G. (2011) Ultra-fast and bias-free All-optical wavelength conversion using III-V on silicon technology. *Opt. Lett.*, **36**, 2450–2452.
54. Kumar, R., Spuesens, T., Mechet, P., Olivier, N., Fedeli, J.-M., Regreny, P., Roelkens, G., Van Thourhout, D., and Morthier, G. (2011) 10 Gbit/s all-optical NRZ-OOK to RZ-OOK format conversion in an ultra-small III-V-on-silicon microdisk fabricated in a CMOS pilot line. *Opt. Express*, **19**, 24647–24656.
55. Morthier, G. and Mechet, P. (2013) Theoretical analysis of unidirectional operation and reflection sensitivity of semiconductor ring or disk lasers. *IEEE Journ. Quant. El.*, **49** (12), 1097–1101.
56. Lloret, J., Kumar, R., Sales, S., Ramos, F., Morthier, G., Mechet, P., Spuesens, T., Van Thourhout, D., Olivier, N., Fedeli, J.-M., and Capmany, J. (2012) Ultra-compact electro-optic phase modulator based on III-V-on-silicon microdisk resonators. *Opt. Lett.*, **37**, 2379–2381.
57. Lloret, J., Morthier, G., Ramos, F., Sales, S., Van Thourhout, D., Spuesens, T., Olivier, N., Fedeli, J.-M., and Capmany, J. (2011) Broadband microwave photonic fully tunable filter using a single heterogeneously integrated III-V/SOI-microdisk-based phase shifter. *Opt. Express*, **19** (18), 17421–17426.

6
Semiconductor Micro-Ring Lasers

Gábor Mezosi and Marc Sorel

6.1
Introduction

This chapter is concerned with the design, fabrication, and characterization of monolithical micro ring lasers on III–V semiconductors. The ring geometry provides these devices with several unique advantages compared to other types of diode lasers.

> *Easy integration into photonic integrated circuits (PICs).* Semiconductor ring lasers (SRLs) do not require cleaved facet or gratings to form a resonant cavity, and thus are particularly easy to integrate with other functional elements in PICs. Waveguide couplers can be used to transfer power from the ring cavity into output waveguides.
>
> *Exact control of cavity size.* Because the cavity size is defined by lithography rather than by cleaving, it can be accurately specified, which allows precise control of longitudinal mode spacing determined by the cavity length.
>
> *High output efficiency.* SRLs can operate in a unidirectional mode, in which practically all the optical energy in the cavity is propagating in one direction (either clock-wise (CW) or counter-clockwise (CCW)). Such characteristics can double the output efficiency compared to equivalent Fabry–Perot (FP) lasers with similar output coupling coefficient (mirror transmittance). Also, the total transmittance of the cavity is defined by the geometry of the waveguide couplers and therefore the desired optical quantum efficiency can be chosen with great flexibility at the lithographic stage.
>
> *Single mode operation.* SRLs can achieve single mode lasing without resorting to the use of gratings and therefore are simple to fabricate, because spatial hole burning, one of the main sources of multi-longitudinal mode behavior in FP lasers, is eliminated when SRLs operate in unidirectional traveling wave mode (see Chapter 7).
>
> *Wafer scale testing.* Because the SRL does not rely on cleaved facets to form the laser cavity, they can be tested on a wafer scale before dicing. Output waveguides can be used as integrated photo-detectors for such tests.

Compact Semiconductor Lasers, First Edition.
Edited by Richard M. De La Rue, Siyuan Yu, and Jean-Michel Lourtioz.
© 2014 Wiley-VCH Verlag GmbH & Co. KGaA. Published 2014 by Wiley-VCH Verlag GmbH & Co. KGaA.

The wave-guiding structures of SRLs largely fall into two types [1]. The first type is a closed loop of single mode waveguide, and hence, known as *ring laser*. The ring cavity can be strictly annular but can often take other shapes. For example, a frequently used shape is a "race-track" consisting of two curved portions connecting two straight waveguides. The second type is whispering gallery mode resonators formed in circular (or near circular) solid disks. The total internal reflection (TIR) along the outer wall of the disk provides the necessary guiding of light, so that an inner wall is not essential. This type is sometimes known as *microdisk lasers*. In essence these two types of lasers share the same characteristics; light circulates the cavity in a closed-loop path, therefore forms traveling wave longitudinal modes according to the round-trip self-consistent phase condition. They can, therefore, be treated under the same framework as SRLs.

The existence of two sets of counter-propagating traveling wave modes in SRLs makes them very distinct from conventional (FP and DFB (distributed feedback)) semiconductor lasers that support standing wave modes. The intricate dynamical interplay between these two sets of traveling wave modes gives rise to very interesting behaviors unseen in other semiconductor lasers, the mechanism of which is treated in detail in the following chapter.

The realization of laser cavities that forces stable unidirectional operation of the SRL has been a particularly active research topic as this is desirable for higher quantum efficiency and more stable $L-I$ characteristics [2–4]. It has been established that unidirectional operation is a natural consequence of the strong cross gain saturation in the semiconductor active medium when optical coupling between the counter-propagating modes is low [5, 6]. Such a type of unidirectional operation leads to bistability between the two counter-propagating directions that can be switched on by externally injected optical trigger pulses. This ability to remain lasing stably in the set direction of operation after each trigger makes the SRL a *self-holding* bistable, which is functionally different from other types of optical bistables that often require a holding optical beam or exhibit switch-on time that is considerably longer than the switch-off time [7]. In particular, SRLs are ideal solutions for developing optical memories that have been elusive to researchers or for the demonstration of optical bistable functionalities on a compact and integrated geometry [8].

6.2
Historical Review of Major Contributions to Research on SRL Devices

The first reported SRL, with four cleaved facets forming the closed loop cavity, dated back to 1976, and was reported by Scifres *et al.* [9]. This was followed by the demonstration of lasing operations on both ring and disk laser geometries by Matsumoto at NTT [1]. In the following two decades, various cavity geometries were proposed and demonstrated, such as pillbox [10], racetrack [2, 11, 12], square [13], and triangular [14] (Figure 6.1). These designs employed various light guiding mechanisms such as the "whispering gallery" effect in the "pillbox" or

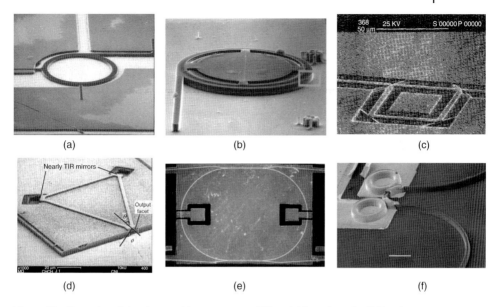

Figure 6.1 Examples of ring lasers with various optical cavities and output coupling geometries. (a) A ring cavity with a Y-junction output coupler [17]; (b) a whispering-gallery pillbox cavity [10]; (c) a square [13] and (d) a triangular [14] cavity with total internal reflection mirrors; (e) a racetrack cavity with multi-mode interference couplers [11]; and (f) a coupled ring geometry with directional couplers [8].

microdisk structure [10], deeply etched or rib-waveguides [2, 12], shallow-etched ridge-waveguides [15], and buried heterostructures [16].

Extensive research in the field of SRL devices was carried out by Hohimer and colleagues in the 1990s – at the Sandia National Laboratories in Albuquerque, New Mexico. The work of this group concentrated on the design of cavities for single mode and high side mode suppression ratio (SMSR) operation [18–20], and on rings with a cross-over waveguide to force unidirectional operation [2]. They also reported on the effect of the feedback from the cleaved output mirrors on the operation and the output power level of an SRL [21]. Furthermore, they were the first researchers to demonstrate a passively mode-locked ring laser and to highlight its potential for accurately defining the repetition rate (i.e., the cavity length) by lithography [22]. The monolithic integration of this mode-locked ring with a millimeter-wave waveguide and a fast detector yielded one of the first PICs capable of generating millimeter-wavelength electromagnetic waves up to 90 GHz frequency [23].

In the same period, forced unidirectional operation was investigated by the photonics group at Cornell University in New York. The approach of Ballantyne and coworkers [4] was to create a loss asymmetry by introducing a tapered waveguide section where the waveguide widens gradually, followed by an abrupt narrowing of the tapered section back to the original width. All the devices fabricated by this group employed a triangular-shaped cavity design with two etched TIR mirrors and

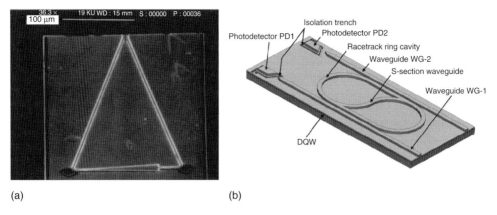

Figure 6.2 Examples of cavity designs that favor unidirectional operation. The losses and coupling between the counter-propagating modes can be unbalanced by introducing (a) a tapered waveguide [4] or (b) an S-shaped waveguide in the center of the cavity [2, 3].

one cleaved mirror [24, 25]. This geometry makes planar integration a cumbersome proposition, but it also provides an easy and straightforward solution for power extraction (Figure 6.2).

On the basis of much the same TIR technology, a rhombus-shaped SRL [26] was introduced into the product portfolio of Binoptics, a spin-off company related to the Cornell group.

In 1994, researchers at Philips demonstrated efficient out-coupling of light from SRL devices using multimode interference (MMI) couplers and MMI combiners. Later, in collaboration with the COBRA institute at Eindhoven in the Netherlands, their attention focused on the investigation of an SRL that included waveguide gratings (AWGs, arrayed waveguide gratings) into the ring, in order to create digitally tunable lasers [27, 28]. Later, the same group produced a passively mode-locked, active–passive integrated extended cavity SRL [29], and a passively mode-locked quantum-dot SRL [30]. In parallel, Hill and coworkers in the same group demonstrated the pulsed operation under cryogenic temperatures of 16 μm diameter coupled rings that could be operated as an optical bistable with a rise time of approximately 20 ps for 13 ps full-width half-maximum (FWHM) input pulses [8]. More recently, in the frame of the European Union (EU) funded research project HISTORIC, a III–V/silicon-on-insulator (SOI) hybrid integration platform was developed and based on this platform 7.5 μm diameter cw-operating room-temperature lasing SRLs that had 60 ps switching times were demonstrated [31].

Research on SRLs at University of Glasgow was initiated in the late 1980s by Jezierski and Laybourn [17]. Krauss and coworkers [10, 32] then demonstrated the first *continuous wave* and room temperature lasing, low threshold, circular ring lasers using deep etching waveguide technology and a pillbox (i.e., microdisk) geometry. Later, using weakly guided shallow-etched waveguides and MMI couplers, they produced highly efficient racetrack geometry SRLs [15, 33, 34], which has become a favorite geometry for SRL because it readily incorporates output couplers (both MMI and directional) that require a straight section in order to couple significant proportions of optical power out of the cavity.

Yu *et al.* [11, 35] investigated passively mode-locked operations of large size SRLs. Analysis of harmonic mode-locked operation of large-cavity SRLs by Avrutin and coworkers [36] was demonstrated to be in accord with experimental observations. Finally, Sorel and coworkers [6, 37], in the late 1990s and early 2000s, demonstrated that SRLs operate naturally in a unidirectional lasing regime when the coupling between the counter-propagating modes is low. They also predicted and experimentally confirmed the different operating regimes of SRLs determined by the bias current level: bidirectional operation, alternate oscillations, and unidirectional operation [38]. Furst demonstrated the monolithic integration of SRLs with externally coupled distributed Bragg reflector (DBR) waveguides [39] and investigated the controlled modal behavior of SRLs [40].

The bistable SRL also shows considerable potential when an external optical injection forces the SRL to operate in a unidirectional regime. In this configuration, the SRL becomes an optical monostable that can be used in the numerous potential applications of all-optical signal processing. For example, if an external holding beam (HB) is injected into the SRL in one direction, the lasing direction of the SRL can only be switched using a counter-directional optical pulse or a continuous wave optical beam that has a higher power level than the HB. After removing the counter-directional excitation, the SRL is restored to the optical state defined by the HB. This kind of operation can be exploited to create various all-optical logic gates, for example, NOT, NOR, or NAND gates, as is illustrated in Figure 6.3. These functions have been demonstrated by Yu and coworkers [41, 42] at the University of Bristol under the general framework of the EU FP6 project IOLOS.

The main thrust of research has been focused on the size reduction of the SRL device to decrease power consumption and improve its potential for integration. This miniaturization process presents major technological challenges that will be discussed in more detail.

6.3
Waveguide Design of Semiconductor Ring Lasers

The reduction of the cavity ring radius and the consequent additional bending loss of the waveguide is a significant issue for small SRLs.

Conventional ridge waveguide FP lasers typically use "shallow etched" ridge waveguide as shown in Figure 6.4a, with the etch depth stopping above the planar waveguide core layer. In large-radius SRLs, shallow etched ridge waveguide is typically used to construct the ring cavity since an appropriate design of the epi-structure can ensure negligible bending until bending radius of about 100–150 μm. Below such values, the shallow-etched waveguide geometry suffers from rapidly increasing bending loss. Therefore smaller device dimensions require different waveguide geometry, with a "deeply etched" profile well below the planar

236 | 6 Semiconductor Micro-Ring Lasers

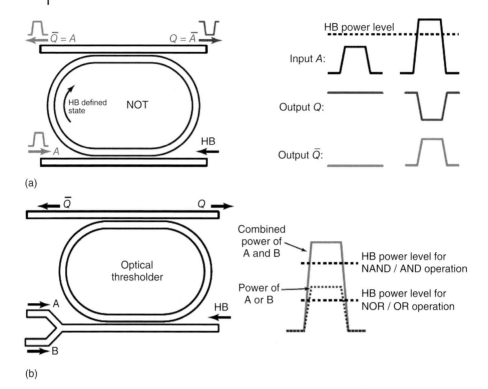

Figure 6.3 Schematic illustration of all-optical logic gates that are based on an SRL monostable. (a) Schematic of an all-optical NOT gate and its principle of operation. (b) Schematic of an all-optical thresholder. The diagram on the right shows its operation in both a NAND and a NOR configuration [42].

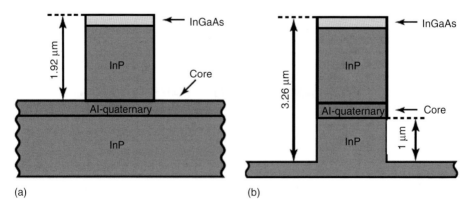

Figure 6.4 Schematic of the cross section of (a) a shallow etched waveguide and (b) a deeply etched waveguide.

waveguide core layers as shown in Figure 6.4b. This offers a much greater optical confinement at the expense of other adverse effects, which would occur if the etch depth penetrates the waveguide core that is also the active gain medium of the semiconductor epi-structure. These adverse effects include surface recombination of carriers and optical scattering loss, both of which will result in increasing laser threshold current density. Moreover, the scattering loss increases the counter-propagating mode scattering, which contributes substantially to the instabilities between the modes and can ultimately suppress the optical bistability. The use of deeply etched waveguides adds several critical challenges to the fabrication process of the structure, some of which are discussed in the following sections.

The increase in the horizontal confinement produced by deeply etched waveguides affects the shape of the modal profile, which becomes more symmetrical, as shown in Figure 6.5a. Figure 6.5b shows plots of the simulated modal effective index of the first three transverse-electric (TE) modes, as a function of the waveguide width (W) for a typical multi-quantum well (MQW) laser material with emission wavelength at 1.55 µm [43].

In comparison with shallow etched waveguide geometry, the corresponding higher-order optical waveguide mode cut-off occurs at larger values of the waveguide width. A simple explanation is that the horizontal confinement is much stronger than the vertical one. As the width of the waveguide is reduced, the guided optical mode is squeezed in the horizontal direction and expands vertically so that radiation into the substrate becomes dominant. Below a width of 2.4 µm, no higher order TE modes are supported and the waveguide becomes single-mode. On the basis of these simulations, the optimum width range for low-loss and single-mode operation is between 2 and 2.4 µm. Because of waveguide imperfections, the cut-off for higher order modes is typically shifted toward wider widths and therefore the range over which single mode operation occurs tends to be larger.

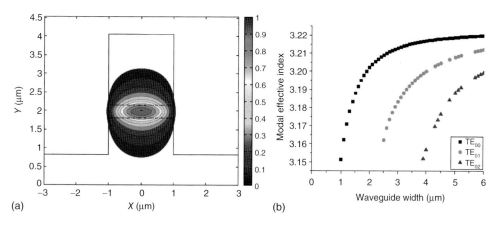

Figure 6.5 (a) Contour map of the TE_{00} mode of a 2 µm wide and 3.2 µm deep waveguide. The straight line plots the waveguide bounds and the dash dotted line indicates the core of the material. (b) Simulated modal effective refractive index of the first three TE modes plotted versus the waveguide width.

6.4
Bending Loss in Semiconductor Ring Lasers

The bending loss of deeply etched waveguide is strongly dependent on both the etch depth and bending radius. Standard 3D BPM (beam propagation method) simulations provide accurate results if the radius is much greater than the waveguide width (e.g., >10 μm). Bends tighter than this can be simulated with other methods that do not rely on the paraxiality approximation, such as the finite-difference time-domain (FDTD) technique.

The three transverse field distribution maps in Figure 6.6a illustrate how the lossy evanescent component of the field diminishes as the slab height increases. Relying only upon the change in shape of the modal cross-section, one can intuitively predict that the bending losses will be completely negligible if the etch depth into the lower cladding is at least 1.5 μm, that is, as much as the height of the upper cladding.

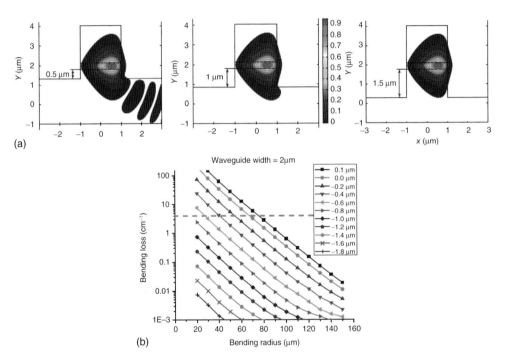

Figure 6.6 (a) Three examples of contour maps of the modal field distribution for a 20 μm radius and 2 μm wide waveguide for relative etch depths from the bottom of the core layer of 0.5, 1.0, and 1.5 μm, respectively. (b) Simulated bending losses of a deeply etched waveguide as a function of the bending radius, with the etch depth as a parameter. The internal propagation loss of a straight waveguide with the same waveguide dimensions is plotted as a reference (i.e., horizontal line at $\sim 4\,cm^{-1}$). (Please find a color version of this figure on the color plates.)

Figure 6.6b plots the bending losses as a function of the bend radius, with the relative depth from the bottom of the planar waveguide core layer as a parameter. The results indicate a 10 times decrease in the bending loss for every 400 nm increase in the waveguide height. By extrapolating the straight lines of the semi-logarithmic plot for radii below 20 μm, it can be predicted that bent waveguides with a etch depth of 1 μm below the core have negligible bending losses down to a radius of a few micrometers.

The bending loss also has dependencies on the width of the waveguide. At the same bend radius, narrower waveguides have higher propagation loss than wider waveguides, as the mode is squeezed more into the substrate and hence more likely to leak, as shown in the contour maps of Figure 6.7.

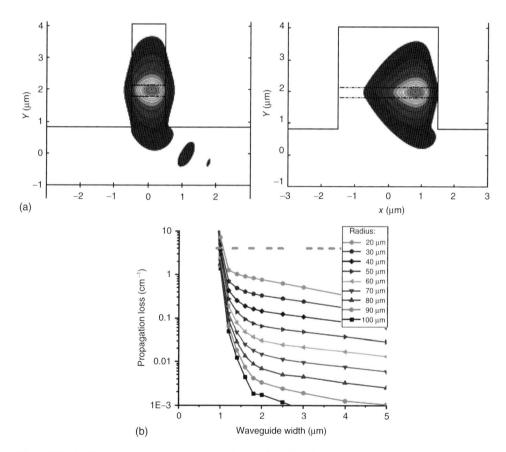

Figure 6.7 (a) The two contour maps illustrate the field distribution of a 20 μm radius curve for waveguide widths of 1 μm (left) and 3 μm (right) with relative etch depths from the bottom of the core layer of 1.0 μm. (b) Simulated bending losses as a function of the width of a deeply etched waveguide, with the radius as a parameter. (Please find a color version of this figure on the color plates.)

These simulation results indicate that single mode propagation with negligible bending losses down to a radius of 5 μm can be obtained if the waveguide width is between 2 and 3 μm and the etching depth is greater than 1 μm below the core layer. Although these figures refer to the aluminum-quaternary material used in this work, they do not vary significantly on most standard epi-layer designs for semiconductor lasers emitting at 1550 nm.

6.5
Nonradiative Carrier Losses

The necessary use of deeply etched waveguide structures for micro-ring semiconductor lasers gives rise to the issue of increased nonradiative recombination of carriers via surface recombination, as the active layer is exposed to the sidewall of the ridge waveguide [44]. It is inevitable that unmatched bonds will occur at the terminations of single crystal material regions that are created by deep etching waveguide structures. These "dangling" bonds can act as traps or recombination centers for carriers and, if present at high density, can form minibands in the bandgap.

The dry etching process used to form the deeply etched ridges typically involves energetic ions created in a plasma, by way of radio frequency (RF) discharge through etchant gases. Such ions (in particular, the lighter ones such as hydrogen) can be implanted into the semiconductor material, damaging the crystal as well as becoming impurities. Both can form recombination centers just below the sidewall that has been exposed to the plasma. The dry etching process may also deposit a layer of reaction by-products on the sidewall.

Therefore deeply etched ridge-waveguides in active semiconductor material may be badly affected by poor sidewall quality and, therefore, by high levels of surface-recombination that deplete the carriers within one diffusion length of the sidewalls. This nonradiative recombination effect is especially detrimental when the surface-area to volume ratio of a device is large (e.g., in narrow waveguides and rings or in very small diameter disks and vertical-cavity surface-emitting lasers (VCSELs)). The effect of the recombination current can be inserted into the phenomenological equations describing the operation of a laser as a reduction of the effective carrier density in the quantum wells. More specifically, there is a contribution to the current that is described by the equation:

$$J_w = \eta_r \eta_i J \tag{6.1}$$

where J is the current density injected into the device, J_w is the current density in the quantum wells (QWs), η_i is the internal efficiency, and η_r is the relative change in internal efficiency because of recombination effects. The latter parameter η_r can be calculated by combining Equation 6.1 with the equation that approximates the current density threshold of a semiconductor laser [44]:

$$J_{th} \approx J_0 \exp\left(\frac{\alpha_i(W) - (1/L)\ln(R)}{n_w \Gamma g_0} - 1\right) \tag{6.2}$$

where W is the waveguide width, α_i is internal waveguide loss, $R \sim 0.3$ is the reflectivity of the cleaved mirror, L is the total length of the cavity, n_w is the number of quantum wells, Γ is the optical confinement factor, and g_0 and J_0 are fitting parameters that denote the point on the gain curve where g/J is maximum [45].

Figure 6.8 shows the threshold current densities of 1 mm long FP lasers as a function of the waveguide width. The first set of data refers to lasers that are etched with a $Cl_2/N_2/Ar$ chemistry and are etched to a depth of 0.8 μm below the core layer. The second set belongs to lasers etched with a $Cl_2/BCl_3/Ar$ chemistry to a depth of over 2 μm below the core layer. Despite the different etches and waveguide depths, the trend of the measured J_{th} on both sets of devices look very similar with an increase of the value of J_{th} for waveguide widths narrower than 3 μm. Below a width of 1 μm, which is the calculated cut-off width of the TE_{00} mode, lasing does not occur anymore.

However, in order to calculate η_r, account must be taken of the increased optical scattering loss of the deeply etched waveguide, which varies with the waveguide width and is highly dependent on the sidewall roughness. The waveguide width dependence of the internal propagation losses α_i were assessed with the standard fringe contrast technique [46] at the wavelength of 1580 nm to minimize the absorption from the semiconductor bandgap. These data indicate a sharp increase

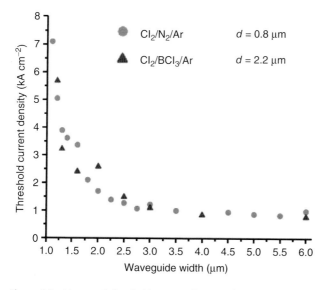

Figure 6.8 Measured threshold current density of deeply etched FP lasers as a function of the waveguide width. Devices etched using different chemistries and different etching depths perform similarly with an increase of J_{th} for widths narrower than 3 μm.

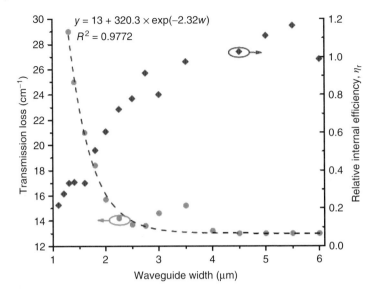

Figure 6.9 Measured transmission loss and calculated relative efficiency as a function of waveguide width of deeply etched FP lasers. The dashed line plots the exponential interpolation of the transmission loss data.

of the internal losses by sidewall scattering for waveguide widths below 2 μm as shown in Figure 6.9.

The internal efficiency results obtained from a set of deeply etched FP lasers are plotted in Figure 6.9 as a function of the waveguide width (denoted by rhombuses). Two regions with different slopes can be distinguished in this plot, with a knee point at a width of around 2.5 μm. The change of slope below the knee point can be explained by the increased recombination rate of the carriers, which occurs as the waveguide width becomes less than twice the diffusion length. The plotted data show clearly that the contribution of surface recombination to the device losses is minimal down to a waveguide width of 2.5 μm, and that, below this width, recombination strongly degrades lasing performance. The degradation of the J_{th} for reducing waveguide width is worsened by the increase of internal propagation loss from sidewall scattering.

It is worth noting that microdisk lasers are more immune to the adverse effects of the deep etching, as only one side of the active volume is exposed to the sidewall. It is therefore expected that they could realize lasing at smaller device sizes than the rings.

6.6
Semiconductor Microring and Microdisk Lasers with Point Couplers

The previous analysis indicates that micro SRLs with a waveguide width of 2.5–3.0 μm and depth of at least 1 μm below the core layer can operate on single

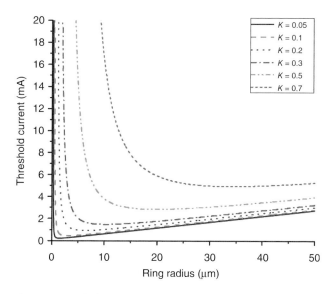

Figure 6.10 Calculated threshold current as a function of the radius of a ring cavity for different values of output coupler $K = 1 - R$. The calculation is based on measured parameters from a 5QW strained Al-quaternary laser material and bending, scattering, and current recombination losses are neglected.

lateral mode with negligible bending losses down to a radius of 5 µm and with low scattering and nonradiative recombination losses. In such a situation, the minimum dimensions of the devices are imposed by the contribution to the loss of the output coupler, which occurs when the $1/L \ln(R)$ term in Equation 6.2 becomes dominant.

In Figure 6.10, the threshold current, I_{th}, of a circular SRL is plotted as a function of the device radius, with the out-coupling strength $(K = 1 - R)$ as a parameter. It is assumed here that the waveguide geometry is designed in such a way that bending loss of deep etched curves is negligible down to a few microns radius and that the internal waveguide loss α_i is comparable to that of shallow etched devices. Each curve follows a similar trend as the device length decreases. Long cavities are dominated by internal losses. In this case, I_{th} decreases proportionally with the radius as the current is proportional to the area of the waveguide to be pumped as threshold current density remains largely constant. After reaching a minimum point, the so-called optimum length (L_{opt}), the threshold current increases sharply as the output coupling loss overtakes the other losses. The SRL radius at which the current threshold is minimum depends strongly on the output-coupling coefficient. Devices with up to 20% output-coupling coefficient can lase down to a radius of 3–4 µm, without requiring a noticeable increase in the threshold current, I_{th}.

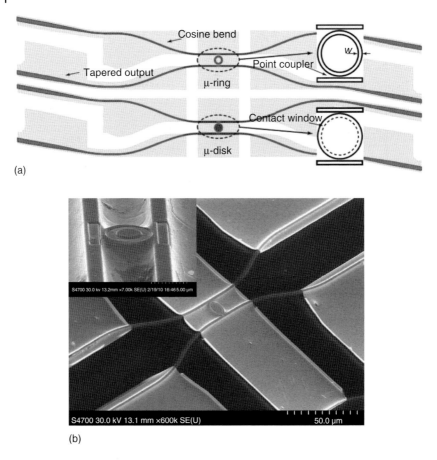

Figure 6.11 (a) Schematic of fabricated ring-shaped (top) and disk-shaped (bottom) laser devices. (b) SEM photograph of a 7 μm radius SDL. The inset shows a close up of a 6 μm diameter microdisk. (Please find a color version of this figure on the color plates.)

Figure 6.11a shows a general layout of SRL and semiconductor disk laser (SDL) devices. Sections colored in red denote pumped waveguides, while blue waveguides are passive and the yellow pads are the metal contacts.

A magnified schematic of the lasing cavity and the coupling region are shown on the right-hand side of the figure. "Point-contact" coupling structures with a narrow gap separation between the ring/disk and the output waveguides are used to tap the light out of the cavity. This coupling geometry is ideal to reduce the geometrical perturbations to the cavity and therefore guarantees minimal counter-propagating mode scattering. As pointed out earlier, the desired unidirectional operation and optical bistability can be severely hampered by the substantial presence of mode coupling. In microdisk lasers, a 1.6 μm wide circular injection window at the outer boundary is defined to restrict the gain to the rim of the disk, avoiding the excitation of higher order modes and excess injection current.

Figure 6.11b shows an SEM image of a fully processed, 7 μm radius, microdisk laser while an even smaller, 6 μm diameter, SDL is shown in the inset of the figure. Two contact windows are opened in the output waveguides, next to the point couplers to ensure the same injection current density in either side of the coupler and prevent potential de-phasing of the coupler when current is injected into the ring or disk.

The $L–I$ characteristics were measured with a large area photodiode, and a thermo-electric cooler (TEC) was used to keep the temperature on the rear of the chip at a constant 20 °C. Figure 6.12 plots the threshold current density J_{th} of devices with 400 nm coupling gap and an etching depth of over 2 μm below the core layer as a function of the ring/disk radius. Room temperature, continuous wave lasing was observed in both rings and disks, down to a radius of 8 μm. Owing to the lack of an inner sidewall, SDLs suffer from a substantial level of current spreading and therefore, only a part of the population of injected carriers contributes to the stimulated emission of the fundamental mode. At larger radii, this effect makes the SDL threshold considerably higher than SRLs. However, at small radii of <10 μm, the lower optical scattering losses make them more favorable than the SRLs with narrow waveguides, while SRLs with wider waveguides perform very similar to SDLs.

For the small radius devices, a strong increase in the threshold can be observed below a radius of 20–25 μm, in the 2 μm wide rings and below 16 μm, for the disks and the 3 μm wide rings. Such an increase is unexpected in the light of the previous loss analysis.

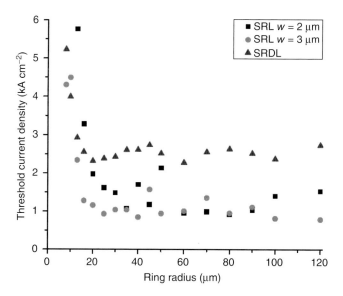

Figure 6.12 Measured threshold current density of micro-SRLs and SDLs – as a function of the device radius. Devices etched with the $Cl_2/Ar/N_2$ chemistry to an etch depth of 3.1 μm show very similar characteristics and are not described in this section. (Please find a color version of this figure on the color plates.)

6.7
Junction Heating in Small SRL Devices

A possible explanation for such a threshold increase may be the greater amount of heating of the junction region that occurs as the device size reduces. In order to confirm this hypothesis, the series resistance (r_s) of the SRLs has been derived from the measured V–I characteristics and plotted in Figure 6.13, as a function of the device radius. The calculated values of the diode resistance (indicated by dashed, dotted, and dash-dotted lines) show good agreement with the experiment, although the resistance values are slightly shifted downward with respect to the measured data (indicated by squares, circles, or triangles) because of the additional resistance of the instrument leads and the n-type contact.

Analysis of the device resistance revealed that the two main contributing factors responsible for the high rs value of a short laser are the resistance of the upper cladding (r_{cl}) and the resistance of the p-type contact (r_c). As a consequence, the greater part of the dissipated heat is generated in the upper cladding and so directly heats the junction region of the diode. When tested under pulsed-current operating conditions, the devices showed the same lasing behavior as when they were biased continuously. This observation indicates that Joule-heating is not responsible for preventing the lasing action of small devices. Moreover, the L–I characteristics of the small radius SRLs were re-measured at a Peltier-controlled temperature of

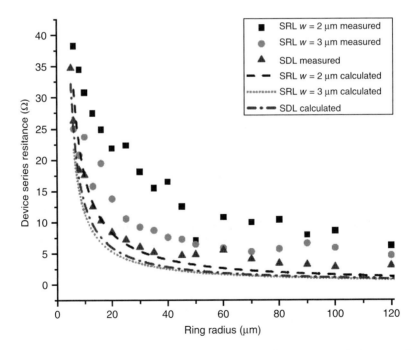

Figure 6.13 Measured and calculated series resistance of micro-SRLs and SDLs, as a function of the device radius.

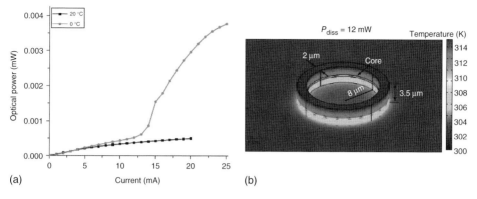

Figure 6.14 (a) L–I characteristic of a 10 μm radius SRL recorded at 20 and 0 °C. (b) Simulated isothermal color-map of the same microring for 12 mW dissipated power.

0 °C and devices showed a 1–2 mA lower value of I_{th}. Furthermore, the minimum radius at which lasing occurs decreased to 6 μm. As an example, Figure 6.14a shows plots of the 20 and 0 °C L–I characteristics of a 10 μm radius and 2 μm wide SRL.

Although the device has a high (8.5 kA cm^{-2}) threshold current density, it exhibits continuous wave lasing with a current threshold of 12 mA. Above a current level of 20 mA (14 kA cm^{-2}), the signs of thermal roll-off can be observed on the slope of the L–I curve. The latter indicates that the thermal degradation of the gain only becomes significant at high device currents (i.e., above 14 kA cm^{-2}), and therefore confirms that thermal gain degradation cannot be responsible for the termination of lasing of the small radius devices.

To validate this observation and to quantify the temperature rise of the junction, thermal calculations based on simple thermal resistance and full numerical simulations were carried out. The thermal resistance calculation results translate into an approximately 0.5° temperature increase in the core layers for every dissipated milliwatt. Figure 6.14b shows a plot of the heat distribution map for the 10 μm radius SRL of the previous example, for a 12 mW dissipated heat level. Good agreement was found between the simple thermal resistance model and the finite element method (FEM) technique. There is only a small 1–2 °C difference between the two results, which is the result of neglecting the thermal resistance of the substrate in the concentrated parameter calculations.

The evaluated thermal properties of SRLs show that material gain degradation due to junction heating becomes observable above a dissipated power level of 20–25 mW. Such dissipation levels are only present in the sub-10 μm diameter devices with high series electrical resistance, or they can be achieved through very strong pumping of larger SRLs (at current levels higher than 18 kA cm^{-2}). Therefore the increased J_{th} of SRLs and SDLs with radii smaller than 20 μm cannot be explained by thermal gain degradation.

6.8
RIE-Lag Effects in Small SRL Devices

A second possible cavity resonance dampening mechanism that scales with the curvature, and is therefore related to the device size, is the bending loss. With deeply etched waveguides, simulation data indicate that bending losses only become visible below radii of 5 µm. Yet so far all SRL bending loss analysis was based on ideally etched ring and disk cavities (i.e., where the etch depth is exactly the same around the entire circumference), but this assumption no longer holds when couplers are included in the geometry. Referring to Figure 6.15, one can observe that the etch depth of the cavity is not constant, since the height of the rings and the disks is reduced in the point-coupler regions, as a result of the so-called "RIE-lag" (RIE, reactive ion etching) effect [47], and that the etch speed is slower in narrow gaps. SEM inspections show that this shallower etched region can comprise as much as 20% of the total cavity circumference in small ring cavities, resulting in higher bending loss in this region. The increased round-trip bending losses of the ring due to nonuniformity in the etch depth provide a valid explanation for the unexpected threshold increase indicated in the experimental data for the devices that have radii smaller than 20 µm.

The presence of the RIE-lag effect also explains a second observation, namely that below a device radius of 20 µm (independent of the etch chemistry used), no unidirectional bistability could be observed, either in the rings or in the disks. A qualitative explanation of this phenomenon is that the modal mismatch caused by the shallower etched parts of the point coupler induces back-reflections. The

Figure 6.15 SEM image of a microdisk after dry etching. The RIE lag effect is clearly visible in the narrow gap forming the output couplers. Not only the RIE lag increases the cavity losses but can also suppress bistability because of the counter-propagating mode coupling that originates from the waveguide profile discontinuity.

slope of the lag becomes steeper with decreasing radius, and the coupling between the counter-propagating modes therefore increases and the directional extinction decreases. Below the critical radius of 20 μm, bistability completely vanishes due to high back-reflection levels, so that only bidirectional lasing is observed.

6.9
Racetrack Geometry Microring Lasers

On the basis of the design considerations set out in the previous sections, racetrack-shape SRLs have been designed for which the general layout is shown in Figure 6.16a. Because of the longer interaction length between the optical cavity and the output bus waveguides, racetrack SRLs can provide substantial

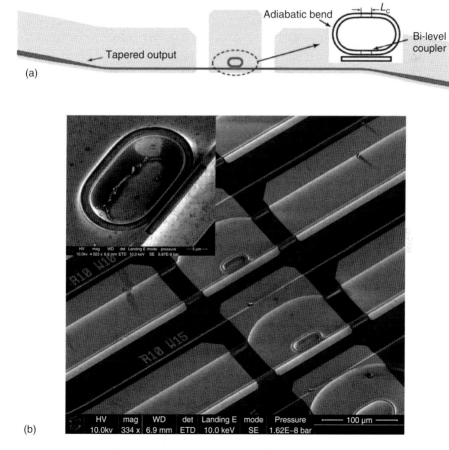

Figure 6.16 (a) Schematic of racetrack shape SRL devices. (b) SEM photograph of racetrack-geometry 10 μm radius SRLs. The inset shows a close up of a 10 μm radius and 5 μm coupler-length SRL.

coupling with a gap between the waveguides that is much larger than in point couplers. If on the one hand, this geometry increases the device dimensions, on the other, it has the potential to decrease the impact of the RIE lag on the modal coupling.

Each racetrack consists of two adiabatic bends and two straight waveguides with a "bi-level" etched coupler [43]. Both the curved and straight sections were designed to have a width of 2 μm. The bending radius of the fabricated SRLs was varied from 50 μm down to 10 μm in steps of 10 μm, and the coupling length (L_c) was designed to be 5, 10, or 15 μm. The SEM image of the fabricated racetrack-shape micro-SRLs is shown in Figure 6.16b.

As an example, Figure 6.17 plots the optical output power levels from the CW direction of a 10 μm radius racetrack with L_c values of 5 and 10 μm, as a function of the injected current level. The output waveguide contacts were left unbiased, which means that the power levels at the device outputs could be further increased by 10–13 dB, by biasing the bus waveguides to transparency. The evaluated threshold current density of the SRLs was between 1.2 and 2.4 kA cm^{-2}. Small radius devices have slightly larger J_{th} values, which is most probably the result of the higher coupling losses per unit length. The large discontinuities in the LI curves indicate the presence of directional switching between the CW and CCW directions (see Figure 6.18).

The emission spectra from a 10 μm radius and 10 μm L_c racetrack SRL, at a 40 mA bias current, show a directional extinction ratio (DER) of 18 dB and an SMSR of 27 dB. It was observed that, above 1.2–1.5 times J_{th}, all devices

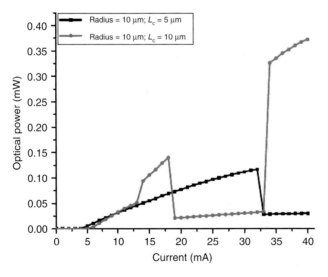

Figure 6.17 L–I characteristics measured on the CW direction for 10 μm radius racetrack SRLs with respective coupling lengths of 5 and 10 μm. The LI characteristics from the CCW direction (not shown in the figure) show an opposite trend that is a typical indication of directional bistability. (Please find a color version of this figure on the color plates.)

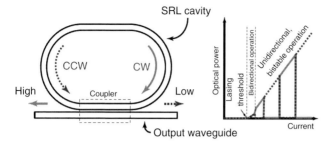

Figure 6.18 Schematic illustration of a bistable SRL. On the right-hand side, a typical L–I curve for an SRL is shown. The LI curve exhibits a narrow region of bidirectional operation just above threshold, followed by an extended unidirectional/bistable operating region. The representations are solid line – clockwise (CW) longitudinal mode and dashed line – counter-clockwise (CCW) longitudinal mode.

show robust unidirectional bistable operation, with ~20 dB DER and ~30 dB SMSR (Figure 6.19). An interesting finding, from the spectral measurements, is that race-track lasers fabricated with a smooth sidewall (etched with, e.g., $Cl_2/Ar/BCl_3$ chemistry) have a higher DER than the devices with a rough sidewall (etched with $Cl_2/Ar/N_2$ chemistry). The reason for the degradation of the unidirectional operation is the coupling of the counter-propagating modes due to the higher backscattering from the rougher waveguide sidewalls. This result confirms once more the significant impact of the waveguide sidewall quality on deeply etched SRL devices and justifies effort on the optimization of the fabrication processes.

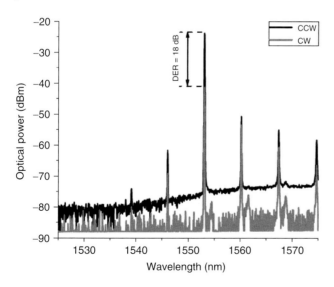

Figure 6.19 Recorded clockwise (CW) and counter-clockwise (CCW) optical spectra of a 10 μm radius racetrack SRL with an L_c of 10 μm.

6.10
Chapter Summary

In this chapter, the design considerations for deeply etched SRLs and their building blocks have been considered in detail.

For a typical MQW semiconductor laser material, a waveguide width of 2–3 μm has been found to be optimal for single-mode operation in deeply etched ridge waveguides.

The bending losses of curved waveguides have been investigated, and it was found that a waveguide etch depth of 1 μm below the waveguide core or deeper is an essential requirement to obtain minimal bending losses down to a radius of ∼5 μm. Furthermore, it was confirmed that a curved waveguide width of 2–3 μm is the optimum dimension for the combination of single transverse-mode operation and low bending losses.

Nonradiative loss processes were reviewed and the effect of surface recombination on the device threshold current has been investigated. The contribution of the surface recombination losses was found to be negligible down to a waveguide width of 2–2.5 μm. If narrower waveguides are required, passivation of the dangling bonds is recommended.

SRLs with various dimensions and evanescent point couplers have been fabricated. These devices lased continuously at room temperature down to a radius of 8 μm. The non-lasing operation below this radius was explained with increased bending losses as a result of the more shallowly etched point-coupler regions. The increasing modal mismatch of these shallower regions with reducing radius also provides the reason for the disappearance of unidirectional operation for devices with a radius smaller than 20 μm.

Analysis of the resistive and thermal material properties show that the considerable dissipation of sub-5 μm radius devices causes early thermal roll-off and prevents SRL size reduction. The material and device design techniques used in VCSEL and high power laser research, such as the reduction of the upper cladding thickness or the use of an n-type doped upper cladding region, together with a buried tunnel junction [48], should allow further miniaturization of the rings.

Another approach, the racetrack-shape SRLs, has been investigated as an alternative to the point coupler with the focus being on decreasing intra-cavity back-reflection levels. Bi-level directional couplers were finally selected as the best candidates for high output coupling ratio (4–10%), minimal intra cavity back-reflection, and short length (10–20 μm). The intra cavity back-reflections, and therefore, the coupling between the counter-propagating modes, was minimized by developing novel straight-to-curved transitions with continuously changing radii. This design effort allows the fabrication of compact racetrack shape SRL devices with output power levels in the milliwatt range. Moreover, thanks to the careful optimal reduction of the intra cavity reflections, a robust unidirectional bistable operation was observed on all the fabricated SRLs, with typical DER and SMSR values of 20 and 30 dB, respectively.

Finally, it should be mentioned that a number of groups have reported *cw*, room temperature lasing SDLs with a diameter of 7.5 μm on a III-V/SOI platform [31, 49]. These devices also showed robust unidirectional operation with excellent SMSR. An important advantage of the III-V/SOI platform is that it can be incorporated into a complementary metal-oxide-semiconductor (CMOS) fabrication flow (see Chapter 5). However, the combination of small device size and the evanescent point-coupling geometry only provide an output power of ~20 μW (at the coupler). Such output power levels are sufficient for demonstrating various optical functionalities, but are inadequate for real telecom applications. The low output power level of these devices illustrates the fundamental problem of SRL size-reduction very well: short cavity SRLs with a small micrometer radius require low mirror losses to allow low-threshold lasing operation (or lasing at all). But this situation decreases the quantum efficiency considerably and hence, also the output power of the device. Therefore, depending on the application, the designer must optimize both the threshold current and the output power level of the SRL by finding a good trade-off between the coupler strength and the size of the SRL device.

References

1. Matsumoto, N. and Kumabe, K. (1977) AlGaAs–GaAs semiconductor ring laser. *Jpn. J. Appl. Phys.*, **16** (8), 1395–1398.
2. Hohimer, P. and Vawter, G.A. (1993) Unidirectional semiconductor ring lasers with racetrack cavities. *Appl. Phys. Lett.*, **63** (18), 2457–2459.
3. Hongjun, C., Hai, L., Liu, C., Hui, D., Benavidez, M., Smagley, V.A., Caldwell, R.B., Peake, G.M., Smolyakov, G.A., Eliseev, P.G., and Osinski, M. (2005) Large S-section-ring-cavity diode lasers: directional switching, electrical diagnostics, and mode beating spectra. *IEEE Photonics Technol. Lett.*, **17**, 282–284.
4. Liang, J.J., Lau, S.T., Leary, M.H., and Ballantyne, J.M. (1997) Unidirectional operation of waveguide diode ring lasers. *Appl. Phys. Lett.*, **70** (10), 1192–1194.
5. Booth, M.F., Schremer, A., and Ballantyne, J.M. (2000) Spatial beam switching and bistability in a diode ring laser. *Appl. Phys. Lett.*, **76** (9), 1095–1097.
6. Sorel, M., Laybourn, P.J.R., Giuliani, G., and Donati, S. (2002) Unidirectional bistability in semiconductor waveguide ring lasers. *Appl. Phys. Lett.*, **80** (17), 3051–3053.
7. Kawaguchi, H. (1997) Bistable laser diodes and their applications: state of the art. *IEEE J. Sel. Top. Quantum Electron.*, **3**, 1254–1270.
8. Hill, M.T., Dorren, H.J.S., de Vries, T., Leijtens, X.J.M., den Besten, J.H., Smalbrugge, B., Oei, Y.-S., Binsma, H., Khoe, G.-D., and Smit, M.K. (2004) A fast low-power optical memory based on coupled micro-ring lasers. *Nature*, **432**, 206–209.
9. Scifres, D.R., Burnham, R.D., and Streifer, W. (1976) Grating-coupled GaAs single heterostructure ring laser. *Appl. Phys. Lett.*, **28** (11), 681–683.
10. Krauss, T., Laybourn, P.J.R., and Roberts, J. (1990) cw operation of semiconductor ring lasers. *Electron. Lett.*, **26**, 2095–2097.
11. Yu, S., Krauss, T.F., and Laybourn, P.J.R. (1998) Mode locking in large monolithic semiconductor ring lasers. *Opt. Eng.*, **37** (4), 1164–1168.
12. Griffel, G., Abeles, J., Menna, R., Braun, A., Connolly, J., and King, M. (2000) Low-threshold InGaAsP ring lasers fabricated using bi-level dry etching. *IEEE Photonics Technol. Lett.*, **12**, 146–148.

13. Han, H., Forbes, D.V., and Coleman, J.J. (1995) InGaAs–AlGaAs–GaAs strained-layer quantum-well heterostructure square ring lasers. *IEEE J. Quantum Electron.*, **31**, 1994–1997.
14. Ji, C., Leary, M., and Ballantyne, J. (1997) Long-wavelength triangular ring laser. *IEEE Photonics Technol. Lett.*, **9**, 1469–1471.
15. Krauss, T., DeLaRue, R.M., Gontijo, I., Laybourn, P.J.R., and Roberts, J.S. (1994) Strip-loaded semiconductor ring lasers employing multimode interference output couplers. *Appl. Phys. Lett.*, **64** (21), 2788–2790.
16. Cockerill, T., Forbes, D., Dantzig, J., and Coleman, J. (1994) Strained-layer InGaAs–GaAs–AlGaAs buried-heterostructure quantum-well lasers by three-step selective-area metalorganic chemical vapor deposition. *IEEE J. Quantum Electron.*, **30**, 441–445.
17. Jezierski, A.F. and Laybourn, P.J.R. (1988) Integrated semiconductor ring laser. *IEE Proc.: Optoelectron.*, **35**, 17–24.
18. Hohimer, J.P., Craft, D.C., Hadley, G.R., Vawter, G.A., and Warren, M.E. (1991) Single-frequency continuous-wave operation of ring resonator diode lasers. *Appl. Phys. Lett.*, **59**, 3360–3362.
19. Hohimer, J.P., Vawter, G.A., Craft, D.C., and Hadley, G.R. (1992) Interferometric ring diode lasers. *Appl. Phys. Lett.*, **61** (12), 1375–1377.
20. Hohimer, J.P., Hadley, G.R., and Vawter, G.A. (1993) Semiconductor ring lasers with reflection output couplers. *Appl. Phys. Lett.*, **63** (3), 278–280.
21. Hohimer, J.P., Vawter, G.A., Craft, D.C., and Hadley, G.R. (1992) Improving the performance of semiconductor ring lasers by controlled reflection feedback. *Appl. Phys. Lett.*, **61** (9), 1013–1015.
22. Hohimer, J.P. and Vawter, G.A. (1993) Passive mode locking of monolithic semiconductor ring lasers at 86 GHz. *Appl. Phys. Lett.*, **63** (12), 1598–1600.
23. Vawter, G., Mar, A., Hietala, V., Zolper, J., and Hohimer, J. (1997) All optical millimeter-wave electrical signal generation using an integrated mode-locked semiconductor ring laser and photodiode. *IEEE Photonics Technol. Lett.*, **9**, 1634–1636.
24. Behfar-Rad, A., Ballantyne, J.M., and Wong, S.S. (1991) Etched-facet AlGaAs triangular-shaped ring lasers with output coupling. *Appl. Phys. Lett.*, **59** (12), 1395–1397.
25. Behfar-Rad, A., Ballantyne, J.M., and Wong, S.S. (1992) AlGaAs/GaAs-based triangular-shaped ring ridge lasers. *Appl. Phys. Lett.*, **60** (14), 1658–1660.
26. Bussjager, R., Erdmann, R., Kovanis, V., McKeon, B., Johns, S., Morrow, A., Green, M., Stoffel, N., Tan, S., Shick, C., Bacon, W., and Beaman, B. (2006) Packaged diamond-shaped ring-laser-diode switch. Avionics Fiber-Optics and Photonics, 2006. IEEE Conference, September 12–14, 2006, pp. 64–65.
27. den Besten, J., Broeke, R., van Geemert, M., Binsma, J., Heinrichsdorff, F., van Dongen, T., de Vries, T., Bente, E., Leijtens, X., and Smit, M. (2002) A compact digitally tunable seven-channel ring laser. *IEEE Photonics Technol. Lett.*, **14**, 753–755.
28. Bente, E., Barbarin, Y., den Besten, J., Smit, M., and Binsma, J. (2004) Wavelength selection in an integrated multiwavelength ring laser. *IEEE J. Quantum Electron.*, **40**, 1208–1216.
29. Barbarin, Y., Bente, E., Heck, M., den Besten, J., Guidi, G., Oei, Y., Binsma, J., and Smit, M. (2005) Realization and modeling of a 27-GHz integrated passively mode-locked ring laser. *IEEE Photonics Technol. Lett.*, **17**, 2277–2279.
30. Heck, M., Tahvili, M., Anantathanasarn, S., Smit, M., Notzel, R., and Bente, E. (2009) Observation of dynamics in a 5 GHz passively mode-locked InAs/InP (100) quantum dot ring laser at 1.55 μm. CLEO Europe – EQEC 2009.
31. Liu, L., Kumar, R., Huybrechts, K., Spuesens, T., Roelkens, G., Geluk, E.-J., de Vries, T., Regreny, P., Van Thourhout, D., Baets, R., and Morthier, G. (2010) An ultra-small, low-power, all-optical flip-flop memory on a silicon chip. *Nat. Photonics*, **4**, 182–187.
32. Krauss, T. and Laybourn, P. (1992) Very low threshold current operation of semiconductor ring lasers. *IEE Proc.: Optoelectron.*, **139**, 383–388.

33. Krauss, T., Laybourn, P., and De La Rue, R. (1994) Improved performance of semiconductor ring lasers with multi-mode interference output couplers. 14th IEEE International Semiconductor Laser Conference, September 19–23, 1994, pp. 87–88.
34. Krauss, T., DeLaRue, R., Laybourn, P., Vogele, B., and Stanley, C. (1995) Efficient semiconductor ring lasers made by a simple self-aligned fabrication process. *IEEE J. Quantum Electron.*, **1**, 757–761.
35. Yu, S., Krauss, T., and Laybourn, P. (1997) Multiple output semiconductor ring lasers with high external quantum efficiency. *IEE Proc.: Optoelectron.*, **144**, 19–22.
36. Avrutin, E.A., Marsh, J.H., Arnold, J.M., Krauss, T.F., Pottinger, H., and De La Rue, R.M. (1999) Analysis of harmonic (sub) THz passive mode-locking in monolithic compound cavity Fabry–Perot and ring laser diodes. *IEE Proc.: Optoelectron.*, **146** (1), 55–61.
37. Sorel, M. and Laybourn, P. (2001) Control of unidirectional operation in semiconductor ring lasers. 14th Annual Meeting of the IEEE Lasers and Electro-Optics Society, 2001, LEOS 2001, Vol. 2, pp. 513–514.
38. Sorel, M., Giuliani, G., Scire, A., Miglierina, R., Donati, S., and Laybourn, P. (2003) Operating regimes of GaAs–AlGaAs semiconductor ring lasers: experiment and model. *IEEE J. Quantum Electron.*, **39**, 1187–1195.
39. Furst, S., Yu, S., and Sorel, M. (2008) Fast and digitally wavelength-tunable semiconductor ring laser using a monolithically integrated distributed Bragg reflector. *IEEE Photonics Technol. Lett.*, **20**, 1926–1928.
40. Furst, S., Perez-Serrano, A., Scire, A., Sorel, M., and Balle, S. (2008) Modal structure, directional and wavelength jumps of integrated semiconductor ring lasers: experiment and theory. *Appl. Phys. Lett.*, **93**, 251109.
41. Thakulsukanant, K., Li, B., Memon, I., Mezosi, G., Wang, Z., Sorel, M., and Yu, S. (2009) All-optical label swapping using bistable semiconductor ring laser in an optical switching node. *J. Lightwave Technol.*, **27**, 631–638.
42. Li, B., Memon, M.I., Mezosi, G., Wang, Z., Sorel, M., and Yu, S. (2009) Characterization of all-optical regeneration potentials of a bistable semiconductor ring laser. *J. Lightwave Technol.*, **27**, 4233–4240.
43. Mezosi, G., Strain, M.J., Furst, S., Wang, Z., Yu, S., and Sorel, M. (2009) Unidirectional bistability in AlGaInAs microring and microdisk semiconductor lasers. *IEEE Photonics Technol. Lett.*, **21**, 88–90.
44. Coldrane, L.A. and Corzine, S.W. (1995) *Diode Lasers and Photonic Integrated Circuits*, Wiley-Interscience.
45. DeTemple, T.A. and Herzinger, C.M. (1993) On the semiconductor laser logarithmic gain-current density relation. *IEEE J. Quantum Electron.*, **29**, 1246–1252.
46. Walker, R.G. (1985) Simple and accurate loss measurement technique for semiconductor optical waveguide. *Electron. Lett.*, **21**, 581–583.
47. Gottscho, R.A., Jurgensen, C.W., and Vitkavage, D.J. (1992) Microscopic uniformity in plasma etching. *J. Vac. Sci. Technol. B*, **10**, 2133–2147.
48. Boucart, J., Starck, C., Gaborit, F., Plais, A., Bouche, N., Derouin, E., Remy, J., Bonnet-Gamard, J., Goldstein, L., Fortin, C., Carpentier, D., Salet, P., Brillouet, F., and Jacquet, J. (1999) Metamorphic DBR and tunnel-junction injection. A CW RT monolithic long-wavelength VCSEL. *IEEE J. Sel. Top. Quantum Electron.*, **5**, 520–529.
49. Campenhout, J.V., Romeo, P.R., Regreny, P., Seassal, C., Thourhout, D.V., Verstuyft, S., Cioccio, L.D., Fedeli, J.-M., Lagahe, C., and Baets, R. (2007) Electrically pumped InP-based microdisk lasers integrated with a nanophotonic silicon-on-insulator waveguide circuit. *Opt. Express*, **15** (11), 6744–6749.

7
Nonlinearity in Semiconductor Micro-Ring Lasers

Xinlun Cai, Siyuan Yu, Yujie Chen, and Yanfeng Zhang

7.1
Introduction

Lasers based on microcavities are attractive for their compactness, low threshold, low power consumption, and potential for ultrafast modulation speed. In the last two decades, much attention has been directed to the ultimate reduction of the laser cavity volume while maintaining a high quality factor. The microsized semiconductor ring laser (SRL) is a promising candidate, due to its many advantages compared to other types of diode laser.

- *Easy integration into photonic integrated circuits (PICs).* SRLs do not require cleaved facet or gratings to form a resonant cavity, and are, therefore, particularly easy to integrate with other functional elements in PICs. Waveguide couplers can be used to transfer power from the ring cavity into output waveguides.
- *Exact control of cavity size.* Because the cavity size is defined by lithography rather than by cleaving, it can be accurately specified, which allows precise control of the longitudinal mode spacing, determined by the cavity length.
- *Single-mode operation.* SRLs can achieve single-mode lasing without resorting to the use of gratings and, therefore, are simple to fabricate, because spatial hole burning, one of the main sources of multi-longitudinal mode behavior in FP lasers, can be eliminated when SRLs are operated in unidirectional mode.
- *Reduced sensitivity to low-level external reflections.* Low-level external reflection is known to severely affect the stable operation of FP and DFB lasers. In SRLs, because external reflections from the output of the clockwise (CW) direction are coupled into the counterclockwise (CCW) direction and vice versa, the sensitivity of the lasing mode to low-level reflections should be reduced, especially when SRLs operate unidirectionally.
- *Wafer scale testing.* Because the output waveguides can be used as integrated photodetectors, SRLs can be tested on a wafer scale, before dicing.

Compact Semiconductor Lasers, First Edition.
Edited by Richard M. De La Rue, Siyuan Yu, and Jean-Michel Lourtioz.
© 2014 Wiley-VCH Verlag GmbH & Co. KGaA. Published 2014 by Wiley-VCH Verlag GmbH & Co. KGaA.

The first demonstration of a ring-shaped semiconductor laser dates back to 1980 [1]. However, it was not until the early 1990s that the idea of integrating SRLs received greater interest. Different cavity geometries were proposed and demonstrated [2–8], utilizing various guiding mechanisms and different output coupling mechanisms [2–4, 9–11]. Although at that time their lasing characteristics were not well understood, it was generally believed that SRLs can only be forced to achieve unidirectional operation. Several forcing mechanisms were introduced to make unidirectional SRLs by providing "nonreciprocal" coupling between the two possible lasing directions, for example, by fabricating a crossover waveguide [5, 12], tapered waveguide inside the cavity [13], or by using the feedback from a cleaved end-facet mirror outside the cavity [6, 14]. Unidirectional SRLs showed improved longitudinal mode purity and enhanced output power over bidirectional ring lasers, but these devices with intracavity structures always suffered from intracavity back-scattering, which introduces complications such as coupled cavity effects and obscures the more fundamental mechanisms behind SRL characteristics. Further experimental studies in triangular SRLs and large diameter SRLs revealed that unidirectional lasing in SRLs can be achieved without any forcing mechanism [15, 16]. However, unlike the forced unidirectional SRLs mentioned above, the lasing direction for these devices is unpredictable. In the triangular lasers of [15], significant intracavity back-scattering may still exist at the corners.

Sorel *et al.*, in particular, experimentally observed several operating regimes in SRLs with minimized intracavity back-scattering, including bidirectional continuous wave (bi-CW), bidirectionality with alternate oscillations (bi-AO), and bistable unidirectional (bis-UNI) (see Figure 7.1). Furthermore they established a two-mode model, which theoretically predicted the various operating regimes [17,

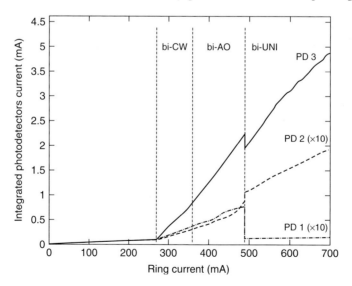

Figure 7.1 Three distinct regimes of operation for semiconductor ring lasers (i) bi-CW, (ii) bi-AO, and (iii) bi-UNI.

18]. The discovery of unidirectional bistability triggered a significant quantity of research in the last decade. When a SRL is working in a bis-UNI region, only one of the possible lasing directions, either CW or CCW, is selected while the other becomes strongly suppressed. In this case, the SRL behaves as an optical bistable between the two directions, and the lasing direction can be controlled by injected optical pulses. This property, while on the one hand can be exploited for applications in all-optical signal processing devices and all-optical flip-flops [19–21], on the other hand can cause problems when the SRLs are designed for the application of a low-cost light source because the lasing directions of SRLs become unpredictable.

The design of SRLs requires a thorough understanding of the mechanisms that determine the peculiar lasing characteristics of the devices. The two-mode model of [17] contains both linear coupling and nonlinear gain suppression between the CW mode and the CCW mode at a single cavity frequency. It succeeds in predicting the existence of various operation regimes, and also shows that the unidirectional bistability stems fundamentally from the fact that the cross-gain suppression coefficient ϵ_c is larger than self-gain suppression coefficient ϵ_s in SRLs. Moreover it reveals that the unidirectional bistability is only possible if the linear coupling due to back-scattering between CW and CCW modes, arising from discontinuities and imperfections in the waveguide such as sidewall roughness, is sufficiently small. Such linear coupling has a significant impact on the performance of SRLs, especially small SRLs with deeply etched waveguide designs because of the higher level of scattering from the roughness on the waveguide sidewalls. Efforts have been made to minimize this roughness and they have led to successful low-threshold room temperature continuous wave (RTCW) lasing and demonstration of bistable operation in semiconductor micro-ring lasers [22, 23].

While the two-mode model can explain some of the lasing characteristics, it cannot explain the periodic switching between the two unidirectional states observed with increasing current or the stability of the lasing direction with decreasing current. Nor does it attempt to explain the influence of various mode coupling mechanisms on the spectral characteristics of SRLs. The explanation of these phenomena needs to be based on the analysis of multimode competition in SRLs. Multimode competition and gain nonlinearities have been intensively studied for FP lasers using density matrix theory [24–32]. Although the modes in FP lasers are standing waves, while in SRLs they are traveling waves, the principles are the same. For the two-mode model, a single pair of self- and cross-gain saturation coefficients is sufficient. As there is no beat frequency between the modes, the photon and carrier populations are also static in time. However, the competition between modes separated in frequency results in population changes that pulsate at the mode beat frequency. Each nonlinear gain mechanism has an associated time constant, which limits the maximum beat frequency to which the respective population pulsation component can respond. The nonlinear gain coefficients are, therefore, different for each combination of longitudinal modes.

Moreover, in multimode operation cases, not only self- and cross-gain suppression but also the contribution of four-wave mixing (FWM), a phase-sensitive phenomenon that leads to power exchange among different modes, needs to be considered. It is well established that the net effect of self- and cross-gain suppression tends to enhance the long wavelength modes and suppress the short wavelength modes (relative to the lasing mode) [27–29]. However, this effect is partly compensated by FWM, which has the opposite asymmetry with respect to wavelength [26, 33], that is, it tends to enhance the modes on the shorter wavelength side. As is discussed in detail later, in the case of SRLs the self- and cross-gain suppression happen among all the modes in both directions, but FWM can only happen among modes propagating in the same direction [33]. Therefore, the existence of an intense dominant lasing mode leads to a relatively symmetric spectrum in the lasing direction (LD) and a highly asymmetric spectrum in the non-lasing direction (NLD), which is a profound difference from FP lasers.

As the injection current is increased, the profile of the gain is shifted toward long wavelength because of joule heating in the active region and so the highest enhanced side mode will be located at the long wavelength side in the NLD. Therefore, this mode is usually selected when the mode hop occurs, resulting in the reversal of lasing direction. As the current is decreased, the gain profile is shifted toward shorter wavelengths. The suppressed short wavelength mode in NLD is unlikely to be selected during the mode hop and so the lasing direction is maintained [34–36].

Therefore, in order to be able to correctly explain the lasing characteristics of SRLs, it is necessary to extend the two mode model to a multimode version. In this chapter, a systematic and comprehensive frequency domain model for SRLs suitable for numerical analysis is presented. The multimode rate equations, including not only third order nonlinearity (χ^3) terms derived from density matrix theory but also terms due to backscattering, are expressed in a concise, matrix form, from which the numerical solutions can be obtained very efficiently. Every aspect of the experimentally observed lasing characteristics of SRLs can be explained by such a model, including the lasing spectra, light–current ($L-I$) characteristics, and lasing direction hysteresis.

7.2
General Formalism

7.2.1
Fundamental Equations for Semiconductor Ring Lasers

The theory behind the nonlinear mode interactions in SRLs is presented in this section. A schematic diagram of the SRL is shown in Figure 7.2. The end-facets of output waveguide are tilted to minimize the reflectivity so that the feedback from these tilted facets may be ignored. The output coupler is assumed to have a low coupling ratio so that the amplitude of the field can be considered to be constant along the ring cavity.

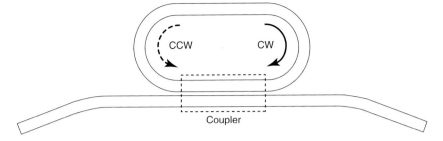

Figure 7.2 Schematic illustration of a SRL. The tilted output waveguides are assumed to have very low reflectivity, and coupling ratio is assumed to be small.

We assume that the laser structure supports only the fundamental TE waveguide mode with the transverse distribution $U(x,y)$. If all fields remain linearly polarized during their interaction, the propagation characteristics can be obtained by solving the scalar wave equation

$$\nabla^2 E - \mu_0 \sigma \frac{\partial E}{\partial t} - \mu_0 \epsilon_0 \epsilon_r \frac{\partial^2 E}{\partial t^2} = \mu_0 \frac{\partial^2 P}{\partial t^2} \tag{7.1}$$

where P is the macroscopic polarization that is assumed to be predominantly decided by the existence of excess electron-hole pairs, ϵ_0 and μ_0 are the dielectric constant and the permeability in vacuum, respectively, ϵ_r is the relative dielectric constant of the laser material without carrier injection, and σ is the conductivity associated with optical loss in the laser cavity, including the equivalent mirror loss. ϵ_r and σ are generally functions of position in the laser cavity.

The following steps are taken to solve Equation 7.1. First, Equation 7.1 is solved with the condition $P=0$ and $\sigma=0$, that is, without polarization by electron-hole pairs and any loss mechanisms. The solutions in this case, which are well known in optical waveguide theory [37], consist of a set of orthogonal eigenfunctions. The solution of the full Equation 7.1 is then expressed in the form of a series of these orthogonal functions, the coefficients of which include the effect of P and σ.

Before going into the details of the theory, it is necessary to define a system for referencing longitudinal modes, as illustrated in Figure 7.3. We consider an odd number $2M+1$ of modes in each direction ($4M+2$ in total) of the SRL. The CW/CCW modes are distinguished by positive/negative numbers, and are labeled in such a way that the index $M+1$ and $-(M+1)$ belong to the central lasing modes at the threshold that coincide with the gain spectrum peak at threshold. The frequencies of the modes are assumed to be equally spaced by $\Delta\omega$, so that $\omega_i = \omega_1 + (i-1)\Delta\omega$, $i=1, 2, \ldots, 2M+1$. Note that this scheme allows an arbitrary number of modes to be modeled.

Assuming in Equation 7.1 that $P=0$, $\sigma=0$, and the time dependent term is $\exp(j\omega_{0i}t)$ with the passive cavity frequency ω_{0i} given below, the spatial distribution of the field is given by the following equation:

$$\nabla^2 U_i(r) + \epsilon_0 \mu_0 \epsilon_r \omega_{0i}^2 U_i(r) = 0 \tag{7.2}$$

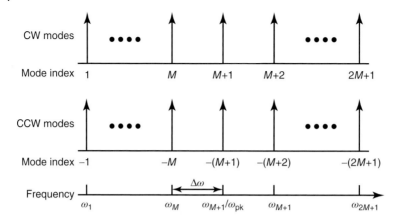

Figure 7.3 Mode referencing system. Modeled lasing spectrum consisting of a large number of discrete modes with mode index of $\pm i$ at equally spaced frequencies ω_i. The central frequency at threshold is also denoted as ω_{pk}.

where $U_i(r)$ is the field distribution with the subscript i being the mode number (integer: $\pm 0, \pm 1, \pm 2, \ldots \pm (2M+1)$), which is separated into transverse and longitudinal distributions ($U_{Ti}(x,y)$ and $U_{Li}(z)$).

$$U_i(r) = U_{Ti}(x, y) U_{Li}(z) \tag{7.3}$$

The transverse distribution $U_{Ti}(x,y)$ is given by a set of characteristic guided modes according to the waveguide structure. The longitudinal distribution is given as:

$$U_{Li}(z) = \sqrt{\frac{1}{L}} \exp\left(-j\frac{2\pi n_{eq}}{\lambda_{0i}} z\right) = \sqrt{\frac{1}{L}} \exp(-jk_i z) \tag{7.4}$$

where L is the cavity circumference, n_{eq} is the equivalent refractive index of the waveguide, and λ_{0i} is the characteristic wavelength corresponding to ω_{0i}. k_i is obtained from the boundary condition of $U_{Li}(z)$ as:

$$k_i = \pm \left|\frac{2m\pi}{L}\right| \quad (m \text{ is the integer } +/- \text{ for CW/CCW}) \tag{7.5}$$

The eigenmodes of the transverse waveguide form a complete orthogonal set:

$$\int_{-\infty}^{\infty} \int_{-\infty}^{\infty} U_{xyi}(x, y) U_{xyj}(x, y) dx dy = \delta_{ij} \tag{7.6}$$

Thus E and P in Equation 7.1 are expanded in terms of the eigenmodes determined by Equation 7.2 as:

$$E(z, t) = \sum_i E_i(t) U_i(r) e^{j\omega_i t} \tag{7.7}$$

$$P(z, t) = \sum_i P_i(t) U_i(r) e^{j\omega_i t} \tag{7.8}$$

where ω_i is the angular frequency of the ith lasing mode, which is normally slightly shifted from the characteristic resonant frequency given by Equation 7.5 due to the nonzero polarization P. $E_i(t)$ and $P_i(t)$ are the complex amplitudes of E and P, which are given by:

$$E_i(t) = \overline{E}_i \exp(j\psi_i(t)) \tag{7.9}$$

$$P_i(t) = \overline{P}_i \exp(j\psi_i(t)) \tag{7.10}$$

$\overline{E}_i$ which is a real number, is the amplitude of E, $\overline{P}_i$, which is a complex number, is the amplitude of P and ψ_i is the phase of E. All quantities are assumed to vary slowly compared with the optical frequency and, therefore, the following inequalities are valid:

$$\left| \frac{\partial \overline{E}_i}{\partial t} \right| \ll |\omega_i \overline{E}_i| \tag{7.11}$$

$$\left| \frac{\partial \overline{P}_i}{\partial t} \right| \ll |\omega_i \overline{P}_i| \tag{7.12}$$

$$\left| \frac{\partial \psi_i}{\partial t} \right| \ll \omega_i \tag{7.13}$$

These equations define the so-called slowly varying amplitude and phase (SVAP) approximation, which plays a central role in laser physics [38]. Physically it means that we consider light waves whose amplitudes and phases vary little within an optical period. The SVAP leads to major mathematical simplifications as can be seen by substituting Equations 7.7–7.10 into Equation 7.1. Using the orthonormal relation (Equation 7.6), and using Equations 7.11–7.13 to eliminate all the small contributions, we obtain

$$\frac{d\overline{E}_i}{dt} = -\frac{\sigma}{2\epsilon_0 n_i n_g} \overline{E}_i - \frac{\omega_i}{2\epsilon_0 n_i n_g} \text{Im}\{\overline{P}_i\} \tag{7.14}$$

$$\frac{d\psi_i}{dt} = \frac{n_i}{n_g}(\omega_{0i} - \omega_i) - \frac{\omega_i}{2\epsilon_0 n_i n_g} \frac{\text{Re}\{\overline{P}_i\}}{\overline{E}_i} \tag{7.15}$$

or equivalently,

$$\frac{dE_i}{dt} = j\frac{\omega_i}{2\epsilon_0 n_i n_g} P_i + j\frac{n_i}{n_g}(\omega_i - \omega_{0i}) E_i - \frac{1}{\tau_p} E_i \tag{7.16}$$

where the photon lifetime τ_p is given by

$$\frac{1}{\tau_p} = \frac{\sigma}{2\epsilon_0 n_i n_g} \tag{7.17}$$

n_i is the modal refractive index, n_g is the modal group refractive index.

Following the treatment of M. Sargent III [32], we write the total induced polarization $P(z,t)$ as

$$P(z,t) = P^{(1)}(z,t) + P^{(3)}(z,t)$$
$$= \epsilon_0 \left[\sum_l \chi_l^1 E_l e^{j(\omega_l t - k_l z)} + \sum_{l,m,n} \chi_{lmn}^3 E_l E_m^* E_n e^{-j(k_l - k_m + k_n)z} e^{j(\omega_l - \omega_m + \omega_n)t} \right] \quad (7.18)$$

χ_l^1 is the first order susceptibility in the presence of pump current and χ_{lmn}^3 is the third order susceptibility. In order to clarify the effect of the induced polarization on individual modes, we project the total polarization onto the ith modes

$$P_i(t) = \frac{1}{L} \int_0^L P(z,t) e^{-j(\omega_i t - k_i z)} dz = P_i^{(1)}(t) + P_i^{(3)}(t) \quad (7.19)$$

where $P_i^{(1)}$ and $P_i^{(3)}$ are the first and third order nonlinear polarization to be evaluated at ω_i. Substituting Equation 7.19 into Equation 7.16 and assuming $\omega_i \approx \omega_{0i}$, the following general expression is obtained for the evolution of the ith mode as it propagates through the SRL:

$$\frac{dE_i}{dt} = \frac{j}{2} \frac{\omega_i}{n_i n_g \epsilon_0} [P_i^{(1)}(t) + P_i^{(3)}(t)] - \frac{1}{2} \frac{E_i}{\tau_p} \quad (7.20)$$

where the photon lifetime in the above equation is given as:

$$\tau_p = \left[\frac{c}{n_g} \left(\alpha_{int} + \frac{1}{L} \ln \frac{1}{T_r} \right) \right]^{-1} \quad (7.21)$$

where α_{int} is the material loss per unit length, T_r is the coupling transmission ratio, and c is the velocity of light.

In semiconductor laser materials, the first order polarization and susceptibility are given by:

$$P_i^{(1)}(t) = \epsilon_0 \chi_i^1 E_i \quad (7.22)$$

$$\chi_i^1 = -\frac{n_i c}{\omega_i} [(\alpha + j) g(N, \omega_i)] \quad (7.23)$$

where α is the linewidth enhancement factor and $g(N, \omega_i)$ is the modal gain, which is a function of carrier density N in the active region and the frequency of the mode.

Both theoretical calculations and experimental measurements show that the linear gain spectrum in semiconductor lasers is asymmetrical with respect to its peak frequency ω_{pk}, and the gain spectrum curve rolls off more quickly on the higher frequency side than on the lower frequency side [38]. In order to account for this asymmetry, a "piecewise-quadratic" frequency dependent relationship is used here:

$$g(N, \omega_i) = \begin{cases} \Gamma \left[a(N - N_0) - \zeta_1(\omega_i - \omega_{pk})^2 \right] & \omega_i \leq \omega_{pk} \\ \Gamma [a(N - N_0) - \zeta_2(\omega_i - \omega_{pk})^2] & \omega_i > \omega_{pk} \end{cases} \quad (7.24)$$

where a is the material gain coefficient, N_0 is the transparency carrier density, and Γ is the confinement factor. ζ_1 and ζ_2 are the quadratic coefficients on the low and high frequency side, respectively, the value of which can be fitted from measured data.

Substituting Equations 7.22 and 7.23 into Equation 7.20, we obtain the rate equations of modal fields in SRL:

$$\frac{dE_i}{dt} = \frac{1}{2}(1 - j\alpha)\frac{c}{n_g}g(N, \omega_i)E_i - \frac{1}{2}\frac{E_i}{\tau_p} + \frac{j}{2}\frac{\omega_i}{n_i n_g \epsilon_0}P_i^{(3)}(t) \quad (7.25)$$

In semiconductor lasers, the gain and phase shift experienced by the optical fields propagating in the device are functions of carrier density N. The carrier density is controlled by the injection current but is also affected by the field intensity in the laser cavity. To analyze the characteristics of SRLs, we need the following carrier density rate equation in addition to Equation 7.25, which describes the relation between the variation of carrier density and injection current.

$$\frac{dN}{dt} = \frac{I}{eV} - \frac{N}{\tau_s} - \sum_i g(N, \omega_i)\left(\frac{2\epsilon_0 n_g n_i}{\hbar \omega_i}|E_i|^2\right) \quad (7.26)$$

where e is the electron charge, V is the volume of the active region, and τ_s is the carrier lifetime given by

$$\tau_s^{-1} = A + BN + CN^2 \quad (7.27)$$

where A, B, and C are the nonradiative, radiative, and Auger recombination coefficients, respectively.

Equations 7.25 and 7.26 form the fundamental equations for SRLs, and they are the basis of the numerical model for the behavior and dynamic characteristics of SRLs. These fundamental equations alone, however, are not sufficient for describing how light propagates through a medium because they only describe how the electromagnetic wave responds to a given induced nonlinear polarization $P_i^{(3)}(t)$. Therefore the third order nonlinear polarization must be determined.

7.2.2
Third Order Susceptibility and Polarization

In general, consider two longitudinal modes propagating in the same direction in an SRL, with complex field amplitudes E_1 and E_2, frequencies ω_1 and ω_2 and wave-vectors k_1 and k_2. The total electric field of the two longitudinal modes at any point in the SRL can be written as:

$$E_{1,2}(z, t) = \text{Re}[E_1 \exp j(\omega_1 t - k_1 z) + E_2 \exp j(\omega_2 t - k_2 z)]$$

The superposition of modes 1 and 2 results in an intensity profile with three components:

1) A DC component proportional to $|E_1|^2$
2) A DC component proportional to $|E_2|^2$

3) An interference pattern proportional to $E_1^* E_2$, which contains both a time-variation at a beat frequency $\Omega_{12} = \omega_2 - \omega_1$, and a spatial variation, with a spatial periodicity given by $2\pi/(k_2 - k_1)$.

The DC components of the intensity, as discussed later, give rise to self-gain/index suppression and the static parts of cross-gain/index suppression. The beating of the two modes leads to the modulation of various parameters of the SRL material at the mode beat frequency $\Omega_{\text{beat}} = \omega_2 - \omega_1$, which gives rise to the dynamic gain/index grating. Diffraction of the mode with frequency ω_2 from the grating leads to the generation of signals at different frequencies $\omega_2 \pm \Omega_{\text{beat}}$, while diffraction of the mode with frequency ω_1 from the grating leads to the generation of signals at different frequencies $\omega_1 \pm \Omega_{\text{beat}}$. In other words, the resulting modulation effects will produce frequency sidebands on the two modes. Basically, any type of nonlinear modulation effects in the SRL material produced by two modes will react back to cross-modulate the two modes with each other and to produce new nonlinear mixing frequencies, and these nonlinear mixing effects can become very complex when a large number of longitudinal modes need to be considered.

The third-order nonlinear processes in SRLs, such as self-suppression, cross-suppression, and FWM, are caused by inter- and intraband effects. The former refer to carrier transitions between the conduction and valence band, which modulate the overall carrier density, whereas the latter are related to carrier–carrier and carrier–phonon scattering and modify the carrier distribution within one band. The inter-band effects change the carrier density because of carrier depletion caused by stimulated emission, which is referred to as *carrier density pulsation* (*CDP*) (see Figure 7.4). The characteristic time of CDP is the effective carrier lifetime τ_s, which is on the order of several hundred picoseconds.

The intraband effects are associated with two ultra-fast phenomena. Firstly, the stimulated emission induced by the intense optical field depletes carriers at the corresponding photon energy level and "burns a hole" in the carrier distribution causing a deviation from the Fermi distribution. This process is referred to as *spectral hole burning* (*SHB*) (see Figure 7.4). The time constant associated with this process is the time needed to restore the Fermi distribution or eliminate the dip caused by SHB. This intraband relaxation time or carrier–carrier scattering time is typically several tens of femtoseconds. Additionally, carriers at low energy levels, which are close to the band-edge, are depleted by stimulated emission or transferred to higher energy levels within the bands because of free carrier absorption. This phenomenon leads to an increase of the temperature of the carrier distributions beyond the lattice temperature and is called *carrier heating* (*CH*) (see Figure 7.4). The characteristic time needed for the "hot"-carrier distributions to relax to the lattice temperature, owing to carrier–phonon scattering, is on the order of several hundred femtoseconds.

CDP, SHB, and CH are the built-in mechanisms in semiconductor laser material that mediate various nonlinear processes. Each nonlinear gain mechanism has an associated time constant, which limits the maximum beat frequency to which the respective population pulsation component can respond. The nonlinear gain coefficients are, therefore, different for different combinations of beating modes.

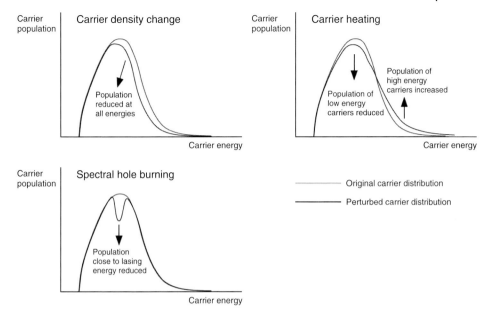

Figure 7.4 Changes to the carrier population caused by carrier density changes, carrier heating, and spectral hole burning.

For beat frequencies in the megahertz–gigahertz range, CDP is the dominant nonlinear mechanism and the modulation of the overall carrier density is very effective. As described above, the modulation of the carrier density leads to the formation of dynamical gain/index gratings, which scatter each of the modes in the SRLs into other modes. The magnitude of the CDP-mediated nonlinear processes quickly drops off for beat frequency exceeding 10 GHz, because of the relatively slow recovery of the carrier density determined by τ_s. Instead, intraband dynamics start to reveal themselves.

The intraband dynamics, including SHB and CH, are the processes that affect the shape of the carrier distributions in the bands, rather than the overall carrier density. Such structural perturbations will also lead to dynamic gain/index gratings in SRLs, which couple the various modes of the SRLs.

In order to derive the nonlinear polarization $P_i^{(3)}$, we consider the interaction between four modes, as shown in Figure 7.5. The figure shows four field components, i, l, m, and n at angular frequencies ω_i, ω_l, ω_m, and ω_n – with corresponding field amplitudes E_i, E_l, E_m, and E_n. Beating between modes m and n can induce oscillations in the carrier distribution, which are responsible for the induced third-order nonlinear polarization oscillations. The lth mode could be scattered into the ith mode by these beating induced oscillations under certain conditions. The total material response can be considered, to a high degree of accuracy, to be a linear combination of each and every beating induced oscillation. Following the commonly employed treatment in the literature, which is supported by more detailed microscopic analysis [39], the contributions from CDP, SHB, and

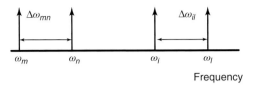

Figure 7.5 Nonlinear interactions of four waves.

CH are considered to be independent. We, therefore, calculate the total material response by adding all of the oscillations. Substituting Equations 7.18 and 7.22 into Equation 7.19 gives:

$$P_i^{(3)}(t) = \epsilon_0 \sum_{l,m,n} \chi_{lmn}^3 E_l E_m^* E_n S_{ilmn} e^{-j(\omega_i - \omega_l + \omega_m - \omega_n)t} \tag{7.28}$$

where

$$S_{ilmn} = \frac{1}{L} \int_0^L e^{j(k_i - k_l + k_m - k_n)z} dz$$

$$= \begin{cases} 1 & k_i - k_l + k_m - k_n = 0 \\ 0 & k_i - k_l + k_m - k_n \neq 0 \end{cases} \tag{7.29}$$

S_{ilmn} is the spatial overlap integration between modes i, l, m, and n. It serves as the selection rule that decides whether effective coupling can occur among the modes. As illustrated in Figure 7.6a, the combinations of i, l, m, and n, leading to a nonzero value of S_{ilmn}, can be divided into two groups: the case $m = n$ yielding the static effects and $m \neq n$ yielding the dynamic effects.

The static effects are not associated with any beating process and they are not dependent on the frequency detuning of the interacting modes. Moreover, only SHB and CH contribute to these nonlinearities. The static effects can be further separated into the case of $m = n = l$ and the case of $m = n \neq l$, which describe the self-gain suppression (the nonlinear refractive index and gain change induced by the mode under consideration itself) and static cross-gain suppression effects (the nonlinear refractive index and gain change induced by all other modes), respectively. From the point of view of physics, the total carrier distribution within the band is changed because of SHB and CH, which causes a change of the gain and the refractive index at the energy level corresponding to the mode under consideration, as well as at all different energy levels.

In contrast to the static effects, the dynamic effects only exist if there is beating between longitudinal modes. Therefore, these effects are dependent on the frequency detuning of various longitudinal modes. All of the three effects – SHB, CH, and CDP – contribute to the dynamic effects. The case $m \neq n = l$ yields the dynamic cross-suppression or Bogatov effects, which is the origin of the asymmetric nature of gain suppression. It works to suppress the lasing gain of the modes on the shorter wavelength side, while enhancing the lasing gain of the modes on the longer wavelength side. The case $m \neq n \neq l$ yields the FWM, which normally exhibits the opposite asymmetry to the dynamic cross-gain suppression. For simplicity, both static and dynamic cross-gain suppression effects are referred to as *cross-gain suppression*.

It should be noted that the total carrier density in a SRL can be divided into a stationary part and an oscillatory part. The stationary part corresponds to the saturated carrier density. Saturation of the carrier density is the strongest nonlinearity in semiconductor optical gain material, the contribution of which is included in χ_l^1. The oscillatory part, which oscillates at the beat frequency, is included in χ_{lmn}^3 – corresponding to the contribution of CDP. Therefore, only SHB and CH contribute to the third-order static effects, whereas SHB, CH, and CDP contribute to the dynamic effects.

In SRLs, gain suppression effects occur between any longitudinal modes, while the FWM effect can only occur among modes propagating in the same direction as a result of the restrictions imposed by the phase matching condition in Equation 7.29. Figure 7.6b shows the interaction paths between the ith mode and its neighborhood modes.

From density matrix theory [34, 36], χ_{lmn}^3 is given by

$$\chi_{lmn}^3 = \chi_{lmn}^N(\Omega_{mn}) + \chi_{lmn}^{ch}(\Omega_{mn}) + \chi_{lmn}^{shb}(\Omega_{mn})$$

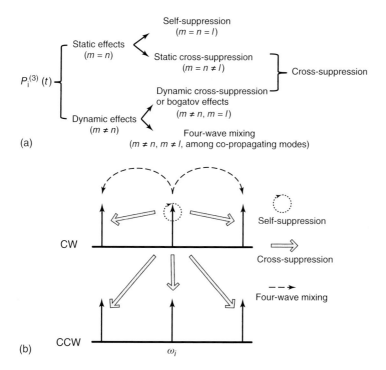

Figure 7.6 (a) Overview of different third order nonlinear effects in SRLs. (b) Interaction paths for ith mode in CW direction.

with

$$\chi_{lmn}^{N} = -\frac{2\epsilon_0 n_{pk} n_g}{\hbar\omega_{pk}} \epsilon_N \eta^N (\alpha + j) \frac{1}{(1 - j\Omega_{mn}\tau_N \eta^N)(1 - j\Omega_{mn}\tau_1 \eta^1)} \chi_{pk}^{1''} \quad (7.30)$$

$$\chi_{lmn}^{ch} = -\frac{2\epsilon_0 n_{pk} n_g}{\hbar\omega_{pk}} \epsilon_{ch} \eta^{ch} (\alpha_{ch} + j) \frac{1}{(1 - j\Omega_{mn}\tau_{ch} \eta^{ch})(1 - j\Omega_{mn}\tau_1 \eta^1)} \chi_{pk}^{1''} \quad (7.31)$$

$$\chi_{lmn}^{shb} = -\frac{2\epsilon_0 n_{pk} n_g}{\hbar\omega_{pk}} \epsilon_{shb} \eta^1 \frac{j}{1 - j\Omega_{mn}\tau_1 \eta^1} \chi_{pk}^{1''} \quad (7.32)$$

ϵ_N, ϵ_{ch}, and ϵ_{shb} give the nonlinear coefficients for CDP, CH, and SHB. α_{ch} is the linewidth enhancement factor for CH. τ_N, τ_{ch}, and τ_1 are differential carrier lifetime, electron CH time, and intraband relaxation time, respectively. $\Omega_{mn} = \omega_n - \omega_m$ is the beat frequency. $\chi_{pk}^{1''}$ is the imaginary part of χ_{pk}^{1}. η^N, η^{cd}, and η^1 are the grating visibilities due to the carrier diffusion. The grating visibilities are equal to 1 for co-propagating modes and less than 1 for counter propagating mode, as the carrier diffusion length is longer than the grating period of $\sim \lambda/2n$, which "washes out" the grating.

It should be noted that the magnitude of CDP may still be comparable to that of the SHB and CH at high beat frequencies beyond the "cut-off" frequency of CDP. Due to the high value of ϵ_N in the DC-low frequency range that is two orders of magnitude higher than ϵ_{ch} and ϵ_{shb}, CDP can still contribute significantly to χ_{lmn}^{3} at Ω_{mn} up to $100 \times (1/\tau_N)$, which is approaching terahertz frequencies. At such beating frequencies, however, χ_{lmn}^{N} is lagging in phase by up to $\pi/2$, relative to its low frequency values, which can have a profound effect on the total χ_{lmn}^{3}, and through it on the nonlinearity. In fact this is the main mechanism contributing to the asymmetry of the gain spectrum mentioned earlier.

7.2.3
Generalized Equations in Matrix Form

Substitution of Equation 7.28 into the electric fields Equation 7.25 gives the differential equations describing the evolution of the ith mode inside the ring cavity in the presence of material oscillations caused by the beating of the mth and nth modes, which scatters the lth mode:

$$\frac{dE_i}{dt} = \frac{1}{2}(1 - j\alpha)\frac{c}{n_g}g(N, \omega_i)E_i - \frac{1}{2}\frac{E_i}{\tau_p} + \frac{1}{2}\frac{\omega_i}{n_i n_g}\sum_{lmn}\chi_{lmn}^{(3)} E_l E_m^* E_n S_{ilmn} - KE_{-i} + k_{fbk}E_{fbk} \quad (7.33)$$

The first term in Equation 7.33 is the linear gain and phase shift experienced by the ith mode. The third term is the nonlinear contribution due to the presence of the mth, nth, and lth modes. The fourth term describes the backscattering between counter-propagating modes, and K is the backscattering rate. For simplicity, the frequency dependence of K has been neglected. The last term accounts for the external feedback or injection. This term is necessary when one wants to simulate SRLs under optical feedback from external facets or optical injection from external laser sources.

The oscillating modes in SRLs can be analyzed by a combination of rate Equation 7.33 and the carrier density rate Equation 7.26. It should be noticed that the carrier density rate Equation 7.26 explicitly takes into account the static change in N due to the overall intensity $\sum |E_i|^2$. Therefore, all terms in Equation 7.33, where $m = n$, are set to zero in order to prevent double counting of this population change component. The SHB and CH terms with $m = n$ contribute to the nonlinear gain compression, which is frequently modeled by introducing a factor of $(1-\epsilon_{p0})$ [40] in the modal gain, where $\epsilon = \epsilon_{ch} + \epsilon_{shb}$.

In Equation 7.33, the number of terms in the summation over modes l, m, and n is equal to the total number of modes *cubed*. This means that the computation will be intensive when a large number of modes needs to be considered. In order to circumvent this problem, we use here a treatment similar to that proposed by M. Summerfield and R. Tucker [41] for simulating multiwave mixing in semiconductor optical amplifiers (SOAs).

Instead of adding up all combinations of l, m, and n that contribute to the ith mode, the summation in Equation 7.33 is rewritten in the form of a matrix operation and Equation 7.33 is extended into a more general vector differential equation from which the numerical solution can be obtained very efficiently.

To perform the analysis, we refer again to the general case represented by Equation 7.18 and shown in Figure 7.2 for a CW mode i, that is, $i > 0$. The summation terms in Equation 7.33 can be rewritten as:

$$\sum_{lmn} \chi^3_{lmn} E_l E_m^* E_n S_{ilmn}$$

$$= \underbrace{\sum_{lmn} \chi^3_{lmn}((n-m)\Delta\omega) E_l E_m^* E_n}_{m \times i < 0, i = l-m+n} + \underbrace{\sum_m [\chi^3_{imm}(0) + \chi^3_{mmi}((i-m)\Delta\omega)] |E_i||E_m|^2}_{m \times i < 0}$$

$$= M^i_{co} + M^i_{counter} \qquad (7.34)$$

M^i_{co} and $M^i_{counter}$ describe the contributions to the ith CW mode from co-propagating and counter-propagating modes, respectively, including all the nonlinear effects, while $M^i_{counter}$ only includes the cross-gain suppression.

χ^3_{lmn} is the function of detuning $\Omega_{mn} = (n-m)\Delta\omega = q\Delta\omega$ (where q is an integer), and so it is rewritten as χ^3_q. For notational convenience, we define five vectors ψ, A_{cw}, A_{ccw}, G, and H as follows:

$$\psi = \{\chi^3_{-2M}, \ldots, \chi^3_0, \ldots, \chi^3_{2M}\}^T \qquad (7.35)$$

$$A^{cw} = \{E_1, \ldots, E_{M+1}, \ldots, E_{2M+1}\}^T \qquad (7.36)$$

$$A^{ccw} = \{E_{-1}, \ldots, E_{-(M+1)}, \ldots, E_{-(2M+1)}\}^T \qquad (7.37)$$

$$G = \left\{ \frac{c}{n_g}(1-j\alpha)g(N,\omega_1) - \frac{1}{\tau_p}, \ldots, \frac{c}{n_g}(1-j\alpha)g(N,\omega_{2M+1}) - \frac{1}{\tau_p} \right\}^T \quad (7.38)$$

$$H = \left\{ \frac{1}{n_g}\frac{\omega_1}{n_1}, \ldots, \frac{1}{n_g}\frac{\omega_{2M+1}}{n_{2M+1}} \right\}^T \quad (7.39)$$

The $(2M+1)$-element vector of all coupling terms in the CW direction, σ^{cw}, can be obtained in the following way:

$$\theta = (\psi \cdot A^{cw} * A^{cw}_{-1}) * A^{cw} \quad (7.40)$$

$$\sigma^{cw} = \{\theta_{2M+1}, \theta_{2M+2}, \ldots, \theta_{4M}, \theta_{4M+1}\} \quad (7.41)$$

where θ is an intermediate vector. $*$ and $\cdot$ indicate the discrete convolution and inner product of two vectors. A^{cw}_{-1} is the order reversed complex-conjugate of A^{cw}.

We also construct a $(2M+1)$ by $(2M+1)$ matrix X in such a way that the element X_{ij} in the ith row and jth column is given as

$$X_{ij} = \chi_0^3 + \chi_{j-i}^3 \quad (7.42)$$

The vector describing the cross-gain suppression experienced by A^{cw} due to the existence of A^{ccw} is given by

$$v^{cw} = [X \times (A^{ccw} \cdot A^{ccw*})] \cdot A^{cw} \quad (7.43)$$

Where $\times$ denotes matrix multiplication. Then the generalized vector differential-equation for A^{cw} can be expressed as

$$\frac{dA^{cw}}{dt} = 0.5G \cdot A^{cw} + 0.5H \cdot \sigma^{cw} + 0.5H \cdot v^{cw} + K_{fbk}E_{fdk} - KA^{ccw} \quad (7.44)$$

Similarly, the equation for A^{ccw} can be expressed as:

$$\frac{dA^{ccw}}{dt} = 0.5G \cdot A^{ccw} + 0.5H \cdot \sigma^{ccw} + 0.5H \cdot v^{ccw} + K_{fbk}E_{fdk} - KA^{cw} \quad (7.45)$$

Equations 7.44 and 7.45 represent a complete and concise description of the set of coupled equations that account for the interaction between all modes, inside the SRLs.

7.3
Numerical Results for Micro-Ring Lasers

We present below the simulation results for SRLs. The device parameters used in the simulations are given in Table 7.1.

Some parameters, including ϵ_N, ϵ_{ch}, ϵ_{shb}, η^N, η^{ch}, η^1, τ_{ch}, and α_{ch}, are obtained by fitting the measurement of the bandwidth and relative frequency detuning of the individual sub-peaks within a SRL longitudinal mode by heterodyne detection methods [33]. Gain-slope and quadratic gain coefficients are measured using the

Table 7.1 SRL parameters used for numerical simulation.

Description	Parameter	Value
SRL cavity length	L (μm)	170
Confinement factor	Γ	0.056
Material loss	α_{int} (m^{-1})	2700
Coupling transmission ratio	T_r	0.995
Nonradiative recombination coefficient	A (s^{-1})	1×10^8
Spontaneous recombination coefficient	B (m^3 s^{-1})	1×10^{-16}
Auger recombination coefficient	C (m^6 s^{-1})	7.5×10^{-41}
Transparency carrier density	N_0 (m^{-3})	2.2×10^{24}
Gain-slope coefficient	a (m^2)	6.35×10^{-20}
Quadratic gain coefficient	ζ_1 (m^{-1} m^{-2})	0.5×10^{18}
Quadratic gain coefficient	ζ_2 (m^{-1} m^{-2})	0.9×10^{18}
Group index	n_g	3.7
Refractive index at gain peak	n_{pk}	3.3
Nonlinear coefficient of CDP	ϵ_N (m^3)	2.5×10^{-21}
Nonlinear coefficient of CH	ϵ_{ch} (m^3)	1.9×10^{-23}
Nonlinear coefficient of SHB	ϵ_{shb} (m^3)	1.7×10^{-23}
Differential carrier lifetime	τ_N (ns)	0.5
Electron carrier heating time	τ_{ch} (ps)	0.6
Intraband relaxation time	τ_1 (ps)	0.1
Linewidth enhancement factor	α	1.57
Linewidth enhancement factor for CH	α_{ch}	2.2
Backscattering rate	K	$1.8 \times 10^7 + 2 \times 10^8$ j
Grating visibility for CDP	η^N	0.0036/1
Grating visibility for CH	η^{ch}	0.73/1
Grating visibility for SHB	η^1	0.95/1

Hakki-Paoli method. The recombination parameters A, B, and C are obtained by measuring the frequency response, via a small modulation signal, of a Fabry-Perot laser fabricated with the same material. The differential carrier lifetime τ_N is found from the 3-dB roll-off of the frequency response. The linewidth enhancement factor α is the calculated value from [36], and it agrees with the measured data [42]. Other parameters are chosen to take the generally accepted values.

The calculated threshold current is $I_{th} = 4.7$ mA. This value is somewhat lower than the experimental values, which may be attributed to underestimated cavity losses or overestimated injection efficiency. However, this discrepancy does not affect the results presented below, because all the values of the injection current in this chapter are normalized with respect to I_{th}. In order to clearly present the simulation results, we modify our mode referencing system presented in Figure 7.3 in such a way that the central modes in both directions are labeled as mode 0. The modes with indices $M > 0$ are assumed to lie on the long-wavelength side of mode 0, and those with indices $M < 0$ are assumed to lie on the short wavelength side.

Suppose a step-current is injected into an SRL, rising abruptly from zero to a stationary value above the threshold, the laser undergoes relaxation oscillations to a steady state. We choose the point of time when the carrier density reaches its threshold value N_{th} for the first time during the relaxation oscillations to be the origin of our time scale. Therefore, the initial condition for the carrier density is $N(0) = N_{th}$. In the following, we shall take 34 modes into account, with 17 modes in each direction and we shall suppose that, at $t = 0$, all modes have the same amplitude $|E_j| = 10^{-3}/3$, but have a statistically independent phase φ_j that follows a uniform distribution in the range $[0, 2\pi]$. The simulation results given here were obtained using a fourth-order Runge–Kutta method. A time step of $\Delta t = 6$ ps was applied to the integration process and the integration was carried out over a time period of $T = 500$ ns, which is sufficiently long to achieve the steady-state in semiconductor lasers.

7.3.1
L–I Characteristics

The steady-state L–I curves for the SRL can be obtained by calculating the time domain response to a fixed current I until steady-state conditions are reached. After recording the steady-state value of the complex amplitude of each mode E_j and the carrier density N for that value of I, the current is increased and the calculation process is repeated using the previously calculated values of E_j and N as initial conditions. The injection current is varied in discrete steps of $(1/50)I_{th}$, ranging from 0 to $4.25 I_{th}$. The thermal shift of the gain profile is assumed to be 13 nm toward the long wavelength from the start to the end of the simulation, corresponding to 3.4 times the free spectral range (FSR) of the SRL.

Figure 7.7a displays the simulated L–I curves in both the CW and CCW directions as the bias current is increased for the SRL. Three distinct operating regimes could be identified. At just above the threshold, the device operates in bi-CW region, where both directions lase with constant output power. From 1.1 to $1.3 I_{th}$, the SRL enters into the bi-AO region, in which the output direction periodically oscillates at a frequency of around 80 MHz. Above $1.3 I_{th}$, unidirectional operation occurs. In this region, the lasing direction does not remain stable for all current values. It is only stable within a particular current range, but as the current is increased further, the lasing direction alternates between the CW and CCW directions.

The device shows single-longitudinal-mode operation with a side-mode suppression-ratio (SMSR) on the order of 20–35 dB, as shown in Figure 7.7b which plots the dominant lasing wavelength and the SMSR at different values of the bias current from both the CW and CCW directions. The dominant lasing wavelength remains constant between directional reversals. However, a large jump to the long wavelength side is observed when the lasing direction reverses, with a sudden drop in the SMSR for the original lasing direction, because the lasing in this direction is now strongly suppressed.

The L–I curves displayed in Figure 7.7a are in very good agreement with the experimental observations [16, 22] on various types of SRL device.

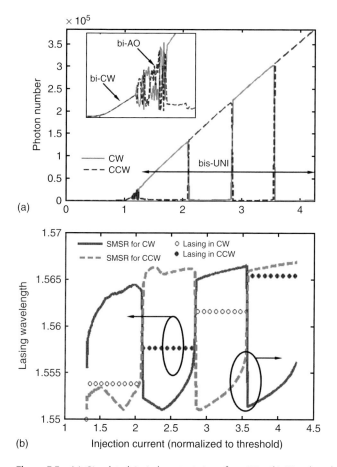

Figure 7.7 (a) Simulated L–I characteristics of an SRL. (b) Wavelength for the dominant mode and SMSR under different injection currents in the bis-UNI region (simulation results).

7.3.2
Temporal Dynamics and Lasing Spectra

Typical examples for the time variation of the total photon number in the two directions of the SRL in the regions of bi-CW, bi-AO, and bis-UNI are plotted in Figure 7.8a–c, which correspond to current values of $I = 1.1$, 1.2, and $1.85 I_{th}$, respectively. The insets show details of the photon number at the beginning of the oscillations.

In Figure 7.8a, the intensities (photon numbers) in the two directions reach almost the same intensity, after relaxation oscillations. The corresponding lasing spectra are depicted in Figure 7.9. In both directions the main mode is mode 0, with

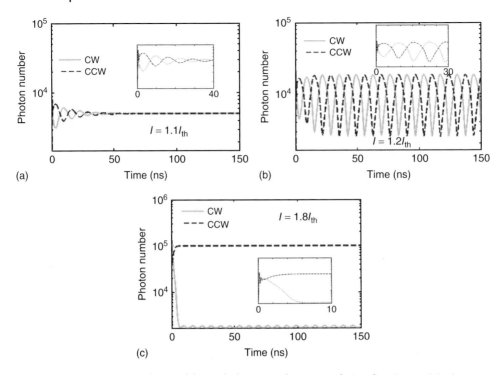

Figure 7.8 Evolution of the total photon numbers in two lasing directions at injection current of (a) $1.1I_{th}$, (b) $1.2I_{th}$, and (c) $1.8I_{th}$. The inset shows the oscillations of the photon numbers at very beginning.

a SMSR of 8.6 dB. The lasing spectra are obtained by calculating the time average for the photon numbers of different modes, after the initial relaxation oscillations die away.

In Figure 7.8b, the intensities in both directions are periodically modulated with complementary phase at a frequency of 86 MHz. The modulation frequency is related to the specific value of the backscattering rate K, and will decrease for increasing pumping current. The oscillation behaviors of different modes are plotted in Figure 7.10.

For higher injection current, unidirectional operation can be obtained. As shown in Figure 7.8c, at $I = 1.85I_{th}$, CCW is the dominant lasing direction with more than 99% of the total power emitted from this direction, while modes in the CW direction are strongly suppressed. Figure 7.11 plots the time evolutions of different modes and the lasing spectra in both directions at $I = 1.85I_{th}$. Here mode 1 in the CCW direction is the dominant mode, and other modes are side modes. It can be noted that the side mode intensities oscillate constantly due to the dynamic effects.

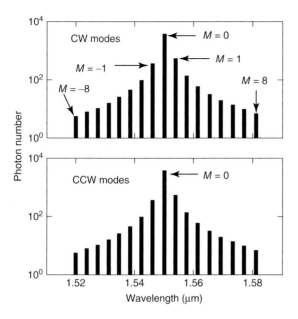

Figure 7.9 Lasing spectrum in CW and CCW directions in bi-CW region at $I = 1.1 I_{th}$.

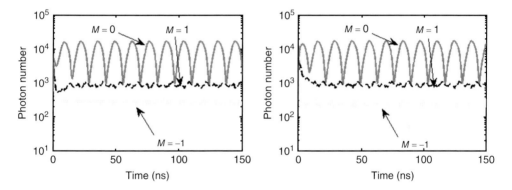

Figure 7.10 (a,b) Evolutions of photon numbers for different modes in CW and CCW directions in bi-AO region at $I = 1.1 I_{th}$.

7.3.3
Lasing Direction Hysteresis

Similarly, the output intensities of the SRL with a decreasing injection current can also be calculated. Figure 7.12 shows the simulated results for the intensities in the CW and CCW directions as functions of increasing and decreasing injection current, respectively. As the current is increased, lasing alternates between the two directions, accompanied by mode hops to the long wavelength. As the current is decreased, however, the lasing direction remains stable until

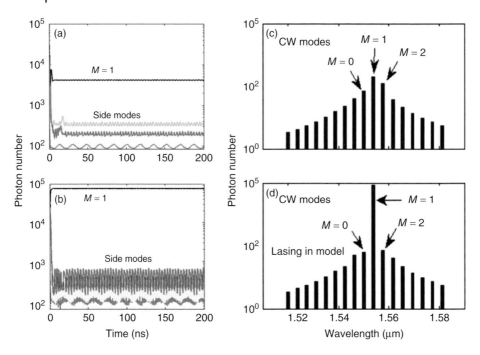

Figure 7.11 Evolutions of photon numbers for different modes in (a) CW and (b) CCW directions in bi-UNI region at $I = 1.8I_{th}$. (b) Lasing spectrum in (c) CW and (d) CCW directions.

the SRL reverts to bidirectionality. During the decrease of the current, the lasing wavelength sequentially steps through each cavity resonance toward shorter wavelengths.

With increasing current, the origins of the periodic switching between two directions can be understood with the help of the lasing spectra shown in Figure 7.11c,d. As discussed in the last section, FWM can only occur in co-propagation modes, while gain-suppression can occur between any modes in both directions. The strong power of CCW mode 1 causes the asymmetric gain suppression to be pronounced, so as to enhance the modes on the long wavelength side and suppress the modes on the short wavelength side, in both directions. This explains the highly asymmetric spectrum in the CW direction shown in Figure 7.11c. In the CCW direction however, the asymmetry caused by gain suppression is compensated by FWM in which CCW mode 1 acts as the pump which is the reason why the CCW lasing spectrum is only slightly asymmetric, in contrast to the CW spectrum. Therefore, CW mode 2 is placed in a more advantageous position with respect to CCW mode 2, in the mode competition. As the injection current is further increased, the joule heating in the active region shifts the gain spectral profile toward longer wavelengths and CW mode 2 has a greater chance to be selected as the dominant lasing mode when the mode hop occurs. It should be pointed out that this process is contributed to by all the possible combinations of mode coupling

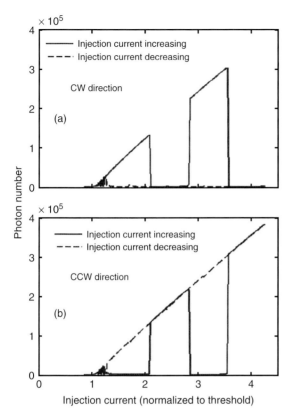

Figure 7.12 Simulated output intensity as the injection current is increased and decreased for (a) CW direction and (b) CCW direction.

included in the model, not just the few modes (Mode 1 and 2 in both direction) directly involved in the transition.

Figure 7.13 plots two examples of the mode dynamics for increasing and decreasing current when the mode hops happen. For the increasing current case, the lasing wavelength hops to longer wavelengths from 1.562 μm (CW mode 3) to 1.565 μm (CCW mode 4). The mode hop is relatively rapid, with a less than 10 ns transition time. Once the transition starts, the intensity of CCW mode 4 increases monotonically until the steady state is achieved. During the transition, very little power appears in other modes. It should be noted that the mode hopping range in the increasing current case is not decided by the FSR of the SRL. It is rather related to the amount of asymmetry in the nonlinear gain spectrum. Hopping of several FSRs was observed when the FSR is small (larger SRL size), and can be readily reproduced using our model [35].

For decreasing current, the lasing wavelength hops to shorter wavelength, from 1.565 μm (CCW mode 4) to 1.562 μm (CCW mode 3). The mode hop takes

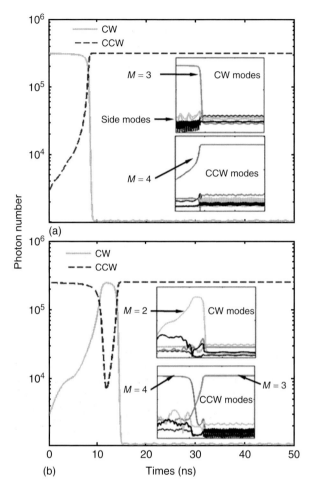

Figure 7.13 Photon number dynamics during mode hops to a longer wavelength and a shorter wavelength. (a) Evolution of total photon numbers in CW and CCW during the mode hop which occurs when the injection current increased from $I = 3.53 I_{th}$ to $I = 3.61 I_{th}$. (b) Evolution of total photon numbers in CW and CCW during the mode hop which occurs when the injection current decreased from $I = 3.13 I_{th}$ to $I = 3.07 I_{th}$. The insets show the photon number in different modes.

around 15 ns. It can be observed that CW mode 2 also takes part in the mode competition, as well as CCW mode 3 and CCW mode 4. At $I = 3.07 I_{th}$, the linear gain peak is at 1.5582 μm, which is very close to the wavelength of mode 2 (1.558 μm), which can explain why mode 2 rises at the beginning of the transition process.

The simulated pattern of behaviors shown in Figure 7.12 agrees very well with the hysteresis pattern from measurements reported on various types of SRL [15, 16, 35].

7.3.4
Experimental Results for Racetrack Shaped SRLs

In order to assess the validity of the model, measurements of fabricated racetrack-shaped SRLs with similar device parameters to those used in the model have been carried out.

The fabricated devices are based on a commercially available, compressively strained, multiple quantum well (MQW) InAlGaAs–InP wafer structure. Above and below the five quantum wells (QWs), two 60 nm thick, graded index separate-confinement heterostructure layers, sandwiched by two wide band-gap layers which prevent electrons and holes from escaping the QW region, are used to ensure better optical confinement. Theoretical and experimental studies show that the conduction band offset of AlGaInAs/InP materials ($\Delta E_c = 0.72 \Delta E_g$) is larger in comparison with that of the InGaAsP/InP materials ($\Delta E_c = 0.4 \Delta E_g$). As a result, devices based on Al-quaternary materials have a lower carrier leakage and a higher characteristic temperature.

Each racetrack consists of two bends and two straight waveguides with an evanescent output coupler. The bending radius varies from 10 to 30 μm and the coupling length L_c varies from 5 to 25 μm. The circumferences of the devices are within the range of 80–240 μm, which is much smaller than all the previous reported unidirectional SRLs [5–7]. Both ends of each output waveguide are 10° tilted with respect to the cleaved facets, in order to minimize optical back-reflections. In addition to the cavity biasing contact, separate contact pads are defined on the output waveguides, which can be pumped as SOAs. An SEM image of the devices is shown in Figure 7.14.

Figure 7.15 shows the experimental setup used to investigate the lasing characteristics of fabricated SRL. Fiber lenses were used to couple the outputs from two directions of SRL into fibres that were fed to a grating based optical spectrum analyzer (OSA) and optical power meters via 3 dB couplers.

Figure 7.16a shows the $L–I$ curves for both the CW and CCW directions of a SRL with a radius of 18 μm and L_c of 24 μm, measured at 293 K. The threshold of this device was measured to be 12 mA and the corresponding threshold current density was 2.16 kA cm^{-2}.

The SRL shows behavior that is commonly observed in similar types of devices. Periodic switching between the CW and CCW modes occurs with increasing biasing current, accompanied by longitudinal mode hops when the lasing direction reverses. It should be noted that the waveguide contacts were left un-pumped. In the unidirectional lasing regime, the device shows single longitudinal mode operation with SMSRs on the order of 20–35 dB (inset of Figure 7.16a).

Figure 7.16b shows the dominant lasing wavelength and the SMSR at different values of the biasing currents for light from both the CW and CCW directions. The dominant lasing wavelength remains constant between directional switching events, except for a small thermal drift. However, a large jump to the long wavelength side is observed when the lasing direction reverses, with a 15 dB drop

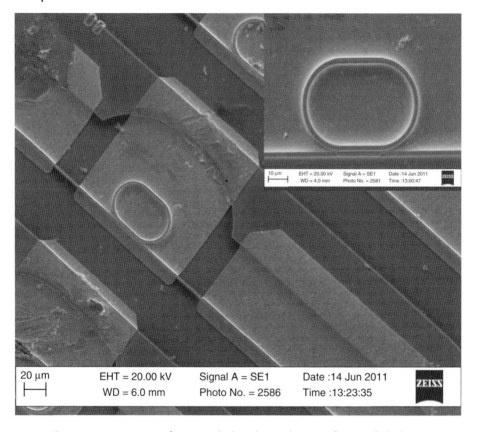

Figure 7.14 SEM image of a racetrack shaped SRL, the circumference of which is 170 μm. The inset shows a zoom-in picture of the racetrack.

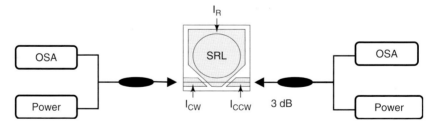

Figure 7.15 Experimental setup for measurement of the L–I curve and the lasing spectra of an SRL.

in the SMSR of the original lasing direction, because the lasing in this direction is now strongly suppressed.

The measured results shown in Figure 7.16 agree very well with the simulated results shown in Figure 7.11, which indicates the effectiveness of the model in describing the lasing characteristics of SRLs.

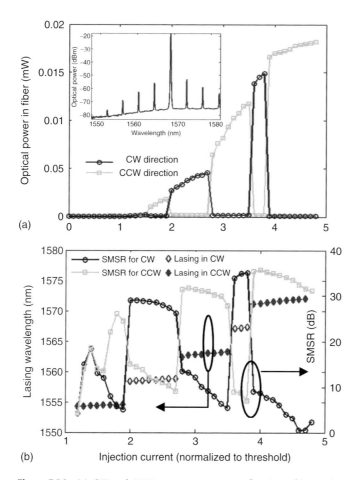

Figure 7.16 (a) CW and CCW output powers as a function of increasing current for a racetrack bistable SRL. The inset shows the CW lasing spectrum at $3.1I_{th}$. (b) Wavelength for the dominant mode and SMSR under different injection current measured from the bistable SRL.

7.4
Numerical Results for Unidirectional Micro-Ring Lasers

The preceding sections demonstrate the effectiveness of the nonlinear multimode rate equation model in describing various lasing characteristics of SRLs. We now present an application of the model in achieving a stable unidirectional SRL by way of precisely controlled external feedback. The model will be used to predict various lasing characteristics of the devices at various different strengths of optical feedback. A high resolution focused ion beam etching (FIBE) process is utilized to fabricate mirror facets that provide such feedback. The angles of the mirror facets can be controlled by the FIBE and help determine the different strengths of the optical feedback.

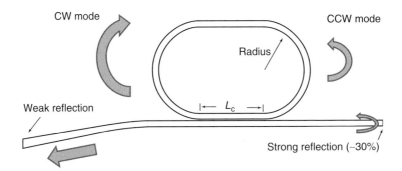

Figure 7.17 Schematic drawing of the unidirectional SRL.

7.4.1
Unidirectional SRL

The structure of the unidirectional SRL is plotted in Figure 7.17. The left-hand end of the output waveguide is tilted relative to the cleaved facet in order to minimize the reflectivity, while the right-hand end is not tilted in order to maximize the reflectivity of the cleaved facet. The SRL would be forced to oscillate in the CW direction, because the light from the CCW direction should be strongly coupled into the CW direction by the reflection from the facet on the right hand side, giving the CW direction an advantage in mode competition.

The rate equations should be revised to take into account the unbalanced coupling introduced by the feedback from the right hand facet.

For the ith mode in the CW direction,

$$E_{\text{fbk}} = E_{-i} \tag{7.46}$$

For the ith mode in the CCW direction,

$$E_{\text{fbk}} = 0 \tag{7.47}$$

The coupling coefficient k_{fbk} is given as:

$$k_{\text{fbk}} = \frac{T_r \sqrt{R_{\text{ref}}}}{\tau_L} \exp(-2jk_i D) \tag{7.48}$$

where T_r is the coupling transmission ratio of the output coupler, R_{ref} is the reflectivity of the right facet, τ_L is the roundtrip time inside the ring cavity, and D is the distance between the output coupler and the facet.

Figure 7.18a plots the simulated $L-I$ curve and Figure 7.18b shows the SMSR and the dominant lasing wavelength for $R_{\text{ref}} = 0.3$ and $D = 100\,\mu\text{m}$. The parameters used in the simulation are the same as those presented in Table 7.1. The threshold current is the same as for the device presented in the last section, but the SRL always lases in the direction predetermined by the position of the mirror (the CW direction in the case of Figure 7.17) and the intensity becomes a monotonically increasing function of the injection current except for several kinks. A comparison

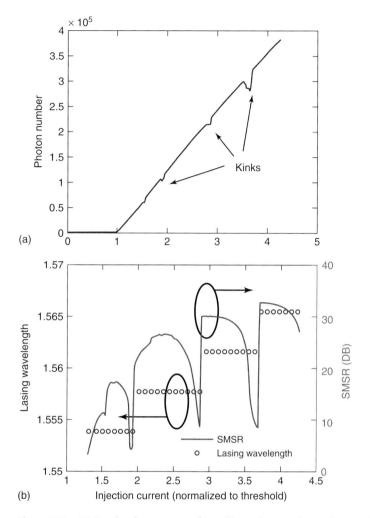

Figure 7.18 (a) Simulated L–I curves of a stable unidirectional SRL. (b) Wavelength for the dominant mode and SMSR under different injection current.

of the L–I curve and the lasing wavelength indicates that those kinks correspond to wavelength hopping between the longitudinal cavity modes, which indicates that, instead of periodic switching between two directions, the mode hops in the device always occur in the same direction as shown in Figure 7.17. Although the SMSR is reduced in the vicinity of mode hops, between them good single longitudinal mode lasing is observed with an SMSR > 25 dB for injection current $I > 2I_{th}$. The simulation results shown in Figure 7.18 indicate that the structure shown in Figure 7.17 can achieve unidirectional lasing.

Based on the fabricated SRL presented in the last section, we use a high-resolution FIBE process to cut one end of the output waveguide, so that a perpendicular mirror

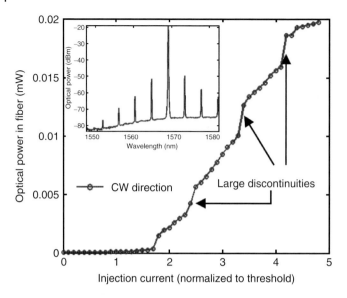

Figure 7.19 CW output powers as a function of increasing current for a racetrack unidirectional SRL. The inset shows the CW lasing spectrum at $3.1 I_{th}$.

facet is formed. It should be noticed that the mirror facets were fabricated using FIBE only for reasons of flexibility and easy comparison. The standard cleaving process can provide equally high, or superior, quality facets.

Figure 7.19 shows the L–I curve for the CW direction of the SRL after FIBE, measured at 293 K. Instead of periodic switching between two directions, the SRL now always lases in the direction predetermined by the position of the mirror (CW direction), and, therefore, it is referred to as a *stable unidirectional SRL*. The threshold current remains the same as for the bistable device (12 mA), but the optical power becomes a monotonically increasing function of the injection current, which is a clear sign of stable unidirectional lasing operation.

Figure 7.20 shows the dominant lasing wavelength and the SMSR, at different values of the biasing current, from the CW direction. A comparison of the L–I curve in Figure 7.19 and the lasing wavelength in Figure 7.20 indicates that the kinks or discontinuities in the L–I curve correspond to wavelength hopping between longitudinal modes in the CW direction. Unlike the bistable SRL (in Figure 7.16), a very good single longitudinal-mode lasing is observed here with SMSR > 25 dB for injection currents $I_{bias} > 2.5 I_{th}$ and reaching 35 dB at $I_{bias} > 4 I_{th}$, which confirms the stability of the unidirectional lasing operation. Again, these experimental results are in good agreement with the simulated results.

Although good agreement between theory and experiment is demonstrated in both bistable SRLs and stable unidirectional SRLs, there are two minor discrepancies between theory and experiment. The first discrepancy is the large abrupt discontinuities in the measured L–I curves displayed in Figure 7.19, while only small kinks are shown in the simulated L–I curves in Figure 7.18a. In both cases

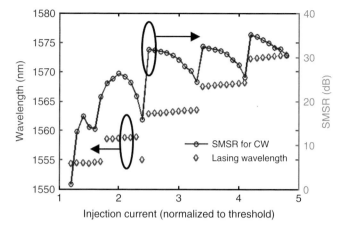

Figure 7.20 Wavelength for the dominant mode and SMSR under different injection current measured from unidirectional SRL.

the discontinuities and the kinks are caused by mode hops. The large discontinuities in the experiment are introduced by the wavelength-dependent absorption in the output waveguide, which leads to an apparently higher output power level at longer wavelengths (due to mode hopping), in addition to the increase due to the SRL's own L–I characteristics. The other discrepancy is in the dominant lasing wavelength and SMSR. It can be observed in the measured results that between mode hops there is a slope in the lasing wavelength, as the injection current is increased. This slope is owing to the change in refractive index with current and temperature. Because this effect is not included in the model, the slope is not present in the simulated results.

7.4.2
Impact of the Feedback Strength on the Operation of Unidirectional SRL

The discussions in the earlier sections have already established that sufficient optical feedback will render a SRL unidirectional, essentially as a single-mode laser. In this section, a detailed investigation of the impact of the feedback strength on the operation of the unidirectional SRL is given.

The feedback strength from the FIB-etched mirror facet can be controlled by the angle θ between the sidewall of the mirror facet and the plane of sample, which can be simply adjusted by tilting the sample when carrying out the FIBE process and provides a very convenient and versatile way to investigate the impact of feedback strength on the operation status of the unidirectional SRL.

Part of the energy couples out of the SRL cavity via the output coupler, which is then partially reflected by the FIB-etched mirror facet, and then couples back into the SRL cavity though the output coupler. Thus we define the feedback strength as

$$S = T_r^2 \cdot R_{\text{ref}} \tag{7.49}$$

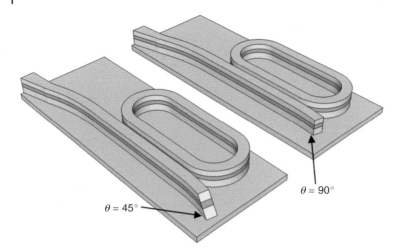

Figure 7.21 Schematic diagram of the unidirectional SRL with different angle θ between the sidewall of the mirror facet and the sample plane.

where T_r is the coupling transmission ratio and R_{ref} is the reflectivity of the FIB etched mirror facet.

Figure 7.21 gives two diagrams for the unidirectional SRLs, with $\theta = 90°$ and 45°. The unidirectional SRL with $\theta = 90°$ is expected to have the strongest feedback strength, while the one with $\theta = 45°$ is expected to have the weakest feedback strength. An SEM image for the device with $\theta = 45°$ is given in Figure 7.22.

By this means, the value of R_{ref} can be changed between 0 and 0.3, determined by the tilt-angle (0 for 45° and 0.3 for 90°). Therefore, the value of S is between 0 and $S_{max} = 0.3T_r^2$. In order to make the range of the S value as large as possible, we use an SRL with multimode interference (MMI) output couplers in which $T_r = 0.5$.

Figure 7.23 gives the calculated $L–I$ curve and SMSR versus the injection current, at different feedback strengths, S. When S is very small ($S = S_{max} \times 10^{-6}$ in Figure 7.23a,b), unidirectional lasing cannot be achieved. It can be seen that, in this case, the bi-AO region still exists, which is a sign of poor unidirectionality. Both the patterns of the $L–I$ curve and the SMSR are very similar to the characteristics for a bistable SRL.

Increasing the feedback strength ($S = S_{max} \times 10^{-4}$ in Figure 7.23c,d), the bi-AO region disappears and unidirectional lasing can be achieved within a range of currents from $I = I_{th}$ to $I = 2.3I_{th}$. Very good single-mode lasing with SMSR > 25 dB can also be achieved for $I > 1.5I_{th}$ except for some points where the SRL goes through mode hops. At higher injection levels ($I > 2.3I_{th}$), the lasing direction of the SRL starts to demonstrate the usual periodical switching, with the SMSR dropping and rising accordingly.

When the feedback strength S is further increased (to $S = S_{max} \times 10^{-2}$, as in Figure 7.23e,f), the range within which the unidirectional lasing can be achieved is extended from $I = I_{th}$ to $I = 3I_{th}$. As shown in Figure 7.23f, except for the mode-hops, the SMSR increases with injection current and an SMSR > 25 dB is observed

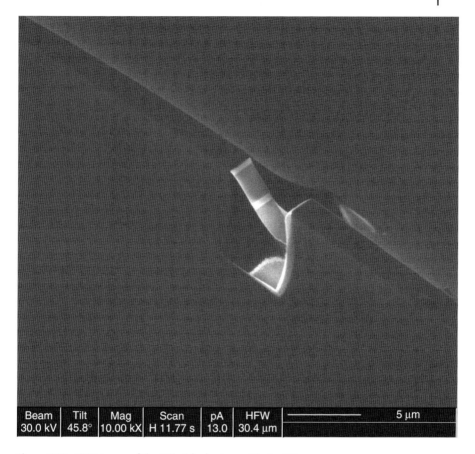

Figure 7.22 SEM image of the FIB etched mirror with $\theta = 45°$.

for injection current $I > 2.5 I_{th}$. However, the SMSR varies between 15 and 20 dB when the injection current is between $I = 1.5 I_{th}$ to $I = 2.5 I_{th}$, which is smaller than the SMSR for $S = S_{max} \times 10^{-4}$ with the same injection level shown in Figure 7.23d. This is an unexpected result that has not been reported previously.

Even more surprising was the fact that, at very strong feedback strength ($S = S_{max}$ in Figure 7.23g,h), the SRL demonstrates unidirectional lasing in the predetermined direction but the SMSR, with maximum value around 15 dB, deteriorates for all injection levels above the threshold.

Figure 7.24 shows the simulated lasing spectra at different strengths of optical feedback at $I = 1.95 I_{th}$. At maximum possible optical feedback, the CW lasing spectrum is a multimode spectrum (in Figure 7.24a). At moderate optical feedback strength ($S = S_{max} \times 10^{-2}$), the CW lasing spectrum becomes a single-mode lasing spectrum with an SMSR of 18.5 dB. At low optical feedback strength ($S = S_{max} \times 10^{-4}$), the CW lasing spectrum becomes a single-mode lasing spectrum, with an improved SMSR of 23.5 dB.

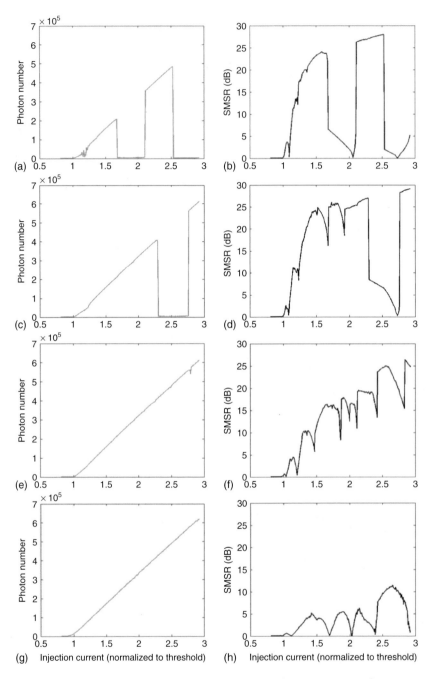

Figure 7.23 Simulated L–I curve and SMSR as functions of injection current after FIBE. (a,b) $S = S_{max} \times 10^{-6}$, (c,d) $S = S_{max} \times 10^{-4}$, (e,f) $S = S_{max} \times 10^{-2}$, and (g,h) $S = S_{max}$.

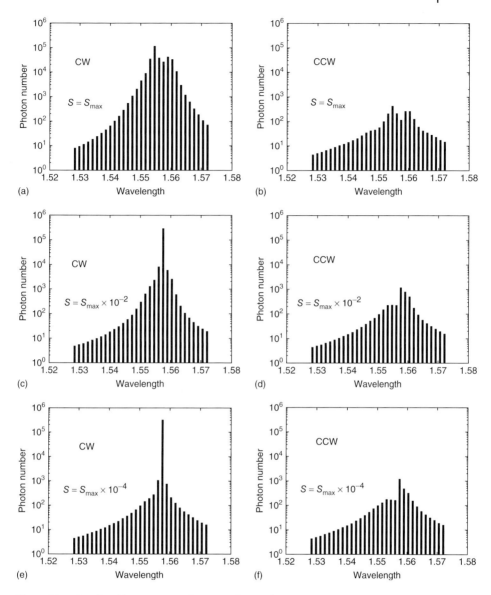

Figure 7.24 Simulated lasing spectra for CW and CCW directions at injection current of $I = 1.95 I_{th}$ for the different strengths of optical feedback. (a,b) $S = S_{max}$, (c,d) $S = S_{max} \times 10^{-2}$, and (e,f) $S = S_{max} \times 10^{-4}$.

The above mentioned phenomena can be understood as follows. The FIB-etched mirror facet reflects part of the energy from the lasing modes in the CCW direction into the lasing modes in the CW direction, which is equivalent to optical injection of multiple optical beams with different wavelengths. The CCW lasing spectrum is highly asymmetric and in many cases the peak wavelength of the CCW direction is different from that of the CW direction. Therefore, there is an offset between the wavelength of maximum optical injection and the wavelength of the main mode in CW direction. That is why the stronger optical feedback results in the reduction of the SMSR for the CW direction. Moreover, very high optical feedback strength will result in strong coupling/injection from the CCW direction into the CW direction and, in turn, results in complicated FWM, which will further degrade the SMSR, because the energy is redistributed into a large number of modes.

The simulation results presented above indicate that the strength of the optical feedback has a huge impact on the operation status of the unidirectional SRL and there exists an optimal feedback strength at which the unidirectional SRL can demonstrate unidirectional lasing with high SMSR. The measured $L-I$ curves and lasing spectra of the unidirectional SRL under different tilted angles fabricated by FIBE are shown in Figures 7.25 and 7.26, and they show features similar to those in the simulated results shown in Figures 7.23 and 7.24, respectively. Again very good agreement between the simulation and measurement is demonstrated.

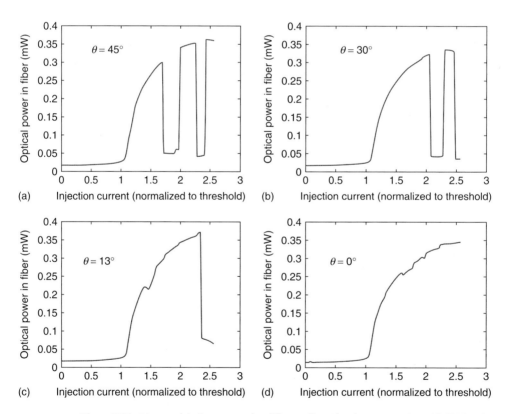

Figure 7.25 Measured $L-I$ curves under different tilt-angle when processing with FIBE. (a) $\theta = 45°$, (b) $\theta = 30°$, (c) $\theta = 13°$, and (d) $\theta = 0°$.

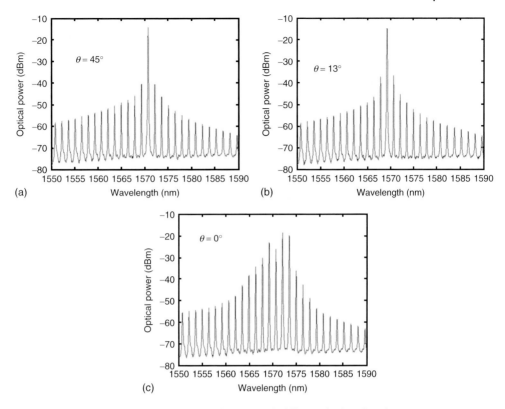

Figure 7.26 Measured lasing spectra in CW direction with different tilted angles when doing FIBE. (a) $\theta = 45°$, (b) $\theta = 13°$, and (c) $\theta = 0°$.

7.5 Summary and Conclusions

In this chapter, a general frequency-domain model of longitudinal mode interactions in SRLs has been presented. The inclusion of nonlinear suppression and wave-mixing terms related to third-order nonlinear susceptibilities χ^3 and also linear terms due to back-scattering between counter-propagating modes are the key to the successful prediction of SRL behavior. The model can handle a large number of modes and complex third-order nonlinear processes such as self-saturation, cross-saturation, and FWM that occur, because of both inter-band and intraband effects. Every aspect of the experimentally observed lasing characteristics of SRLs, including lasing spectra, light–current curves, and lasing direction hysteresis, can be reproduced by the model.

To further assess the performance and validity of the model, several miniaturized SRLs are tested, which show very good agreement with the numerical results. Stable unidirectional lasing in SRLs is also predicted and experimentally demonstrated by introducing asymmetric feedback from an external facet that is fabricated using a high-resolution FIBE process. The model is used to investigate the lasing characteristics of the unidirectional SRL, in particular, the effect of the strength of external optical feedback on the operation of the devices. Again, very good agreement between theoretical and experimental results is achieved in predicting and experimentally demonstrating optimal feedback strength for high-performance single-mode unidirectional operation of the SRL. More recently, wavelength tunable lasers and multi-wavelength lasers have been developed, based on SRLs under optical feedback [43, 44].

It can be convincingly concluded that the central role of the interplay between the linear and nonlinear mode coupling processes that drives the behavior of the SRL is now well understood, and can be readily applied to other semiconductor lasers where multimode interaction in a shared gain medium is important.

The model used, which is constructed under the SVAP assumption, may pose a limitation on how small the semiconductor laser size can be. A rough estimate would be an optical round-trip length of about 10 wavelengths. For SRLs, this would mean a radius on the order of 1.5 wavelengths or about 725 nm. Submicron lasers of such sizes must be treated differently by including the two-dimensional (2D) modal field and carrier distributions in a group of coupled equations, in which 2D carrier diffusion must be considered in addition to the effects already included here. For electrically pumped semiconductor microlasers operating at room temperature, their sizes are still in the order of 5-μm radius and the dominant optical confinement mechanism is still the whispering gallery mode. In such devices the model presented here is valid.

References

1. Liao, A.S.-H. and Wang, S. (1980) Semiconductor injection lasers with a circular resonator. *Appl. Phys. Lett.*, **36**, 801–803.
2. Krauss, T.F., Laybourn, P.J.R., and Roberts, J.S. (1990) CW operation of semi- conductor ring lasers. *Electron. Lett.*, **26**, 2095–2097.
3. Hohimer, J.P., Craft, D.C., Hadley, G.R., and Vawter, G.A. (1992) CW room temperature operation of y-junction semiconductor ring lasers. *Electron. Lett.*, **28**, 374–375.
4. Griffel, G., Abeles, J.H., Menna, R.J., Braun, A.M., Connolly, J.C., and King, M. (2000) Low threshold InGaAsP ring lasers fabricated using bi-level dry etching. *IEEE Photonics Technol. Lett.*, **12**, 146–148.
5. Hoimer, J.P. and Vawter, G.A. (1993) Unidirectional semiconductor ring laser with racetrack cavities. *Appl. Phys. Lett.*, **63**, 2457–2459.
6. Oku, S., Okayasu, M., and Ikeda, M. (1991) Low-threshold operation of square-shaped semiconductor ring lasers (orbiter lasers). *IEEE Photonics Technol. Lett.*, **3**, 588–590.
7. Han, H., Forbes, D.V., and Coleman, J.J. (1995) InGaAs-AlGaAs-GaAs strained-layer quantum-well heterostructure square ring lasers. *IEEE J. Quantum Electron.*, **31**, 1994–1997.
8. Ji, C., Leary, M.H., and Ballantyne, J.M. (1997) Long-wavelength triangular ring laser. *IEEE Photonics Technol. Lett.*, **9**, 1469–1471.
9. Krauss, T.F., De La Rue, R.M., Laybourn, P.J.R., Vogele, B., and Stanley, C.R. (1995) Efficient semiconductor ring lasers made by a simple self- aligned fabrication process. *IEEE J. Sel. Top. Quantum Electron.*, **1**, 757–761.
10. Cockerill, T.M., Forbes, D.V., Dantzig, J.A., and Coleman, J.J. (1994) A

strained-layer InGaAs-GaAs-AlGaAs buried heterostructure quantum-well lasers by three step selective area metalorganic chemical vapor deposition. *IEEE J. Quantum Electron.*, **30**, 441–445.
11. Zhang, J.P., Chu, D.Y., Wu, S.L., Bi, W.G., Tiberio, R.C., Tu, C.W., and Ho, S.T. (1996) Directional light output from photonic-wire microcavity semiconductor lasers. *IEEE Photonics Technol. Lett.*, **8**, 968–970.
12. Hohimer, J.P., Vawter, G.A., and Craft, D.C. (1993) Unidirectional operation in a semiconductor ring diode laser. *Appl. Phys. Lett.*, **62**, 1185–1187.
13. Liang, J.J., Lau, S.T., Leary, M.H., and Ballantyne, J.M. (1997) Unidirectional operation of waveguide diode ring lasers. *Appl. Phys. Lett.*, **70**, 1192–1194.
14. Hohimer, J.P., Vawter, G.A., Craft, D.C., and Hadley, G.R. (1992) Improving the performance of semiconductor ring lasers by controlled reflection feedback. *Appl. Phys. Lett.*, **61**, 1013–1015.
15. Booth, M.F., Schremer, A., and Ballantyne, J.M. (2000) Spatial beam switching and bistability in a diode ring laser. *Appl. Phys. Lett.*, **76**, 1095–1097.
16. Sorel, M., Laybourn, P.J.R., Giuliani, G., and Donati, S. (2002) Unidirectional bistability in semiconductor waveguide ring lasers. *Appl. Phys. Lett.*, **80**, 3051–3053.
17. Sorel, M., Laybourn, P.J.R., Scire, A., Balle, S., Giuliani, G., Miglierina, R., and Donati, S. (2002) Alternate oscillations in semiconductor ring lasers. *Opt. Lett.*, **27**, 1992–1994.
18. Sorel, M., Giuliani, G., Scire, A., Miglierina, R., Donati, S., and Laybourn, P.J.R. (2003) operating regimes of GaAs-AlGaAs semiconductor ring lasers: experiment and model. *IEEE J. Quantum Electron.*, **39**, 1187–1195.
19. Thakulsukanant, K., Bei, L., Memon, I., Mezosi, G., Wang, Z., Sorel, M., and Yu, S. (2009) All-optical label swapping using bistable semiconductor ring laser in an optical switching node. *IEEE J. Lightwave Technol.*, **27**, 632–638.
20. Li, B., Memon, M.I., Mezosi, G., Yuan, G., Wang, Z., Sorel, M., and Yu, S. (2008) All optical response of semiconductor ring laser to dual-optical injections. *IEEE Photonics Technol. Lett.*, **20**, 770–772.
21. Yuan, G. and Yu, S. (2007) Analysis of dynamic switching behavior of bistable semiconductor ring lasers triggered by resonant optical pulse injection. *IEEE J. Sel. Top. Quantum Electron.*, **13**, 1227–1234.
22. Mezosi, G., Strain, M.J., Furst, S., Wang, Z., Yu, S., and Sorel, M. (2009) Unidirectional bistability in AlGaInAs microring and micro-disc semiconductor lasers. *IEEE Photonics Technol. Lett.*, **21** (2), 88–90.
23. Liu, L., Kumar, R., Huybrechts, K., Spuesens, T., Roelkens, G., Geluk, E.-J., de Vries, T., Regreny, P., Van Thourhout, D., Baets, R., and Morthier, G. (2010) An ultra-small, low-power, all-optical flip-flop memory on a silicon chip. *Nat. Photonics*, **4**, 182–187.
24. Bogatov, A.P., Eliseev, P.G., and Sverdlov, B.N. (1975) Anomalous interaction of spectral modes in a semiconductor laser. *IEEE J. Quantum Electron.*, **11**, 510–515.
25. Yamada, M. (1989) Theoretical-analysis of nonlinear optical phenomena taking into account the beating vibration of the electron-density in semiconductor lasers. *J. Appl. Phys.*, **66**, 81–89.
26. Kazarinov, R.F., Henry, C.H., and Logan, R.A. (1982) Longitudinal mode self-stabilization in semiconductor lasers. *J. Appl. Phys.*, **53**, 4631–4644.
27. Ogasawara, N. and Ito, R. (1988) Longitudinal mode competition and asymmetric gain saturation in semiconductor injection lasers. II. Theory. *Jpn. J. Appl. Phys.*, **27**, 615–626.
28. Alalusi, M.R. and Darling, R.B. (1995) Effects of nonlinear gain on mode-hopping in semiconductor-laser diodes. *IEEE J. Quantum Electron.*, **31**, 1181–1192.
29. Ahmed, M. and Yamada, M. (2002) Influence of instantaneous mode competition on the dynamics of semiconductor lasers. *IEEE J. Quantum Electron.*, **38**, 682–693.
30. Herzog, U. (1991) Longitudinal mode interaction in semiconductor lasers due to nonlinear gain suppression and

four wave mixing. *Opt. Commun.*, **82**, 390–405.
31. Zhou, F.L., Sargent, M. III,, Koch, S.W., and Chow, W. (1990) Population pulsations and sidemode generation in semiconductors. *Phys. Rev. A*, **40**, 463–474.
32. Sargent, M. III, (1993) Theory of a multimode quasiequilibrium semiconductor laser. *Phys. Rev. A*, **48**, 771–726.
33. Born, C., Sorel, M., and Yu, S. (2005) Linear and nonlinear mode interactions in a semiconductor ring laser. *IEEE J. Quantum Electron.*, **41**, 261–271.
34. Born, C., Yuan, G., Wang, Z., and Yu, S. (2008) Nonlinear gain in semiconductor ring lasers. *IEEE J. Quantum Electron.*, **44**, 1055–1064.
35. Born, C., Yuan, G., Wang, Z., Sorel, M., and Yu, S. (2008) Laing mode hysteresis characteristics in semiconductor ring lasers. *IEEE J. Quantum Electron.*, **44**, 1171–1179.
36. Born, C. (2006) Nonlinear mode interactions in semiconductor ring lasers. PhD thesis. University of Bristol, Bristol.
37. Okamoto, K. (2006) *Fundamentals of Optical Waveguides*, Academic Press.
38. Agrawl, G.P. and Dutta, N.K. (1986) *Long Wavelength Semiconductor Lasers*, Van Nostrand, New York.
39. Uskov, A., Mørk, J., and Mark, J. (1994) Wave mixing in semiconductor laser amplifiers due to carrier heating and spectral-hole burning. *IEEE J. Quantum Electron.*, **30**, 1769–1781.
40. Willatzen, M., Uskov, A., Mørk, J., Olesen, H., Tromborg, B., and Jauho, A.P. (1991) Nonlinear gain suppression in semiconductor lasers due to carrier heating. *IEEE Photonics Technol. Lett.*, **3**, 606–609.
41. Summerfield, M.A. and Tucker, R.S. (1999) Frequency-domain model of multiwave mixing in bulk semiconductor optical amplifiers. *IEEE J. Quantum Electron.*, **5**, 839–850.
42. Giuliani, G. and Yu, S. (2008) Integrated Optical Logic and Memory Using Ultrafast Micro-Ring Bistable Semiconductor Lasers. D6.3 Report on Characteristics of All-Optical Functional Devices Optical Signal Regeneration Functions. Report No. D6.3-034743, European Commission.
43. Ermakov, I.V., Beri, S., Ashour, M., Danckaert, J., Docter, B., Bolk, J., Leijtens, X., and Verschaffelt, G. (2012) Semiconductor ring laser with on-chip filtered optical feedback for discrete wavelength tuning. *IEEE J. Quantum Electron.*, **48**, 129–136.
44. Khoder, M., Verschaffelt, G., Nguimdo, R., Leijtens, X., Bolk, J., and Danckaert, J. (2013) Controlled multiwavelength emission using semiconductor ring lasers with on-chip filtered optical feedback. *Opt. Lett.*, **38**, 2608–2610.

Index

a
α DFB lasers   102–103
amplified spontaneous emission (ASE)   11, 12, 24

b
Bloch mode device   70, 71
Bragg mirrors   33

c
carrier density pulsation (CDP)   266
carrier heating (CH)   266
chemically assisted ion-beam etching (CAIBE)   111
III–V compact lasers integrated onto silicon (SOI)   195–197
– heterogeneously integrated edge-emitting laser diodes   204–205
– – DFB and tunable lasers   209–211
– – Fabry-Perot lasers   205–207
– – heat sinking strategies for heterogeneously integrated lasers   212–213
– – mode-locked lasers   207
– – proposed novel laser architectures   211–212
– – racetrack resonator lasers   207, 208
– membrane bonding on SOI   197–199
– – adhesive bonding   199–201
– – direct bonding   201, 202–203
– – substrate removal   203, 204
– microdisk and microring lasers   213, 214–215
– – dynamic operation and switching   219–225
– – microdisk lasers design   215–217
– – static operation   217–219
composite gain waveguide (CGW)   4
composite metallo-dielectric-gain resonators   4–5
– composite gain-dielectric-metal 3D resonators   7–9
– composite gain-dielectric-metal waveguides   5–7
continuous wave (CW) operation via nonradiative recombination reduction   66–74
coupled-cavity waveguide (CCW)   101–102
coupled mode theory (CMT)   78

d
density matrix   259, 260, 269
distributed Bragg reflector lasers   209
distributed feedback (DFB)-like photonic crystal waveguide lasers   100–102
distributed feedback lasers   209
DVS-BCB (divinylsiloxane-bis-benzocyclobutene)   199–200
dynamic operation and switching
– all-optical set-reset flip-flop   219–220
– direct modulation   219
– gating and wavelength conversion   221–222
– microwave photonic filter   224, 225
– narrowband optical isolation and reflection sensitivity of microdisk lasers   222, 223–224
– phase modulation   224

e
edge-emitting PhC lasers   77
electrical injection   93, 94, 96, 97, 99, 100, 102, 125, 135
electrically pumped subwavelength metallo-dielectric lasers   13
– cavity design and modeling   13–17

Compact Semiconductor Lasers, First Edition.
Edited by Richard M. De La Rue, Siyuan Yu, and Jean-Michel Lourtioz.
© 2014 Wiley-VCH Verlag GmbH & Co. KGaA. Published 2014 by Wiley-VCH Verlag GmbH & Co. KGaA.

electrically pumped subwavelength metallo-dielectric lasers (*contd.*)
– fabrication   17–18
– measurements and discussion   18–20
emission statistics   58
etching   10, 13, 17, 18, 20, 234, 240–242, 245, 248. *See also* vertical cavity surface-emitting lasers (VCSELs)
exchange Bragg coupling laser structure   211

f
Fabry-Perot lasers   205–207
3D finite-difference time-domain (FDTD) calculations   157
Finite Element Method   157
focused ion beam etching (FIBE)   283 285–287, 290, 292, 294
four-wave mixing (FWM)   259–260

g
graded photonic heterostructure (GPH)   129

h
heat sinking
– comparison between membrane and bonded PhC laser   69, 70–71
– higher thermal conductivity   73–74
– through substrate   72–73
heterogeneously integrated edge-emitting laser diodes   204–205
– DFB and tunable lasers   209–211
– Fabry-Perot lasers   205–207
– heat sinking strategies for heterogeneously integrated lasers   212–213
– mode-locked lasers   207
– proposed novel laser architectures   211–212
– racetrack resonator lasers   207, 208
heterogeneously integrated multiwavelength laser   210, 211

i
inductively coupled plasma (ICP) etching   111
injection-locked ring lasers   219
InP-Based PhC fine processing   68

l
laser diodes and quantum cascade lasers and photonic crystal lasers   91–92
– mid-infrared and terahertz (THz) quantum cascade lasers   112–114
– – microdisk quantum cascade lasers   114–116

– – surface emission and small modal volumes   116, 117–131
– – THz QC lasers with truly subwavelength dimensions   131–135
– near-infrared and visible laser diodes   93
– – microcavity lasers   93–98
– – nonradiative carrier recombination in photonic crystal laser diodes   109–112
– – photonic crystal surface-emitting lasers (PCSELs)   106–109
– – waveguide lasers in substrate approach   98–106
lasing region   24
light cone   38
light–current (L–I) curves   274–275
linear rate equation model   52–53

m
membrane device   70, 71
micro/nano cavity based PhC lasers   36
– lasers based on 2D PhC cavities   36–44
– lasers based on 3D PhC cavities   44–46
microcavity lasers   93–98
microdisk   232
– and microring lasers   213, 214–215
– – dynamic operation and switching   219–225
– – microdisk lasers design   215–217
– – static operation   217–219
microdisk quantum cascade lasers
– at mid-infrared wavelength   114–115
– THz waves   115–116
microwave photonic filter   224 225
mid-infrared and terahertz (THz) quantum cascade lasers   112–114
– microdisk quantum cascade lasers   114–116
– surface emission and small modal volumes   116, 117–131
– THz QC lasers with truly subwavelength dimensions   131–135
mid-infrared quantum cascade lasers
– with deeply etched photonic crystal structures   116, 117–121
– terahertz (THz) 2D photonic crystal quantum-cascade (QC) lasers   123–131
– with thin metallic photonic crystal layer   121–123
mode coupling   259, 278, 294
mode-locked lasers   207
multimode analysis   259, 260

n

nanoscale metallo-dielectric coherent light sources 1–4
- composite metallo-dielectric-gain resonators 4–5
- – composite gain-dielectric-metal 3D resonators 7–9
- – composite gain-dielectric-metal waveguides 5–7
- electrically pumped subwavelength metallo-dielectric lasers 13
- – cavity design and modeling 13–17
- – fabrication 17–18
- – measurements and discussion 18–20
- subwavelength metallo-dielectric lasers experimental validations for operation at room-temperature 9–10
- – characterization and testing 11–13
- – fabrication processes 10–11
- thresholdless nanoscale coaxial lasers 20–22
- – characterization and discussion 23, 24–27
- – design and fabrication 22–23

near-infrared and visible laser diodes 93
- microcavity lasers 93–98
- nonradiative carrier recombination in photonic crystal laser diodes 109–112
- photonic crystal surface-emitting lasers (PCSELs) 106–109
- waveguide lasers in substrate approach 98–106

nonlinear gain 259, 266, 271, 279
nonlinearity, in semiconductor microring lasers 257–260
- general formalism
- – fundamental equations 260–265
- – generalized equations in matrix form 270–272
- – third order susceptibility and polarization 265–270
- numerical results, for microring lasers 272–274
- – experimental results for racetrack shaped SRLs 281–283
- – lasing direction hysteresis 277, 278–280
- – L–I characteristics 274–275
- – temporal dynamics and lasing spectra 275–277
- numerical results, for unidirectional microring lasers 283
- – feedback strength impact on unidirectional SRL operation 287–293
- – unidirectional SRL 284–287

nonradiative carrier losses 240–242
nonradiative carrier recombination in photonic crystal laser diodes 109–112
non-radiative lifetime 55

o

optical bistability 259
optical feedback 271, 283 287, 289 291, 292 294
out-coupling, through surface emission directionality control 75–76

p

passive optical waveguides through epitaxial regrowth 78–80
III–V PhC lasers heterogeneously integrated onto SOI waveguides circuitry 80–82
photoluminescence region 24
photonic band-gaps (PBGs) 33, 35, 181–185
photonic crystal lasers. *See also* individual entries 33–34
- characteristics 52
- – dynamics 60–65
- – rate equation model and PhC laser parameters 52–55
- – stationary regime in PhC lasers 55–60
- design and fabrication 35–36
- – micro/nano cavity based PhC lasers 36–46
- – slow-light based PhC lasers 46–52
- isssues to be solved and partially solved 65
- – continuous wave (CW) operation via nonradiative recombination reduction 66–74
- – interfacing and power issues 74–82

photonic crystal surface-emitting lasers (PCSELs) 106–109
photonic integrated circuits (PICs) 195, 196, 226, 231, 257
photonic modes 3
photon lifetime and active volume 53–54
plane-wave admittance method (PWAM) 158–166
plane-wave effective-index method 157
plasma-enhanced chemical vapor deposition (PECVD) 10, 17
pump–probe spatiotemporally resolved thermo-reflectance technique 70
Purcell factor 21, 29, 54

q

quantum dots   66–67
quantum electrodynamic (QED) effects   21–22
quantum wells (QWs)   67, 68–69, 149, 151, 168, 182–184
quantum wires   67

r

racetrack geometry microring lasers   249–251
racetrack resonator lasers   207, 208
radiative lifetime and spontaneous emission factor   54–55
rate-equation model   25–27
– and PhC laser parameters   52–55
reactive ion etching (RIE)-lag effects in small SRL devices   248–249
resonator mirrors   211–212
ridge waveguide lasers with photonic crystal mirrors   103–106
ring laser   232–50. *See also* individual entries
ring-resonator based tunable laser   210

s

sampled grating DBR lasers   209, 210
scanning electron microscope (SEM)   13, 17–21, 23, 94, 97, 101, 104, 107, 109, 115–118, 121, 125, 126, 132, 134, 218, 220, 244, 245, 248–250, 281, 282, 288, 289. *See also* photonic crystal lasers
Schawlow–Townes formula (modified)   60
semiconductor ring lasers (SRLs)   231. *See also* nonlinearity, in semiconductor microring lasers
– bending loss   238–240
– historical review of contributions to research on devices   232–235
– junction heating in small devices   246–247
– microring and microdisk lasers with point couplers   242–245
– nonradiative carrier losses   240–242
– racetrack geometry microring lasers   249–251
– reactive ion etching (RIE)-lag effects in small devices   248–249
– waveguide design   235, 236–237
single mode and high side mode suppression ratio (SMSR)   233
slow-light based PhC lasers   46
– 2D PhC DFB-like lasers
– – for in-plane emission   47–50
– – for surface emission   50–52
spatially diagonal transition   113
spectral hole burning (SHB)   266
spin-on-glasses (SOGs)   199
split-ring resonator (SRR)   133, 134
stable unidirectional SRL   286
surface emission and small modal volumes
– mid-infrared quantum cascade lasers
– – with deeply etched photonic crystal structures   116, 117–121
– – terahertz (THz) 2D photonic crystal quantum-cascade (QC) lasers   123, 125–131
– – with thin metallic photonic crystal layer   121–123
surface-emitting QC lasers   120
surface passivation of improved efficiency, in PhC lasers   68, 69
surface plasmon polariton (SPP)   3

t

tapered fibers   76–77
terahertz (THz)
– 2D photonic crystal quantum-cascade (QC) lasers   123, 125–131
– QC lasers with truly subwavelength dimensions   131–135
threshold current   237, 241, 243, 245, 247, 250, 252, 253
threshold gain for 3D resonators   5
thresholdless lasers   57
thresholdless nanoscale coaxial lasers   20–22
– characterization and discussion   23–27
– design and fabrication   22–23
top and bottom-emitting photonic-crystal VCSELs   170–173
transverse modes   149, 151, 152, 154, 156, 185–187
two-mode model   258–259

v

Vernier effect   104
vertical cavity surface-emitting lasers (VCSELs)   92, 106, 149–155
– enhanced fundamental mode operation   173–177
– highly birefringent and dichroic   177–181
– numerical methods for modeling   156–157
– photonic crystal depth impact on threshold characteristics   166–169
– plane-wave admittance method (PWAM)   158–166
– top and bottom-emitting   170–173
– with true photonic bandgap   181–185
vertical outgassing channels (VOCs)   203

w

wafer scale testing 231, 257
waveguide lasers in substrate approach 98–100
– α DFB lasers 102–103
– distributed feedback (DFB)-like photonic crystal waveguide lasers 100–102
– ridge waveguide lasers with photonic crystal mirrors 103–106
wire cavity laser 41–42